SCHAUM'S OUTLINE OF

THEORY AND PROBLEMS

OF

DIGITAL PRINCIPLES

Third Edition

•

ROGER L. TOKHEIM, M.S.

SCHAUM'S OUTLINE SERIES
McGraw-Hill
New York   San Francisco   Washington, D.C.   Auckland   Bogotá
Caracas   Lisbon   London   Madrid   Mexico City   Milan
Montreal   New Delhi   San Juan   Singapore
Sydney   Tokyo   Toronto

ROGER L. TOKHEIM holds B.S., M.S., and Ed.S. degrees from St. Cloud State University and the University of Wisconsin-Stout. He is the author of *Digital Electronics* and its companion *Activities Manual for Digital Electronics*, *Schaum's Outline of Microprocessor Fundamentals*, and numerous other instructional materials on science and technology. An experienced educator at the secondary and college levels, he is presently an instructor of Technology Education and Computer Science at Henry Sibley High School, Mendota Heights, Minnesota.

Schaum's Outline of Theory and Problems of
DIGITAL PRINCIPLES

11 12 13 14 15 VFM VFM 05 04 03

ISBN 0-07-065050-0

Sponsoring Editor: John Aliano
Production Supervisor: Denise Puryear
Editing Supervisor: Patty Andrews

Library of Congress Cataloging–in–Publication Data
Tokheim, Roger L.
    Schaum's outline of theory and problems of digital prinicples/by
  Roger L. Tokheim—3rd ed.
      p. cm.—(Schaum's outline series)
    Includes index.
    ISBN 0-07-065050-0
    1. Digital electronics. I. Title. II. Series.
TK7868.D5T66 1994   93-64
621.3815—dc20   CIP

McGraw-Hill

A Division of The McGraw-Hill Companies

Preface

Digital electronics is a rapidly growing technology. Digital circuits are used in most new consumer products, industrial equipment and controls, and office, medical, military, and communications equipment. This expanding use of digital circuits is the result of the development of inexpensive integrated circuits and the application of display, memory, and computer technology.

Schaum's Outline of Digital Principles provides information necessary to lead the reader through the solution of those problems in digital electronics one might encounter as a student, technician, engineer, or hobbyist. While the principles of the subject are necessary, the Schaum's Outline philosophy is dedicated to showing the student how to apply the principles of digital electronics through practical solved problems. This new edition now contains over 1000 solved and supplementary problems.

The third edition of *Schaum's Outline of Digital Principles* contains many of the same topics which made the first two editions great successes. Slight changes have been made in many of the traditional topics to reflect the technological trend toward using more CMOS, NMOS, and PMOS integrated circuits. Several microprocessor/microcomputer-related topics have been included, reflecting the current practice of teaching a microprocessor course after or with digital electronics. A chapter detailing the characteristics of TTL and CMOS devices along with several interfacing topics has been added. Other display technologies such as liquid-crystal displays (LCDs) and vacuum fluorescent (VF) displays have been given expanded coverage. The chapter on microcomputer memory has been revised with added coverage of hard and optical disks. Sections on programmable logic arrays (PLA), magnitude comparators, demultiplexers, and Schmitt trigger devices have been added.

The topics outlined in this book were carefully selected to coincide with courses taught at the upper high school, vocational-technical school, technical college, and beginning college level. Several of the most widely used textbooks in digital electronics were analyzed. The topics and problems included in this Schaum's Outline reflect those encountered in standard textbooks.

Schaum's Outline of Digital Principles, Third Edition, begins with number systems and digital codes and continues with logic gates and combinational logic circuits. It then details the characteristics of both TTL and CMOS ICs, along with various interfacing topics. Next encoders, decoders, and display drivers are explored, along with LED, LCD, and VF seven-segment displays. Various arithmetic circuits are examined. It then covers flip-flops, other multivibrators, and sequential logic, followed by counters and shift registers. Next semiconductor and bulk storage memories are explored. Finally, multiplexers, demultiplexers, latches and buffers, digital data transmission, magnitude comparators, Schmitt trigger devices, and programmable logic arrays are investigated. The book stresses the use of industry-standard digital ICs (both TTL and CMOS) so that the reader becomes familiar with the practical hardware aspects of digital electronics. Most circuits in this Schaum's Outline can be wired using standard digital ICs.

I wish to thank my son Marshall for his many hours of typing, proofreading, and testing circuits to make this book as accurate as possible. Finally, I extend my appreciation to other family members Daniel and Carrie for their help and patience.

ROGER L. TOKHEIM

Contents

Chapter 1

Numbers Used in Digital Electronics

1-1 INTRODUCTION

The decimal number system is familiar to everyone. This system uses the symbols 0, 1, 2, 3, 4, 5, 6, 7, 8, and 9. The decimal system also has a place-value characteristic. Consider the decimal number 238. The 8 is in the 1s position or place. The 3 is in the 10s position, and therefore the three 10s stand for 30 units. The 2 is in the 100s position and means two 100s, or 200 units. Adding $200 + 30 + 8$ gives the total decimal number of 238. The decimal number system is also called the *base 10 system*. It is referred to as base 10 because it has 10 different symbols. The base 10 system is also said to have a *radix* of 10. "Radix" and "base" are terms that mean exactly the same thing.

Binary numbers (base 2) are used extensively in digital electronics and computers. Both hexadecimal (base 16) and octal (base 8) numbers are used to represent groups of binary digits. Binary and hexadecimal numbers find wide use in modern microcomputers.

All the number systems mentioned (decimal, binary, octal, and hexadecimal) can be used for counting. All these number systems also have the place-value characteristic.

1-2 BINARY NUMBERS

The binary number system uses only two symbols $(0, 1)$. It is said to have a radix of 2 and is commonly called the *base 2 number system*. Each *binary digit* is called a *bit*.

Counting in binary is illustrated in Fig. 1-1. The binary number is shown on the right with its decimal equivalent. Notice that the *least significant bit* (LSB) is the 1s place. In other words, if a 1 appears in the right column, a 1 is added to the binary count. The second place over from the right is the 2s place. A 1 appearing in this column (as in decimal 2 row) means that 2 is added to the count. Three other binary place values also are shown in Fig. 1-1 (4s, 8s, and 16s places). Note that each larger place value is an added power of 2. The 1s place is really 2^0, the 2s place 2^1, the 4s place 2^2, the 8s place 2^3, and the 16s place 2^4. It is customary in digital electronics to memorize at least the binary counting sequence from 0000 to 1111 (say: one, one, one, one) or decimal 15.

Consider the number shown in Fig. 1-2a. This figure shows how to convert the binary 10011 (say: one, zero, zero, one, one) to its decimal equivalent. Note that, for each 1 bit in the binary number, the decimal equivalent for that place value is written below. The decimal numbers are then added $(16 + 2 + 1 = 19)$ to yield the decimal equivalent. Binary 10011 then equals a decimal 19.

Consider the binary number 101110 in Fig. 1-2b. Using the same procedure, each 1 bit in the binary number generates a decimal equivalent for that place value. The *most significant bit* (MSB) of the binary number is equal to 32. Add 8 plus 4 plus 2 to the 32 for a total of 46. Binary 101110 then equals decimal 46. Figure 1-2b also identifies the binary point (similar to the decimal point in decimal numbers). It is customary to omit the binary point when working with whole binary numbers.

What is the value of the number 111? It could be one hundred and eleven in decimal or one, one, one in binary. Some books use the system shown in Fig. 1-2c to designate the base, or radix, of a number. In this case 10011 is a base 2 number as shown by the small subscript 2 after the number. The number 19 is a base 10 number as shown by the subscript 10 after the number. Figure 1-2c is a summary of the binary-to-decimal conversions in Fig. 1-2a and b.

How about converting fractional numbers? Figure 1-3 illustrates the binary number 1110.101 being converted to its decimal equivalent. The place values are given across the top. Note the value of each position to the right of the binary point. The procedure for making the conversion is the same as with whole numbers. The place value of each 1 bit in the binary number is added to form the decimal number. In this problem $8 + 4 + 2 + 0.5 + 0.125 = 14.625$ in decimal.

Decimal count	Binary count				
	16s	8s	4s	2s	1s
0					0
1					1
2				1	0
3				1	1
4			1	0	0
5			1	0	1
6			1	1	0
7			1	1	1
8		1	0	0	0
9		1	0	0	1
10		1	0	1	0
11		1	0	1	1
12		1	1	0	0
13		1	1	0	1
14		1	1	1	0
15		1	1	1	1
16	1	0	0	0	0
17	1	0	0	0	1
18	1	0	0	1	0
19	1	0	0	1	1
	2^4	2^3	2^2	2^1	2^0
	Powers of 2				

Fig. 1-1 Counting in binary and decimal

Powers of 2	2^4	2^3	2^2	2^1	2^0
Place value	16s	8s	4s	2s	1s
Binary	**1**	**0**	**0**	**1**	**1** . ←——Binary point
Decimal	16	+		2	+ 1 = 19

(a) Binary-to-decimal conversion

Powers of 2	2^5	2^4	2^3	2^2	2^1	2^0
Place value	32s	16s	8s	4s	2s	1s
Binary	**1**	**0**	**1**	**1**	**1**	**0** . ←——Binary point
Decimal	32	+	8	+ 4	+ 2	= 46

(b) Binary-to-decimal conversion

$$10011_2 = 19_{10} \qquad 101110_2 = 46_{10}$$

(c) Summary of conversions and use of small subscripts to indicate radix of number

Fig. 1-2

Powers of 2	2^3	2^2	2^1	2^0	$1/2^1$	$1/2^2$	$1/2^3$
Place value	8s	4s	2s	1s	0.5s	0.25s	0.125s
Binary	1	1	1	0 .	1	0	1
Decimal	8 +	4 +	2	+	0.5	+	0.125 = 14.625

Fig. 1-3 Binary-to-decimal conversion

Convert the decimal number 87 to a binary number. Figure 1-4 shows a convenient method for making this conversion. The decimal number 87 is first divided by 2, leaving 43 with a remainder of 1. The remainder is important and is recorded at the right. It becomes the LSB in the binary number. The quotient (43) then is transferred as shown by the arrow and becomes the dividend. The quotients are repeatedly divided by 2 until the quotient becomes 0 with a remainder of 1, as in the last line of Fig. 1-4. Near the bottom the figure shows that decimal 87 equals binary 1010111.

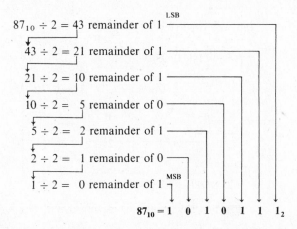

Fig. 1-4 Decimal-to-binary conversion

Convert the decimal number 0.375 to a binary number. Figure 1-5a illustrates one method of performing this task. Note that the decimal number (0.375) is being *multiplied* by 2. This leaves a product of 0.75. The 0 from the integer place (1s place) becomes the bit nearest the binary point. The 0.75 is then multiplied by 2, yielding 1.50. The carry of 1 to the integer (1s place) is the next bit in the binary number. The 0.50 is then multiplied by 2, yielding a product of 1.00. The carry of 1 in the integer place is the final 1 in the binary number. When the product is 1.00, the conversion process is complete. Figure 1-5a shows a decimal 0.375 being converted into a binary equivalent of 0.011.

Figure 1-5b shows the decimal number 0.84375 being converted into binary. Again note that 0.84375 is multiplied by 2. The integer of each product is placed below, forming the binary number. When the product reaches 1.00, the conversion is complete. This problem shows a decimal 0.84375 being converted to binary 0.11011.

Consider the decimal number 5.625. Converting this number to binary involves two processes. The integer part of the number (5) is processed by *repeated division* near the top in Fig. 1-6. Decimal 5 is converted to a binary 101. The fractional part of the decimal number (.625) is converted to binary .101 at the bottom in Fig. 1-6. The fractional part is converted to binary through the *repeated multiplication* process. The integer and fractional sections are then combined to show that decimal 5.625 equals binary 101.101.

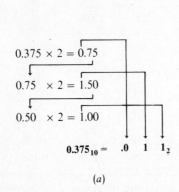

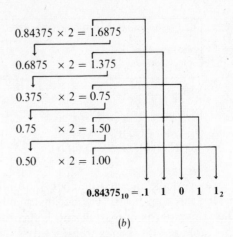

(a)   (b)

Fig. 1-5 Fractional decimal-to-binary conversions

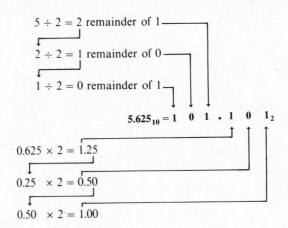

Fig. 1-6 Decimal-to-binary conversion

SOLVED PROBLEMS

1.1   The binary number system is the base _____ system and has a radix of _____ .

Solution:

     The binary number system is the base 2 system and has a radix of 2.

1.2   The term bit means _____ _____ when dealing with binary numbers.

Solution:

     Bit means binary digit.

1.3   How would you say the number 1001 in (a) binary and (b) decimal?

Solution:

     The number 1001 is pronounced as follows: (a) one, zero, zero, one;   (b) one thousand and one.

1.4   The number 110_{10} is a base _____ number.

Solution:

     The number 110_{10} is a base 10 number, as indicated by the small 10 after the number.

1.5   Write the base 2 number one, one, zero, zero, one.

Solution:

11001_2

1.6   Convert the following binary numbers to their decimal equivalents:
(*a*)   001100   (*b*)   000011   (*c*)   011100   (*d*)   111100   (*e*)   101010   (*f*)   111111
(*g*)   100001   (*h*)   111000

Solution:

Follow the procedure shown in Fig. 1-2. The decimal equivalents of the binary numbers are as follows:
(*a*) $001100_2 = 12_{10}$ (*c*) $011100_2 = 28_{10}$ (*e*) $101010_2 = 42_{10}$ (*g*) $100001_2 = 33_{10}$
(*b*) $000011_2 = 3_{10}$ (*d*) $111100_2 = 60_{10}$ (*f*) $111111_2 = 63_{10}$ (*h*) $111000_2 = 56_{10}$

1.7 $11110001111_2 = \underline{\hspace{1cm}}_{10}$

Solution:

Follow the procedure shown in Fig. 1-2. $11110001111_2 = 1935_{10}$.

1.8 $11100.011_2 = \underline{\hspace{1cm}}_{10}$

Solution:

Follow the procedure shown in Fig. 1-3. $11100.011_2 = 28.375_{10}$.

1.9 $110011.100\,11_2 = \underline{\hspace{1cm}}_{10}$

Solution:

Follow the procedure shown in Fig. 1-3. $110011.10011_2 = 51.593\,75_{10}$.

1.10 $1010101010.1_2 = \underline{\hspace{1cm}}_{10}$

Solution:

Follow the procedure shown in Fig. 1-3. $1010101010.1_2 = 682.5_{10}$.

1.11   Convert the following decimal numbers to their binary equivalents:
(*a*)   64, (*b*)   100, (*c*)   111, (*d*)   145, (*e*)   255, (*f*)   500.

Solution:

Follow the procedure shown in Fig. 1-4. The binary equivalents of the decimal numbers are as follows:
(*a*) $64_{10} = 1000000_2$ (*c*) $111_{10} = 1101111_2$ (*e*) $255_{10} = 11111111_2$
(*b*) $100_{10} = 1100100_2$ (*d*) $145_{10} = 10010001_2$ (*f*) $500_{10} = 111110100_2$

1.12 $34.75_{10} = \underline{\hspace{1cm}}_2$

Solution:

Follow the procedure shown in Fig. 1-6. $34.75_{10} = 100010.11_2$.

1.13 $25.25_{10} = $ _____$_2$

Solution:

Follow the procedure shown in Fig. 1-6. $25.25_{10} = 11001.01_2$.

1.14 $27.1875_{10} = $ _____$_2$

Solution:

Follow the procedure shown in Fig. 1-6. $27.1875_{10} = 11011.0011_2$.

1-3   HEXADECIMAL NUMBERS

The hexadecimal number system has a radix of 16. It is referred to as the *base 16 number system*. It uses the symbols 0-9, A, B, C, D, E, and F as shown in the hexadecimal column of the table in Fig. 1-7. The letter A stands for a count of 10, B for 11, C for 12, D for 13, E for 14, and F for 15. The advantage of the hexadecimal system is its usefulness in converting directly from a 4-bit binary number. Note in the shaded section of Fig. 1-7 that each 4-bit binary number from 0000 to 1111 can be represented by a unique hexadecimal digit.

Decimal	Binary	Hexadecimal	Decimal	Binary	Hexadecimal
0	0000	0	16	10000	10
1	0001	1	17	10001	11
2	0010	2	18	10010	12
3	0011	3	19	10011	13
4	0100	4	20	10100	14
5	0101	5	21	10101	15
6	0110	6	22	10110	16
7	0111	7	23	10111	17
8	1000	8	24	11000	18
9	1001	9	25	11001	19
10	1010	A	26	11010	1A
11	1011	B	27	11011	1B
12	1100	C	28	11100	1C
13	1101	D	29	11101	1D
14	1110	E	30	11110	1E
15	1111	F	31	11111	1F

Fig. 1-7   Counting in decimal, binary, and hexadecimal number systems

Look at the line labeled 16 in the decimal column in Fig. 1-7. The hexadecimal equivalent is 10. This shows that the hexadecimal number system uses the place-value idea. The 1 (in 10_{16}) stands for 16 units, while the 0 stands for zero units.

Convert the hexadecimal number 2B6 into a decimal number. Figure 1-8a shows the familiar process. The 2 is in the 256s place so $2 \times 256 = 512$, which is written in the decimal line. The hexadecimal digit B appears in the 16s column. Note in Fig. 1-8 that hexadecimal B corresponds to decimal 11. This means that there are eleven 16s (16×11), yielding 176. The 176 is added into the decimal total near the bottom in Fig. 1-8a. The 1s column shows six 1s. The 6 is added into the decimal line. The decimal values are added ($512 + 176 + 6 = 694$), yielding 694_{10}. Figure 1-8a shows that $2B6_{16}$ equals 694_{10}.

Powers of 16	16^2	16^1	16^0
Place value	256s	16s	1s

Hexadecimal number	**2**	**B**	**6**

Decimal

$$\begin{array}{ccccccc} 256 & & 16 & & 1 & & \\ \times\ 2 & & \times\ 11 & & \times\ 6 & & \\ \hline 512 & + & 176 & + & 6 & = & 694_{10} \end{array}$$

(a) Hexadecimal-to-decimal conversion

Powers of 16	16^2	16^1	16^0	$1/16^1$
Place value	256s	16s	1s	.0625s

Hexadecimal number	**A**	**3**	**F** ·	**C**

Decimal

$$\begin{array}{ccccccccc} 256 & & 16 & & 1 & & .0625 & & \\ \times\ 10 & & \times\ 3 & & \times\ 15 & & \times\ 12 & & \\ \hline 2560 & + & 48 & + & 15 & + & 0.75 & = & 2623.75_{10} \end{array}$$

(b) Fractional hexadecimal-to-decimal conversion

Fig. 1-8

Convert the hexadecimal number A3F.C to its decimal equivalent. Figure 1-8b details this problem. First consider the 256s column. The hexadecimal digit A means that 256 must be multiplied by 10, resulting in a product of 2560. The hexadecimal number shows that it contains three 16s, and therefore $16 \times 3 = 48$, which is added to the decimal line. The 1s column contains the hexadecimal digit F, which means $1 \times 15 = 15$. The 15 is added to the decimal line. The 0.0625s column contains the hexadecimal digit C, which means $12 \times 0.0625 = 0.75$. The 0.75 is added to the decimal line. Adding the contents of the decimal line ($2560 + 48 + 15 + 0.75 = 2623.75$) gives the decimal number 2623.75. Figure 1-8b converts $A3F.C_{16}$ to 2623.75_{10}.

Now reverse the process and convert the decimal number 45 to its hexadecimal equivalent. Figure 1-9a details the familiar repeated divide-by-16 process. The decimal number 45 is first divided by 16, resulting in a quotient of 2 with a remainder of 13. The remainder of 13 (D in hexadecimal) becomes the LSD of the hexadecimal number. The quotient (2) is transferred to the dividend position and divided by 16. This results in a quotient of 0 with a remainder of 2. The 2 becomes the next digit in the

$250 \div 16 = 15$ remainder of 10 ⎯⎯⎯⎯⎯⎯⎯⎯⎯

$15 \div 16 = \ 0$ remainder of 15 ⎯⎯⎯⎯⎯⎯⎯

$45 \div 16 = 2$ remainder of 13 ⎯⎯⎯⎯⎯⎯⎯

$\mathbf{250.25_{10} = F \quad A \cdot 4_{16}}$

$2 \div 16 = 0$ remainder of $\ 2$ ⎯⎯⎯⎯

$0.25 \times 16 = 4.00$

$\mathbf{45_{10} = 2 \quad D_{16}}$

$0.00 \times 16 = 0.00$

(a) Decimal-to-hexadecimal conversion

(b) Fractional decimal-to-hexadecimal conversion

Fig. 1-9

hexadecimal number. The process is complete because the integer part of the quotient is 0. The process in Fig. 1-9a converts the decimal number 45 to the hexadecimal number 2D.

Convert the decimal number 250.25 to a hexadecimal number. The conversion must be done by using two processes as shown in Fig. 1-9b. The integer part of the decimal number (250) is converted to hexadecimal by using the repeated divide-by-16 process. The remainders of 10 (A in hexadecimal) and 15 (F in hexadecimal) form the hexadecimal whole number FA. The fractional part of the 250.25 is multiplied by 16 (0.25×16). The result is 4.00. The integer 4 is transferred to the position shown in Fig. 1-9b. The completed conversion shows the decimal number 250.25 equaling the hexadecimal number FA.4.

The prime advantage of the hexadecimal system is its easy conversion to binary. Figure 1-10a shows the hexadecimal number 3B9 being converted to binary. Note that each hexadecimal digit forms a group of four binary digits, or bits. The groups of bits are then combined to form the binary number. In this case $3B9_{16}$ equals 1110111001_2.

$$
\begin{array}{ccc}
3 & B & 9_{16} \\
\downarrow & \downarrow & \downarrow \\
0011 & 1011 & 1001
\end{array}
\qquad 3B9_{16} = 1110111001_2
$$

(a) Hexadecimal-to-binary conversion

$$
\begin{array}{cccc}
4 & 7 & .\ F & E \\
\downarrow & \downarrow & \downarrow & \downarrow \\
0100 & 0111 & .\ 1111 & 1110
\end{array}
\qquad 47.FE_{16} = 1000111.1111111_2
$$

(b) Fractional hexadecimal-to-binary conversion

$$
\begin{array}{ccc}
1010 & 1000 & 0101\ . \\
\downarrow & \downarrow & \downarrow \\
A & 8 & 5
\end{array}
\qquad 101010000101_2 = A85_{16}
$$

(c) Binary-to-hexadecimal conversion

$$
\begin{array}{cccc}
0001 & 0010 & .\ 0110 & 1100 \\
\downarrow & \downarrow & \downarrow & \downarrow \\
1 & 2 & .\ 6 & C
\end{array}
\qquad 10010.011011_2 = 12.6C_{16}
$$

(d) Fractional binary-to-hexadecimal conversion

Fig. 1-10

Another hexadecimal-to-binary conversion is detailed in Fig. 1-10b. Again each hexadecimal digit forms a 4-bit group in the binary number. The hexadecimal point is dropped straight down to form the binary point. The hexadecimal number 47.FE is converted to the binary number 1000111.1111111. It is apparent that hexadecimal numbers, because of their compactness, are much easier to write down than the long strings of 1s and 0s in binary. The hexadecimal system can be thought of as a shorthand method of writing binary numbers.

Figure 1-10c shows the binary number 101010000101 being converted to hexadecimal. First divide the binary number into 4-bit groups *starting at the binary point*. Each group of four bits is then translated into an equivalent hexadecimal digit. Figure 1-10c shows that binary 101010000101 equals hexadecimal A85.

Another binary-to-hexadecimal conversion is illustrated in Fig. 1-10d. Here binary 10010.011011 is to be translated into hexadecimal. First the binary number is divided into groups of four bits, starting at the binary point. Three 0s are added in the leftmost group, forming 0001. Two 0s are added to the rightmost group, forming 1100. Each group now has 4 bits and is translated into a hexadecimal digit as shown in Fig. 1-10d. The binary number 10010.011011 then equals $12.6C_{16}$.

As a practical matter, many modern hand-held calculators perform number base conversions. Most can convert between decimal, hexadecimal, octal, and binary. These calculators can also perform arithmetic operations in various bases (such as hexadecimal).

SOLVED PROBLEMS

1.15   The hexadecimal number system is sometimes called the base _____ system.

Solution:

The hexadecimal number system is sometimes called the base 16 system.

1.16   List the 16 symbols used in the hexadecimal number system.

Solution:

Refer to Fig. 1-7. The 16 symbols used in the hexadecimal number system are 0, 1, 2, 3, 4, 5, 6, 7, 8, 9, A, B, C, D, E, and F.

1.17   Convert the following whole hexadecimal numbers to their decimal equivalents:
(*a*) C, (*b*) 9F, (*c*) D52, (*d*) 67E, (*e*) ABCD.

Solution:

Follow the procedure shown in Fig. 1-8*a*. Refer also to Fig. 1-7. The decimal equivalents of the hexadecimal numbers are as follows:
(*a*) $C_{16} = 12_{10}$ (*c*) $D52_{16} = 3410_{10}$ (*e*) $ABCD_{16} = 43981_{10}$
(*b*) $9F_{16} = 159_{10}$ (*d*) $67E_{16} = 1662_{10}$

1.18   Convert the following hexadecimal numbers to their decimal equivalents:
(*a*) F.4, (*b*) D3.E, (*c*) 1111.1, (*d*) 888.8, (*e*) EBA.C.

Solution:

Follow the procedure shown in Fig. 1-8*b*. Refer also to Fig. 1-7. The decimal equivalents of the hexadecimal numbers are as follows:
(*a*) $F.4_{16} = 15.25_{10}$ (*c*) $1111.1_{16} = 4369.0625_{10}$ (*e*) $EBA.C_{16} = 3770.75_{10}$
(*b*) $D3.E_{16} = 211.875_{10}$ (*d*) $888.8_{16} = 2184.5_{10}$

1.19   Convert the following whole decimal numbers to their hexadecimal equivalents:
(*a*) 8, (*b*) 10, (*c*) 14, (*d*) 16, (*e*) 80, (*f*) 2560, (*g*) 3000, (*h*) 62,500.

Solution:

Follow the procedure shown in Fig. 1-9*a*. Refer also to Fig. 1-7. The hexadecimal equivalents of the decimal numbers are as follows:
(*a*) $8_{10} = 8_{16}$ (*c*) $14_{10} = E_{16}$ (*e*) $80_{10} = 50_{16}$ (*g*) $3000_{10} = BB8_{16}$
(*b*) $10_{10} = A_{16}$ (*d*) $16_{10} = 10_{16}$ (*f*) $2560_{10} = A00_{16}$ (*h*) $62\,500_{10} = F424_{16}$

1.20   Convert the following decimal numbers to their hexadecimal equivalents:
(*a*) 204.125, (*b*) 255.875, (*c*) 631.25, (*d*) 10 000.003 906 25.

Solution:

Follow the procedure shown in Fig. 1-9*b*. Refer also to Fig. 1-7. The hexadecimal equivalents of the decimal numbers are as follows:
(*a*) $204.125_{10} = CC.2_{16}$ (*c*) $631.25_{10} = 277.4_{16}$
(*b*) $255.875_{10} = FF.E_{16}$ (*d*) $10\,000.003\,906\,25_{10} = 2710.01_{16}$

1.21   Convert the following hexadecimal numbers to their binary equivalents:
(*a*) B, (*b*) E, (*c*) 1C, (*d*) A64, (*e*) 1F.C, (*f*) 239.4

Solution:

Follow the procedure shown in Fig. 1-10a and b. Refer also to Fig. 1-7. The binary equivalents of the hexadecimal numbers are as follows:

(a) $B_{16} = 1011_2$ (c) $1C_{16} = 11100_2$ (e) $1F.C_{16} = 11111.11_2$

(b) $E_{16} = 1110_2$ (d) $A64_{16} = 101001100100_2$ (f) $239.4_{16} = 1000111001.01_2$

1.22 Convert the following binary numbers to their hexadecimal equivalents:

  (a) 1001.1111   (c) 110101.011001   (e) 10100111.111011

  (b) 10000001.1101   (d) 10000.1   (f) 1000000.0000111

Solution:

Follow the procedure shown in Fig. 1-10c and d. Refer also to Fig. 1-7. The hexadecimal equivalents of the binary numbers are as follows:

(a) $1001.1111_2 = 9.F_{16}$ (c) $110101.011001_2 = 35.64_{16}$ (e) $10100111.111011_2 = A7.EC_{16}$

(b) $10000001.1101_2 = 81.D_{16}$ (d) $10000.1_2 = 10.8_{16}$ (f) $1000000.0000111_2 = 40.0E_{16}$

1-4   2s COMPLEMENT NUMBERS

The 2s complement method of representing numbers is widely used in microprocessor-based equipment. Until now, we have assumed that all numbers are positive. However, microprocessors must process both positive and negative numbers. By using *2s complement representation*, the *sign as well as the magnitude* of a number can be determined.

Assume a microprocessor register 8 bits wide such as that shown in Fig. 1-11a. The most-significant bit (MSB) is the *sign bit*. If this bit is 0, then the number is $(+)$ positive. However, if the sign bit is 1, then the number is $(-)$ negative. The other 7 bits in this 8-bit register represent the magnitude of the number.

The table in Fig. 1-11b shows the 2s complement representations for some positive and negative numbers. For instance, a $+127$ is represented by the 2s complement number 01111111. A decimal -128 is represented by the 2s complement number 10000000. Note that *the 2s complement representations for all positive values are the same as the binary equivalents* for that decimal number.

Convert the signed decimal -1 to a 2s complement number. Follow Fig. 1-12 as you make the conversion in the next five steps.

Step 1.   Separate the sign and magnitude part of -1. The negative sign means the sign bit will be 1 in the 2s complement representation.

Step 2.   Convert decimal 1 to its 7-bit binary equivalent. In this example decimal 1 equals 0000001 in binary.

Step 3.   Convert binary 0000001 to its 1s complement form. In this example binary 0000001 equals 1111110 in 1s complement. Note that each 0 is changed to a 1 and each 1 to a 0.

Step 4.   Convert the 1s complement to its 2s complement form. In this example 1s complement 1111110 equals 1111111 in 2s complement. Add $+1$ to the 1s complement to get the 2s complement number.

Step 5.   The 7-bit 2s complement number (1111111 in this example) becomes the magnitude part of the entire 8-bit 2s complement number.

The result is that the signed decimal -1 equals 11111111 in 2s complement notation. The 2s complement number is shown in the register near the top of Fig. 1-12.

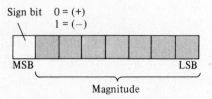

(a) The MSB of an 8-bit register is the sign bit

Signed decimal	8-bit 2s complement representation		
+127	0	111	1111
+126	0	111	1110
+125	0	111	1101
+124	0	111	1100
⋮	⋮	⋮	⋮
+5	0	000	0101
+4	0	000	0100
+3	0	000	0011
+2	0	000	0010
+1	0	000	0001
+0	0	000	0000
−1	1	111	1111
−2	1	111	1110
−3	1	111	1101
−4	1	111	1100
−5	1	111	1011
⋮	⋮	⋮	⋮
−125	1	000	0011
−126	1	000	0010
−127	1	000	0001
−128	1	000	0000
	Sign	Magnitude	

same as binary numbers

(b) 2s complement representations of positive and negative numbers

Fig. 1-11

Reverse the process and convert the 2s complement 11111000 to a signed decimal number. Follow Fig. 1-13 as the conversion is made in the following four steps.

Step 1. Separate the sign bit from the magnitude part of the 2s complement number. The MSB is a 1; therefore, the sign of decimal number will be (−) negative.

Step 2. Take the 1s complement of the magnitude part. The 7-bit magnitude 1111000 equals 0000111 in 1s complement notation.

Step 3. Add +1 to the 1s complement number. Adding 0000111 to 1 gives us 0001000. The 7-bit number 0001000 is now in binary.

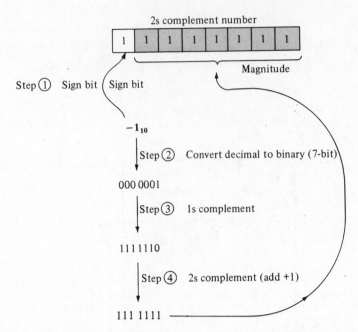

Fig. 1-12 Converting a signed decimal number to a 2s complement number

Step 4. Convert the binary number to its decimal equivalent. In this example, binary 0001000 equals 8 in decimal notation. The magnitude part of the number is 8.

The procedure in Fig. 1-13 shows how to convert 2s complement notation to negative signed decimal numbers. In this example, 2s complement 11111000 equals −8 in decimal notation.

Regular binary-to-decimal conversion (see Fig. 1-4) is used to convert 2s complements that equal positive decimal numbers. Remember that, for positive decimal numbers, the binary and 2s complement equivalents are the same.

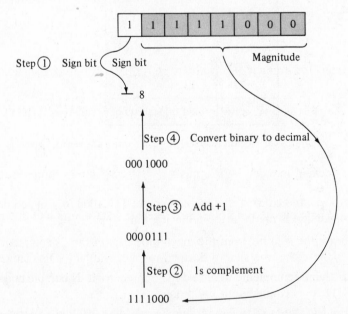

Fig. 1-13 Converting a 2s complement number to a signed decimal number

SOLVED PROBLEMS

1.23 The _____ (LSB, MSB) of a 2s complement number is the sign bit.

The MSB (most-significant bit) of a 2s complement number is the sign bit.

1.24 The 2s complement number 10000000 is equal to _____ in signed decimal.

Solution:

Follow the procedure shown in Fig. 1-13. The 2s complement number 10000000 equals −128 in decimal.

1.25 The number 01110000 is equal to _____ in signed decimal.

Solution:

The 0 in the MSB position means this is a positive number, and conversion to decimal follows the rules used in binary-to-decimal conversion. The number 01110000 is equal to +112 in signed decimal.

1.26 The signed decimal number +75 equals _____ in 8-bit 2s complement.

Solution:

Follow the procedure shown in Fig. 1-4. Decimal +75 equals 01001011 in 2s complement and binary.

1.27 The 2s complement number 11110001 is equal to _____ in signed decimal.

Solution:

Follow the procedure shown in Fig. 1-13. The 2s complement number 11110001 is equal to −15 in signed decimal.

1.28 The signed decimal number −35 equals _____ in 8-bit 2s complement.

Solution:

Follow the procedure shown in Fig. 1-12. Decimal −35 equals 11011101 in 2s complement.

1.29 The signed decimal number −100 equals _____ in 8-bit 2s complement.

Solution:

Follow the procedure shown in Fig. 1-12. Decimal −100 equals 10011100 in 2s complement.

1.40 The signed decimal number +20 equals _____ in 8-bit 2s complement.

Solution:

Follow the procedure shown in Fig. 1-4. Decimal +20 equals 00010100 in 2s complement and binary

Supplementary Problems

1.31 The number system with a radix of 2 is called the _____ number system.   *Ans.* binary

1.32 The number system with a radix of 10 is called the _____ number system.   *Ans.* decimal

1.33 The number system with a radix of 8 is called the _____ number system.   *Ans.* octal

1.34 The number system with a radix of 16 is called the _____ number system.   *Ans.* hexadecimal

1.35 A *binary digit* is sometimes shortened and called a(n) _____.   *Ans.* bit

1.36 How would you pronounce the number 1101 in (*a*) binary and (*b*) decimal?
Ans. (*a*) one, one, zero, one   (*b*) one thousand one hundred and one

1.37 The number 1010_2 is a base ___(*a*)___ number and is pronounced ___(*b*)___ .
Ans. (*a*) 2   (*b*) one, zero, one, zero

1.38 Convert the following binary numbers to their decimal equivalents:
(*a*) 00001110, (*b*) 11100000, (*c*) 10000011, (*d*) 10011010.
Ans. (*a*) $00001110_2 = 14_{10}$ (*c*) $10000011_2 = 131_{10}$
       (*b*) $11100000_2 = 224_{10}$ (*d*) $10011010_2 = 154_{10}$

1.39 $110011.11_2 =$ _____ $_{10}$ *Ans.* 51.75

1.40 $11110000.0011_2 =$ _____ $_{10}$ *Ans.* 240.1875

1.41 Convert the following decimal numbers to their binary equivalents:
(*a*) 32, (*b*) 200, (*c*) 170, (*d*) 258.
Ans. (*a*) $32_{10} = 100000_2$ (*c*) $170_{10} = 10101010_2$
       (*b*) $200_{10} = 11001000_2$ (*d*) $258_{10} = 100000010_2$

1.42 $40.875_{10} =$ _____ $_2$ *Ans.* 101000.111

1.43 $999.125_{10} =$ _____ $_2$ *Ans.* 1111100111.001

1.44 Convert the following hexadecimal numbers to their decimal equivalents:
(*a*) 13AF, (*b*) 25E6, (*c*) B4.C9, (*d*) 78.D3.
Ans. (*a*) $13AF_{16} = 5039_{10}$ (*c*) $B4.C9_{16} = 180.78515_{10}$
       (*b*) $25E6_{16} = 9702_{10}$ (*d*) $78.D3_{16} = 120.82421_{10}$

1.45 Convert the following decimal numbers to their hexadecimal equivalents:
(*a*) 3016, (*b*) 64881, (*c*) 17386.75, (*d*) 9817.625.
Ans. (*a*) $3016_{10} = BC8_{16}$ (*c*) $17386.75_{10} = 43EA.C_{16}$
       (*b*) $64881_{10} = FD71_{16}$ (*d*) $9817.625_{10} = 2659.A_{16}$

1.46 Convert the following hexadecimal numbers to their binary equivalents:
(*a*) A6, (*b*) 19, (*c*) E5.04, (*d*) 1B.78.
Ans. (*a*) $A6_{16} = 10100110_2$ (*c*) $E5.04_{16} = 11100101.000001_2$
       (*b*) $19_{16} = 11001_2$ (*d*) $1B.78_{16} = 11011.01111_2$

1.47 Convert the following binary numbers to their hexadecimal equivalents:
(*a*) 11110010, (*b*) 11011001, (*c*) 111110.000011, (*d*) 10001.11111.
Ans. (*a*) $11110010_2 = F2_{16}$ (*c*) $111110.000011_2 = 3E.0C_{16}$
       (*b*) $11011001_2 = D9_{16}$ (*d*) $10001.11111_2 = 11.F8_{16}$

1.48   When 2s complement notation is used, the MSB is the _____ bit.   *Ans.*   sign

1.49   Convert the following signed decimal numbers to their 8-bit 2s complement equivalents:
(*a*)   +13,   (*b*)   +110,   (*c*)   −25,   (*d*)   −90.
Ans.   (*a*)   00001101   (*b*)   01101110   (*c*)   11100111   (*d*)   10100110

1.50   Convert the following 2s complement numbers to their signed decimal equivalents:
(*a*)   01110000,   (*b*)   00011111,   (*c*)   11011001,   (*d*)   11001000.
Ans.   (*a*)   +112   (*b*)   +31   (*c*)   −39   (*d*)   −56

Chapter 2

Binary Codes

2-1 INTRODUCTION

Digital systems process only codes consisting of 0s and 1s (binary codes). That is due to the bistable nature of digital electronic circuits. The straight binary code was discussed in Chap. 1. Several other, special binary codes have evolved over the years to perform specific functions in digital equipment. All those codes use 0s and 1s, but their meanings may vary. Several binary codes will be detailed here, along with the methods used to translate them into decimal form. In a digital system, electronic translators (called *encoders* and *decoders*) are used for converting from code to code. The following sections will detail the process of conversion from one code to another.

2-2 WEIGHTED BINARY CODES

Straight binary numbers are somewhat difficult for people to understand. For instance, try to convert the binary number 10010110_2 to a decimal number. It turns out that $10010110_2 = 150_{10}$, but it takes quite a lot of time and effort to make this conversion without a calculator.

The *binary-coded decimal* (BCD) *code* makes conversion to decimals much easier. Figure 2-1 shows the 4-bit BCD code for the decimal digits 0-9. Note that the BCD code is a weighted code. The most significant bit has a weight of 8, and the least significant bit has a weight of only 1. This code is more precisely known as the *8421 BCD code*. The 8421 part of the name gives the weighting of each place in the 4-bit code. There are several other BCD codes that have other weights for the four place values. Because the 8421 BCD code is most popular, it is customary to refer to it simply as the BCD code.

Decimal	BCD 8s 4s 2s 1s
0	0 0 0 0
1	0 0 0 1
2	0 0 1 0
3	0 0 1 1
4	0 1 0 0
5	0 1 0 1
6	0 1 1 0
7	0 1 1 1
8	1 0 0 0
9	1 0 0 1

Fig. 2-1 The 8421 BCD code

How is the decimal number 150 expressed as a BCD number? Figure 2-2*a* shows the very simple technique for converting decimal numbers to BCD (8421) numbers. Each decimal digit is converted to its 4-bit BCD equivalent (see Fig. 2-1). The decimal number 150 then equals the BCD number 000101010000.

Converting BCD numbers to decimal numbers also is quite simple. Figure 2-2*b* shows the technique. The BCD number 10010110 is first divided into groups of 4 bits starting at the binary point. Each group of 4 bits is then converted to its equivalent decimal digit, which is recorded below. The BCD number 10010110 then equals decimal 96.

16

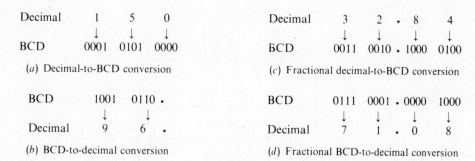

(a) Decimal-to-BCD conversion

(c) Fractional decimal-to-BCD conversion

(b) BCD-to-decimal conversion

(d) Fractional BCD-to-decimal conversion

Fig. 2-2

Figure 2-2c illustrates a fractional decimal number being converted to its BCD equivalent. Each decimal digit is converted to its BCD equivalent. The decimal point is dropped down and becomes the binary point. Figure 2-2c shows that decimal 32.84 equals the BCD number 00110010.10000100.

Convert the fractional BCD number 01110001.00001000 to its decimal equivalent. Figure 2-2d shows the procedure. The BCD number is first divided into groups of 4 bits starting at the binary point. Each group of four bits is then converted to its decimal equivalent. The binary point becomes the decimal point in the decimal number. Figure 2-2d shows the BCD number 01110001.00001000 being translated into its decimal equivalent of 71.08.

Consider converting a BCD number to its straight binary equivalent. Figure 2-3 shows the three-step procedure. Step 1 shows the BCD number being divided into 4-bit groups starting from the binary point. Each 4-bit group is translated into its decimal equivalent. Step 1 in Fig. 2-3 shows the BCD number 000100000011.0101 being translated into the decimal number 103.5.

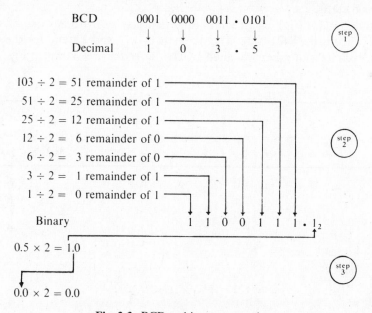

Fig. 2-3 BCD-to-binary conversion

Step 2 in Fig. 2-3 shows the integer part of the decimal number being translated into binary. The 103_{10} is converted into 1100111_2 in step 2 by the repeated divide-by-2 procedure.

Step 3 in Fig. 2-3 illustrates the fractional part of the decimal number being translated into binary. The 0.5_{10} is converted into 0.1_2 in step 3 by the repeated multiply-by-2 procedure. The integer and fractional parts of the binary number are joined. The BCD number 000100000011.0101 then equals the binary number 1100111.1.

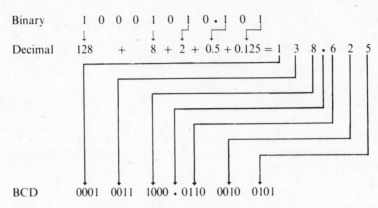

Fig. 2-4 Binary-to-BCD conversion

Note that it is usually more efficient to write down a figure in straight binary numbers than in BCD numbers. Binary numbers usually contain fewer 1s and 0s, as seen in the conversion in Fig. 2-3. Although longer, BCD numbers are used in digital systems when numbers must be easily converted to decimals.

Translate the binary number 10001010.101 into its BCD (8421) equivalent. The procedure is shown in Fig. 2-4. The binary number is first converted to its decimal equivalent. The binary number 10001010.101 then equals 138.625_{10}. Each decimal digit is then translated into its BCD equivalent. Figure 2-4 shows decimal 138.625 being converted into the BCD number 000100111000.011000100101. The entire conversion, then, translates binary 10001010.101_2 into the BCD number 000100111000.011000100101.

"Binary-coded decimal (BCD)" is a general term that may apply to any one of several codes. The most popular BCD code is the 8421 code. The numbers 8, 4, 2, and 1 stand for the weight of each bit in the 4-bit group. Examples of other weighted BCD 4-bit codes are shown in Fig. 2-5.

Decimal	8421 BCD		4221 BCD		5421 BCD	
	8s 4s 2s 1s	8s 4s 2s 1s	4s 2s 2s 1s	4s 2s 2s 1s	5s 4s 2s 1s	5s 4s 2s 1s
0		0 0 0 0		0 0 0 0		0 0 0 0
1		0 0 0 1		0 0 0 1		0 0 0 1
2		0 0 1 0		0 0 1 0		0 0 1 0
3		0 0 1 1		0 0 1 1		0 0 1 1
4		0 1 0 0		1 0 0 0		0 1 0 0
5		0 1 0 1		0 1 1 1		1 0 0 0
6		0 1 1 0		1 1 0 0		1 0 0 1
7		0 1 1 1		1 1 0 1		1 0 1 0
8		1 0 0 0		1 1 1 0		1 0 1 1
9		1 0 0 1		1 1 1 1		1 1 0 0
10	0 0 0 1	0 0 0 0	0 0 0 1	0 0 0 0	0 0 0 1	0 0 0 0
11	0 0 0 1	0 0 0 1	0 0 0 1	0 0 0 1	0 0 0 1	0 0 0 1
12	0 0 0 1	0 0 1 0	0 0 0 1	0 0 1 0	0 0 0 1	0 0 1 0
13	0 0 0 1	0 0 1 1	0 0 0 1	0 0 1 1	0 0 0 1	0 0 1 1

Fig. 2-5 Three weighted BCD codes

SOLVED PROBLEMS

2.1   The letters BCD stand for _____ - _____ _____.

Solution:

The letters BCD stand for binary-coded decimal.

2.2   Convert the following 8421 BCD numbers to their decimal equivalents:
(*a*)   1010   (*b*)   00010111   (*c*)   10000110   (*d*)   010101000011   (*e*)   00110010.10010100
(*f*)   0001000000000000.0101

Solution:

The decimal equivalents of the BCD numbers are as follows:
(*a*)   1010 = ERROR (no such BCD number)   (*d*)   010101000011 = 543
(*b*)   00010111 = 17   (*e*)   00110010.10010100 = 32.94
(*c*)   10000110 = 86   (*f*)   0001000000000000.0101 = 1000.5

2.3   Convert the following decimal numbers to their 8421 BCD equivalents:
(*a*)   6,   (*b*)   13,   (*c*)   99.9,   (*d*)   872.8,   (*e*)   145.6   (*f*)   21.001.

Solution:

The BCD equivalents of the decimal numbers are as follows:
(*a*)   6 = 0110   (*c*)   99.9 = 10011001.1001   (*e*)   145.6 = 000101000101.0110
(*b*)   13 = 00010011   (*d*)   872.8 = 100001110010.1000   (*f*)   21.001 = 00100001.000000000001

2.4   Convert the following binary numbers to their 8421 BCD equivalents:
(*a*)   10000,   (*b*)   11100.1,   (*c*)   101011.01,   (*d*)   100111.11,   (*e*)   1010.001,
(*f*)   1111110001.

Solution:

The BCD equivalents of the binary numbers are as follows:
(*a*)   10000 = 00010110   (*d*)   100111.11 = 00111001.01110101
(*b*)   11100.1 = 00101000.0101   (*e*)   1010.001 = 00010000.000100100101
(*c*)   101011.01 = 01000011.00100101   (*f*)   1111110001 = 0001000000001001

2.5   Convert the following 8421 BCD numbers to their binary equivalents:
(*a*)   00011000   (*b*)   01001001   (*c*)   0110.01110101   (*d*)   00110111.0101
(*e*)   01100000.00100101   (*f*)   0001.001101110101

Solution:

The binary equivalents of the BCD numbers are as follows:
(*a*)   00011000 = 10010   (*d*)   00110111.0101 = 100101.1
(*b*)   01001001 = 110001   (*e*)   01100000.00100101 = 111100.01
(*c*)   0110.01110101 = 110.11   (*f*)   0001.001101110101 = 1.011

2.6   List three weighted BCD codes.

Solution:

Three BCD codes are: (*a*)   8421 BCD code,   (*b*)   4221 BCD code,   (*c*)   5421 BCD code.

2.7   The 4221 BCD equivalent of decimal 98 is _____ .

Solution:

The 4221 BCD equivalent of decimal 98 is 11111110.

2.8   The 5421 BCD equivalent of decimal 75 is _____ .

Solution:

The 5421 BCD equivalent of decimal 75 is 10101000.

2.9   What kind of number (BCD or binary) would be easier for a worker to translate to decimal?

Solution:

BCD numbers are easiest to translate to their decimal equivalents.

2-3   NONWEIGHTED BINARY CODES

Some binary codes are nonweighted. Each bit therefore has no special weighting. Two such nonweighted codes are the excess-3 and Gray codes.

The *excess-3* (XS3) code is related to the 8421 BCD code because of its binary-coded-decimal nature. In other words, each 4-bit group in the XS3 code equals a specific decimal digit. Figure 2-6 shows the XS3 code along with its 8421 BCD and decimal equivalents. Note that the XS3 number is always *3 more* than the 8421 BCD number.

Decimal	8421 BCD		XS3 BCD	
	10s	1s	10s	1s
0		0000	0011	0011
1		0001	0011	0100
2		0010	0011	0101
3		0011	0011	0110
4		0100	0011	0111
5		0101	0011	1000
6		0110	0011	1001
7		0111	0011	1010
8		1000	0011	1011
9		1001	0011	1100
10	0001	0000	0100	0011
11	0001	0001	0100	0100

Fig. 2-6 The excess-3 (XS3) code

Consider changing the decimal number 62 to an equivalent XS3 number. Step 1 in Fig. 2-7*a* shows 3 being added to each decimal digit. Step 2 shows 9 and 5 being converted to their 8421 BCD equivalents. The decimal number 62 then equals the BCD XS3 number 10010101.

Convert the 8421 BCD number 01000000 to its XS3 equivalent. Figure 2-7*b* shows the simple procedure. The BCD number is divided into 4-bit groups starting at the binary point. Step 1 shows 3 (binary 0011) being added to each 4-bit group. The sum is the resulting XS3 number. Figure 2-7*b* shows the 8421 BCD number 01000000 being converted to its equivalent BCD XS3 number, which is 01110011.

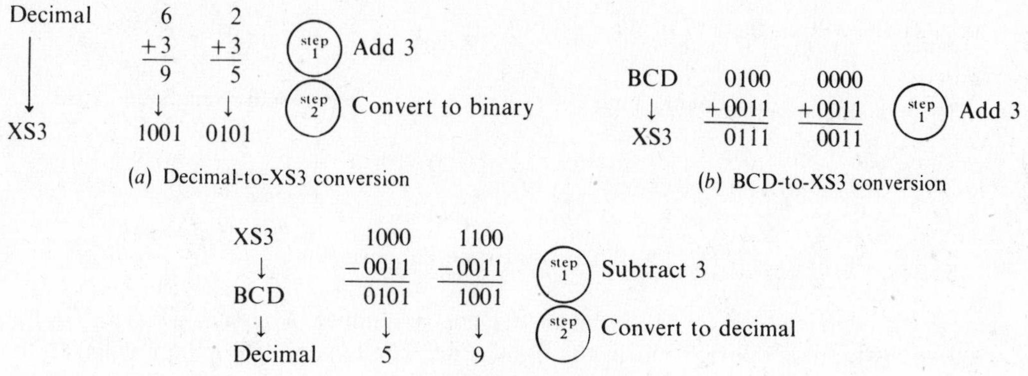

Decimal   6   2   (step 1) Add 3
  | +3   +3
  ↓   9   5   (step 2) Convert to binary
 XS3   1001 0101

(a) Decimal-to-XS3 conversion

BCD   0100   0000
 ↓   +0011   +0011   (step 1) Add 3
XS3   0111   0011

(b) BCD-to-XS3 conversion

XS3   1000   1100
 ↓   -0011   -0011   (step 1) Subtract 3
BCD   0101   1001
 ↓   (step 2) Convert to decimal
Decimal   5   9

(c) XS3-to-decimal conversion

Fig. 2-7

Consider the conversion from XS3 code to decimal. Figure 2-7c shows the XS3 number 10001100 being converted to its decimal equivalent. The XS3 number is divided into 4-bit groups starting at the binary point. Step 1 shows 3 (binary 0011) being subtracted from each 4-bit group. An 8421 BCD number results. Step 2 shows each 4-bit group in the 8421 BCD number being translated into its decimal equivalent. The XS3 number 10001100 is equal to decimal 59 according to the procedure in Fig. 2-7c.

The XS3 code has significant value in arithmetic circuits. The value of the code lies in its ease of complementing. If each bit is complemented (0s to 1s and 1s to 0s), the resulting 4-bit word will be the 9s complement of the number. Adders can use 9s complement numbers to perform subtraction.

The *Gray code* is another nonweighted binary code. The Gray code is not a BCD-type code. Figure 2-8 compares the Gray code with equivalent binary and decimal numbers. Look carefully at the Gray code. Note that each increase in count (increment) is accompanied by *only 1 bit changing state*. Look at the change from the decimal 7 line to the decimal 8 line. In binary all four bits change state (from 0111 to 1000). In this same line the Gray code has only the left bit changing state (0100 to 1100). This change of a single bit in the code group per increment characteristic is important in some applications in digital electronics.

Decimal	Binary	Gray code	Decimal	Binary	Gray code
0	0000	0000	8	1000	1100
1	0001	0001	9	1001	1101
2	0010	0011	10	1010	1111
3	0011	0010	11	1011	1110
4	0100	0110	12	1100	1010
5	0101	0111	13	1101	1011
6	0110	0101	14	1110	1001
7	0111	0100	15	1111	1000

Fig. 2-8 The Gray code

Consider converting a binary number to its Gray code equivalent. Figure 2-9a shows the binary number 0010 being translated into its Gray code equivalent. Start at the MSB of the binary number. Transfer this to the left position in the Gray code as shown by the downward arrow. Now *add* the 8s bit to the next bit over (4s bit). The sum is 0 (0 + 0 = 0), which is transferred down and written as the second bit from the left in the Gray code. The 4s bit is now added to the 2s bit of the binary number. The sum is 1 (0 + 1 = 1) and is transferred down and written as the third bit from the left in the Gray

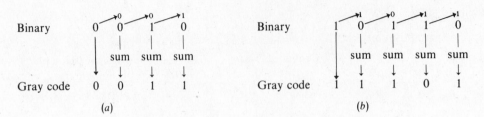

Fig. 2-9 Binary-to-Gray code conversion

code. The 2s bit is now added to the 1s bit of the binary number. The sum is 1 $(1 + 0 = 1)$ and is transferred and written as the right bit in the Gray code. The binary 0010 is then equal to the Gray code number 0011. This can be verified in the decimal 2 line of the table in Fig. 2-8.

The rules for converting from any binary number to its equivalent Gray code number are as follows:

1. The left bit is the same in the Gray code as in the binary number.
2. Add the MSB to the bit on its immediate right and record the sum (neglect any carry) below in the Gray code line.
3. Continue adding bits to the bits on their right and recording sums until the LSB is reached.
4. The Gray code number will always have the same number of bits as the binary number.

Try these rules when converting binary 10110 to its Gray code equivalent. Figure 2-9b shows the MSB(1) in the binary number being transferred down and written as part of the Gray code number. The 16s bit is then added to the 8s bit of the binary number. The sum is 1 $(1 + 0 = 1)$, which is recorded in the Gray code (second bit from left). Next the 8s bit is added to the 4s bit of the binary number. The sum is 1 $(0 + 1 = 1)$, which is recorded in the Gray code (third bit from the left). Next the 4s bit is added to the 2s bit of the binary number. The sum is 0 $(1 + 1 = 10)$ because the carry is dropped. The 0 is recorded in the second position from the right in the Gray code. Next the 2s bit is added to the 1s bit of the binary number. The sum is 1 $(1 + 0 = 1)$, which is recorded in the Gray code (right bit). The process is complete. Figure 2-9b shows the binary number 10110 being translated into the Gray code number 11101.

Convert the Gray code number 1001 to its equivalent binary number. Figure 2-10a details the procedure. First the left bit (1) is transferred down to the binary line to form the 8s bit. The 8s bit of the binary number is transferred (see arrow) up above the next Gray code bit, and the two are added. The sum is 1 $(1 + 0 = 1)$, which is written in the 4s bit place in the binary number. The 4s bit (1) is then added to the next Gray code bit. The sum is 1 $(1 + 0 = 1)$. This 1 is written in the 2s place of the binary number. The binary 2s bit (1) is added to the right Gray code bit. The sum is 0 $(1 + 1 = 10)$ because the carry is neglected. This 0 is written in the 1s place of the binary number. Figure 2-10a shows the Gray code number 1001 being translated into its equivalent binary number 1110. This conversion can be verified by looking at decimal line 14 in Fig. 2-8.

Convert the 6-bit Gray code number 011011 to its 6-bit binary equivalent. Start at the left and follow the arrows in Fig. 2-10b. Follow the procedure, remembering that a $1 + 1 = 10$. The carry of 1 is neglected, and the 0 is recorded on the binary line. Figure 2-10b shows that the Gray code number 011011 is equal to the binary number 010010.

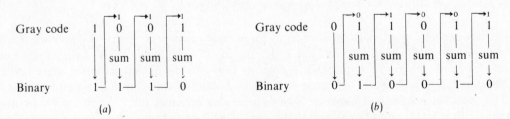

Fig. 2-10 Gray code-to-binary conversions

SOLVED PROBLEMS

2.10 The letters and numbers XS3 stand for _____ – _____ code.

Solution:

    XS3 stands for the excess-3 code.

2.11 The _____ (8421, XS3) BCD code is an example of a nonweighted code.

Solution:

    The XS3 BCD code is an example of a nonweighted code.

2.12 The _____ (Gray, XS3) code is a BCD code.

Solution:

    The XS3 code is a BCD code.

2.13 Convert the following decimal numbers to their XS3 code equivalents:
(*a*) 9, (*b*) 18, (*c*) 37, (*d*) 42, (*e*) 650.

Solution:

    The XS3 equivalents of the decimal numbers are as follows:
(*a*) $9 = 1100$ (*c*) $37 = 01101010$ (*e*) $650 = 100110000011$
(*b*) $18 = 01001011$ (*d*) $42 = 01110101$

2.14 Convert the following 8421 BCD numbers to their XS3 code equivalents:
(*a*) 0001, (*b*) 0111, (*c*) 01100000, (*d*) 00101001, (*e*) 10000100.

Solution:

    The XS3 equivalents of the 8421 BCD numbers are as follows:
(*a*) $0001 = 0100$ (*c*) $01100000 = 10010011$ (*e*) $10000100 = 10110111$
(*b*) $0111 = 1010$ (*d*) $00101001 = 01011100$

2.15 Convert the following XS3 numbers to their decimal equivalents:
(*a*) 0011, (*b*) 01100100, (*c*) 11001011, (*d*) 10011010, (*e*) 10000101.

Solution:

    The decimal equivalents of the XS3 numbers are as follows:
(*a*) $0011 = 0$ (*c*) $11001011 = 98$ (*e*) $10000101 = 52$
(*b*) $01100100 = 31$ (*d*) $10011010 = 67$

2.16 The _____ (Gray, XS3) code is usually used in arithmetic applications in digital circuits.

Solution:

    The XS3 code is usually used in arithmetic applications.

2.17 Convert the following straight binary numbers to their Gray code equivalents:
(*a*) 1010, (*b*) 10000, (*c*) 10001, (*d*) 10010, (*e*) 10011.

Solution:

    The Gray code equivalents of the binary numbers are as follows:
(*a*) $1010 = 1111$ (*c*) $10001 = 11001$ (*e*) $10011 = 11010$
(*b*) $10000 = 11000$ (*d*) $10010 = 11011$

2.18 Convert the following Gray code numbers to their straight binary equivalents:
(a) 0100, (b) 11111, (c) 10101, (d) 110011, (e) 011100.

Solution:

The binary equivalents of the Gray code numbers are as follows:
(a) 0100 = 0111   (c) 10101 = 11001   (e) 011100 = 010111
(b) 11111 = 10101   (d) 110011 = 100010

2.19 The Gray code's most important characteristic is that, when the count is incremented by 1, _____ (more than, only) 1 bit will change state.

Solution:

The Gray code's most important characteristic is that, when the count is incremented by 1, only 1 bit will change state.

2-4 ALPHANUMERIC CODES

Binary 0s and 1s have been used to represent various numbers to this point. Bits can also be coded to represent letters of the alphabet, numbers, and punctuation marks. One such 7-bit code is the *American Standard Code for Information Interchange* (ASCII, pronounced "ask-ee"), shown in Fig. 2-11. Note that the letter A is represented by 1000001, whereas B in the ASCII code is 1000010.

Character	ASCII		EBCDIC		Character	ASCII		EBCDIC	
Space	010	0000	0100	0000	A	100	0001	1100	0001
!	010	0001	0101	1010	B	100	0010	1100	0010
"	010	0010	0111	1111	C	100	0011	1100	0011
#	010	0011	0111	1011	D	100	0100	1100	0100
$	010	0100	0101	1011	E	100	0101	1100	0101
%	010	0101	0110	1100	F	100	0110	1100	0110
&	010	0110	0101	0000	G	100	0111	1100	0111
'	010	0111	0111	1101	H	100	1000	1100	1000
(	010	1000	0100	1101	I	100	1001	1100	1001
)	010	1001	0101	1101	J	100	1010	1101	0001
*	010	1010	0101	1100	K	100	1011	1101	0010
+	010	1011	0100	1110	L	100	1100	1101	0011
,	010	1100	0110	1011	M	100	1101	1101	0100
-	010	1101	0110	0000	N	100	1110	1101	0101
.	010	1110	0100	1011	O	100	1111	1101	0110
/	010	1111	0110	0001	P	101	0000	1101	0111
0	011	0000	1111	0000	Q	101	0001	1101	1000
1	011	0001	1111	0001	R	101	0010	1101	1001
2	011	0010	1111	0010	S	101	0011	1110	0010
3	011	0011	1111	0011	T	101	0100	1110	0011
4	011	0100	1111	0100	U	101	0101	1110	0100
5	011	0101	1111	0101	V	101	0110	1110	0101
6	011	0110	1111	0110	W	101	0111	1110	0110
7	011	0111	1111	0111	X	101	1000	1110	0111
8	011	1000	1111	1000	Y	101	1001	1110	1000
9	011	1001	1111	1001	Z	101	1010	1110	1001

Fig. 2-11 Alphanumeric codes

The ASCII code is used extensively in small computer systems to translate from the keyboard characters to computer language. The chart in Fig. 2-11 is not a complete list of all the combinations in the ASCII code.

Codes that can represent both letters and numbers are called *alphanumeric codes*. Another alphanumeric code that is widely used is the *Extended Binary-Coded Decimal Interchange Code* (EBCDIC, pronounced "eb-si-dik"). Part of the EBCDIC code is shown in Fig. 2-11. Note that the EBCDIC code is an 8-bit code and therefore can have more variations and characters than the ASCII code can have. The EBCDIC code is used in many larger computer systems.

The alphanumeric ASCII code is the modern code for getting information into and out of microcomputers. ASCII is used when interfacing computer keyboards, printers, and video displays. ASCII has become the standard input/output code for microcomputers.

Other alphanumeric codes that you may encounter are:

1. 7-bit BCDIC (*Binary-Coded Decimal Interchange Code*).

2. 8-bit EBCDIC (*Extended Binary-Coded Decimal Interchange Code*). Used on some IBM equipment.

3. 7-bit Selectric. Used to control the spinning ball on IBM Selectric typewriters.

4. 12-bit Hollerith. Used on punched paper cards.

SOLVED PROBLEMS

2.20 Binary codes that can represent both numbers and letters are called _____ codes.

Solution:

Alphanumeric codes can represent both numbers and letters.

2.21 The following are abbreviations for what?
(*a*) ASCII (*b*) EBCDIC

Solution:

(*a*) ASCII = American Standard Code for Information Interchange
(*b*) EBCDIC = Extended Binary-Coded Decimal Interchange Code

2.22 Refer to Fig. 2-12. The ASCII keyboard-encoder output would be _____ if the K on the typewriter-like keyboard were pressed.

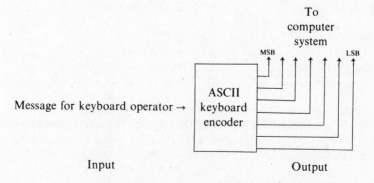

Fig. 2-12 ASCII keyboard-encoder system

Solution:

The ASCII output would be 1001011 if the K on the keyboard were pressed.

2.23 Refer to Fig. 2-12. List the 12 ASCII keyboard-encoder outputs for entering the message "pay $1000.00."

Solution:

The ASCII codes for the characters in the message are as follows:
(a) P = 1010000   (d) Space = 0100000   (g) 0 = 0110000   (j) . = 0101110
(b) A = 1000001   (e) $ = 0100100   (h) 0 = 0110000   (k) 0 = 0110000
(c) Y = 1011001   (f) 1 = 0110001   (i) 0 = 0110000   (l) 0 = 0110000

2.24 The _____ code is a 12-bit alphanumeric code used on punched paper cards.

Solution:

The 12-bit Hollerith code is used on punched cards.

2.25 The 7-bit _____ code is considered the industry standard for input/output on microcomputers.

Solution:

The 7-bit ASCII (American Standard Code for Information Interchange) code is considered the industry standard for microcomputer inputs and outputs.

Supplementary Problems

2.26 Electronic devices that translate from one code to another are called __(a)__ or __(b)__.
Ans. (a) encoders (b) decoders

2.27 Convert the following 8421 BCD numbers to their decimal equivalents:
(a) 10010000, (b) 11111111, (c) 0111.0011, (d) 01100001.00000101.
Ans. (a) 10010000 = 90   (c) 0111.0011 = 7.3
       (b) 11111111 = ERROR (no such BCD number)   (d) 01100001.00000101 = 61.05

2.28 Convert the following decimal numbers to their 8421 BCD equivalents:
(a) 10, (b) 342, (c) 679.8, (d) 500.6.
Ans. (a) 10 = 00010000   (c) 679.8 = 011001111001.1000
       (b) 342 = 001101000010   (d) 500.6 = 010100000000.0110

2.29 Convert the following binary numbers to their 8421 BCD equivalents:
(a) 10100, (b) 11011.1, (c) 100000.01, (d) 111011.11.
Ans. (a) 10100 = 00100000   (c) 100000.01 = 00110010.00100101
       (b) 11011.1 = 00100111.0101   (d) 111011.11 = 01011001.01110101

2.30 Convert the following 8421 BCD numbers to their binary equivalents:
(a) 01011000, (b) 000100000000, (c) 1001.01110101, (d) 0011.0000011000100101.
Ans. (a) 01011000 = 111010   (c) 1001.01110101 = 1001.11
       (b) 000100000000 = 1100100   (d) 0011.0000011000100101 = 11.0001

2.31 The 4221 BCD equivalent of decimal 74 is _____.   *Ans.* 11011000

2.32 The 5421 BCD equivalent of decimal 3210 is _____.   *Ans.* 0011001000010000

2.33 The BCD code is convenient when translations must be made to _____ (binary, decimal) numbers.
Ans. decimal

2.34   "Excess-3" code is often shortened to _____.   *Ans.*   XS3

2.35   Convert the following decimal numbers to their XS3 code equivalents:
(*a*)   7,   (*b*)   16,   (*c*)   32,   (*d*)   4089.
Ans.   (*a*)   7 = 1010   (*c*)   32 = 01100101
         (*b*)   16 = 01001001   (*d*)   4089 = 0111001110111100

2.36   Convert the following XS3 numbers to their decimal equivalents:
(*a*)   1100,   (*b*)   10101000,   (*c*)   100001110011,   (*d*)   0100101101100101.
Ans.   (*a*)   1100 = 9   (*c*)   100001110011 = 540
         (*b*)   10101000 = 75   (*d*)   0100101101100101 = 1832

2.37   Convert the following straight binary numbers to their Gray code equivalents:
(*a*)   0110,   (*b*)   10100,   (*c*)   10101,   (*d*)   10110.
Ans.   (*a*)   0110 = 0101   (*b*)   10100 = 11110   (*c*)   10101 = 11111   (*d*)   10110 = 11101

2.38   Convert the following Gray code numbers to their straight binary equivalents:
(*a*)   0001,   (*b*)   11100,   (*c*)   10100,   (*d*)   10101.
Ans.   (*a*)   0001 = 0001   (*b*)   11100 = 10111   (*c*)   10100 = 11000   (*d*)   10101 = 11001

2.39   EBCDIC is a(n) _____ -bit alphanumeric code used in some IBM equipment.
Ans.   8

2.40   The 7-bit alphanumeric _____ code serves as the industry standard for input/output on micro-computers.
Ans.   ASCII

Chapter 3

Basic Logic Gates

3-1 INTRODUCTION

The *logic gate* is the basic building block in digital systems. Logic gates operate with binary numbers. Gates are therefore referred to as binary logic gates. All voltages used with logic gates will be either HIGH or LOW. In this book, a HIGH voltage will mean a binary 1. A LOW voltage will mean a binary 0. Remember that logic gates are electronic circuits. These circuits will respond only to HIGH voltages (called 1s) or LOW (ground) voltages (called 0s).

All digital systems are constructed by using only three basic logic gates. These basic gates are called the AND gate, the OR gate, and the NOT gate. This chapter deals with these very important basic logic gates, or functions.

3-2 THE AND GATE

The AND gate is called the "all or nothing" gate. The schematic in Fig. 3-1a shows the idea of the AND gate. The lamp (Y) will light only when *both* input switches (A and B) are closed. All the possible combinations for switches A and B are shown in Fig. 3-1b. The table in this figure is called a *truth table*. The truth table shows that the output (Y) is *enabled* (lit) only when both inputs are closed.

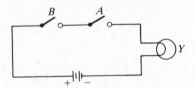

(a) AND circuit using switches

Input switches		Output light
B	A	Y
open	open	no
open	closed	no
closed	open	no
closed	closed	yes

(b) Truth table

Fig. 3-1

The standard *logic symbol* for the AND gate is drawn in Fig. 3-2a. This symbol shows the inputs as A and B. The output is shown as Y. This is the symbol for a 2-input AND gate. The truth table for the 2-input AND gate is shown in Fig. 3-2b. The inputs are shown as *binary* digits (bits). Note that only when both input A and input B are 1 will the output be 1. Binary 0 is defined as a LOW, or ground, voltage. Binary 1 is defined as a HIGH voltage. In this book, a HIGH voltage will mean about +5 volts (V) if the integrated circuits (ICs) being used are from the TTL family.

Boolean algebra is a form of symbolic logic that shows how logic gates operate. A *Boolean expression* is a "shorthand" method of showing what is happening in a logic circuit. The Boolean expression for the circuit in Fig. 3-2 is

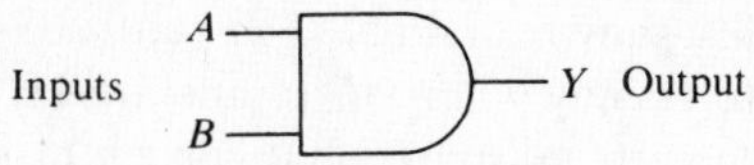

(a) AND-gate symbol

Inputs		Output
B	A	Y
0	0	0
0	1	0
1	0	0
1	1	1

0 = low voltage
1 = high voltage

(b) AND truth table

Fig. 3-2

$$A \cdot B = Y$$

The Boolean expression is read as A AND ($\cdot$ means AND) B equals the output Y. The dot ($\cdot$) means the logic function AND in Boolean algebra, *not* multiply as in regular algebra.

Sometimes the dot ($\cdot$) is left out of the Boolean expression. The Boolean expression for the 2-input AND gate is then:

$$AB = Y$$

The Boolean expression reads A AND B equals the output Y.

A logic circuit will often have three *variables*. Figure 3-3a shows the Boolean expression for a 3-input AND gate. The input variables are A, B, and C. The output is shown as Y. The logic symbol for this 3-input AND expression is drawn in Fig. 3-3b. The three inputs (A, B, C) are on the left of the symbol. The single output (Y) is on the right of the symbol. The truth table in Fig. 3-3c shows the eight possible combinations of the variables A, B, and C. Note that the top line in the table is the binary count 000. The binary count then proceeds upward to 001, 010, 011, 100, 101, 110, and finally 111. Note that *only* when all inputs are 1 is the output of the AND gate enabled with a 1.

$$A \cdot B \cdot C = Y$$

(a) Three-variable Boolean expression

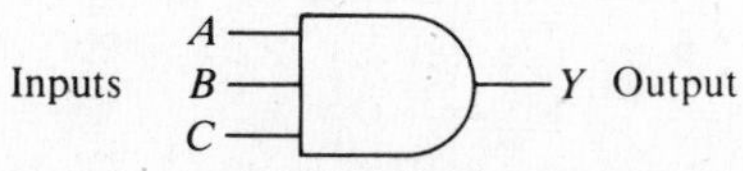

(b) 3-input AND gate symbol

Inputs			Output
C	B	A	Y
0	0	0	0
0	0	1	0
0	1	0	0
0	1	1	0
1	0	0	0
1	0	1	0
1	1	0	0
1	1	1	1

(c) Truth table with three variables

Fig. 3-3

Consider the AND truth tables shown in Figs. 3-2b and 3-3c. In each truth table the *unique output* from the AND gate is a HIGH only when *all* inputs are HIGH. Designers look at each gate's unique output when deciding which gate will perform a certain task.

The laws of Boolean algebra govern how AND gates operate. The formal laws for the *AND function* are:

$$A \cdot 0 = 0$$
$$A \cdot 1 = A$$
$$A \cdot A = A$$
$$A \cdot \overline{A} = 0$$

You can prove the accuracy of these laws by referring back to the truth table in Fig. 3-2. These are general statements that are always true about the AND function. AND gates must follow these laws. Note the bar over the variable in the last law. The bar over the variable means *not A*, or the opposite of *A*.

SOLVED PROBLEMS

3.1 Write the Boolean expression for a 4-input AND gate.

Solution:

$A \cdot B \cdot C \cdot D = Y$ or $ABCD = Y$

3.2 Draw the logic symbol for a 4-input AND gate.

Solution:

See Fig. 3-4.

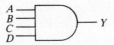

Fig. 3-4 Symbol for a 4-input AND gate

3.3 Draw a truth table for a 4-input AND gate.

Solution:

Inputs				Output	Inputs				Output
D	C	B	A	Y	D	C	B	A	Y
0	0	0	0	0	1	0	0	0	0
0	0	0	1	0	1	0	0	1	0
0	0	1	0	0	1	0	1	0	0
0	0	1	1	0	1	0	1	1	0
0	1	0	0	0	1	1	0	0	0
0	1	0	1	0	1	1	0	1	0
0	1	1	0	0	1	1	1	0	0
0	1	1	1	0	1	1	1	1	1

3.4   In Fig. 3-5, what would the output pulse train look like?

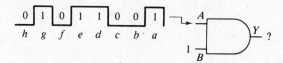

Fig. 3-5 Pulse-train problem

Solution:

In Fig. 3-5, the output waveform would look exactly like the input waveform at input A.

pulse $a = 1$ pulse $c = 0$ pulse $e = 1$ pulse $g = 1$
pulse $b = 0$ pulse $d = 1$ pulse $f = 0$ pulse $h = 0$

3.5   In Fig. 3-6, what would the output pulse train look like? Note that two pulse trains are being ANDed.

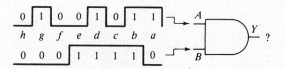

Fig. 3-6 Pulse-train problem

Solution:

In Fig. 3-6, the output pulses would be as follows:
pulse $a = 0$ pulse $c = 0$ pulse $e = 0$ pulse $g = 0$
pulse $b = 1$ pulse $d = 1$ pulse $f = 0$ pulse $h = 0$

3-3 THE OR GATE

The OR gate is called the "any or all" gate. The schematic in Fig. 3-7a shows the idea of the OR gate. The lamp (Y) will glow when either switch A or switch B is closed. The lamp will also glow when both switches A and B are closed. The lamp (Y) will *not* glow when both switches (A and B) are open. All the possible switch combinations are shown in Fig. 3-7b. The truth table details the *OR function* of the switch and lamp circuit. The output of the OR circuit will be enabled (lamp lit) when any or all input switches are closed.

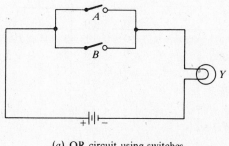

Input switches		Output light
B	A	Y
open	open	no
open	closed	yes
closed	open	yes
closed	closed	yes

(a) OR circuit using switches   (b) Truth table

Fig. 3-7

The standard logic symbol for an OR gate is drawn in Fig. 3-8a. Note the different shape of the OR gate. The OR gate has two inputs labeled A and B. The output is labeled Y. The shorthand Boolean expression for this OR function is given as $A + B = Y$. Note that the plus ($+$) symbol means OR in Boolean algebra. The expression ($A + B = Y$) is read as A OR ($+$ means OR) B equals output Y. You will note that the plus sign does *not* mean to add as it does in regular algebra.

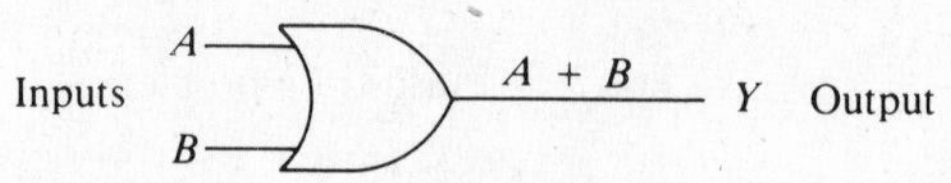

(a) OR-gate symbol

Inputs	Output
B A	Y
0 0	0
0 1	1
1 0	1
1 1	1

0 = low voltage
1 = high voltage

(b) OR truth table

Fig. 3-8

The truth table for the 2-input OR gate is drawn in Fig. 3-8b. The input *variables* (A and B) are given on the left. The resulting output (Y) is shown in the right column of the table. The OR gate is *enabled* (output is 1) anytime a 1 appears at any or all of the inputs. As before, a 0 is defined as a LOW (ground) voltage. A 1 in the truth table represents a HIGH ($+5$ V) voltage.

The Boolean expression for a 3-input OR gate is written in Fig. 3-9a. The expression reads A OR B OR C equals output Y. The plus sign again signifies the OR function.

A logic symbol for the 3-input OR gate is drawn in Fig. 3-9b. Inputs A, B, and C are shown on the left of the symbol. Output Y is shown on the right of the OR symbol. This symbol represents some circuit that will perform the OR function.

$$A + B + C = Y$$

(a) Three-variable Boolean expression

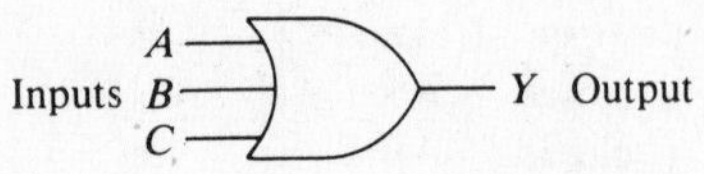

(b) 3-input OR gate symbol

Inputs			Output
C	B	A	Y
0	0	0	0
0	0	1	1
0	1	0	1
0	1	1	1
1	0	0	1
1	0	1	1
1	1	0	1
1	1	1	1

(c) Truth table with three variables

Fig. 3-9

A truth table for the 3-input OR gate is shown in Fig. 3-9c. The variables (A, B, and C) are shown on the left side of the table. The output (Y) is listed in the right column. Anytime a 1 appears at any input, the output will be 1.

Consider the OR truth tables in Figs. 3-8b and 3-9c. In each truth table the *unique output* from the OR gate is a LOW only when *all* inputs are LOW. Designers look at each gate's unique output when deciding which gate will perform a certain task.

The laws of Boolean algebra govern how an OR gate will operate. The formal laws for the OR function are:

$$A + 0 = A$$
$$A + 1 = 1$$
$$A + A = A$$
$$A + \overline{A} = 1$$

Looking at the truth table in Fig. 3-8 will help you check these laws. These general statements are always true of the OR function. The bar over the last variable means *not A*, or the opposite of *A*.

SOLVED PROBLEMS

3.6 Write the Boolean expression for a 4-input OR gate.

  Solution:

   $A + B + C + D = Y$

3.7 Draw the logic symbol for a 4-input OR gate.

  Solution:

   See Fig. 3-10.

Fig. 3-10 Symbol used for a 4-input OR gate

3.8 Draw a truth table for a 4-input OR gate.

  Solution:

Inputs				Output	Inputs				Output
D	C	B	A	Y	D	C	B	A	Y
0	0	0	0	0	1	0	0	0	1
0	0	0	1	1	1	0	0	1	1
0	0	1	0	1	1	0	1	0	1
0	0	1	1	1	1	0	1	1	1
0	1	0	0	1	1	1	0	0	1
0	1	0	1	1	1	1	0	1	1
0	1	1	0	1	1	1	1	0	1
0	1	1	1	1	1	1	1	1	1

3.9   In Fig. 3-11, what would the output pulse train look like?

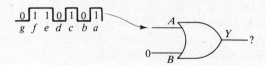

Fig. 3-11 Pulse-train problem

Solution:

In Fig. 3-11, the output waveform would look exactly like the input waveform at input A.

pulse $a = 1$ pulse $c = 1$ pulse $e = 1$ pulse $g = 0$
pulse $b = 0$ pulse $d = 0$ pulse $f = 1$

3.10   In Fig. 3-12, what would the output pulse train look like? Note that two pulse trains are being ORed together.

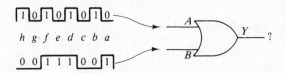

Fig. 3-12 Pulse-train problem

Solution:

In Fig. 3-12, the output pulses would be as follows:

pulse $a = 1$ pulse $c = 0$ pulse $e = 1$ pulse $g = 0$
pulse $b = 1$ pulse $d = 1$ pulse $f = 1$ pulse $h = 1$

3-4 THE NOT GATE

A NOT gate is also called an *inverter*. A NOT gate, or inverter, is an unusual gate. The NOT gate has only one input and one output. Figure 3-13a illustrates the logic symbol for the inverter, or NOT gate.

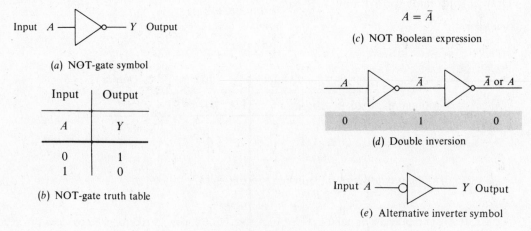

Fig. 3-13

The process of inverting is simple. Figure 3-13b is the truth table for the NOT gate. The input is always changed to its opposite. If the input is 0, the NOT gate will give its *complement*, or opposite, which is 1. If the input to the NOT gate is a 1, the circuit will complement it to give a 0. This inverting is also called *complementing* or *negating*. The terms negating, complementing, and inverting all mean the same thing.

The Boolean expression for inverting is shown in Fig. 3-13c. The expression $A = \overline{A}$ reads as A equals the output *not A*. The bar over the A means to complement A. Figure 3-13d illustrates what would happen if two inverters were used. The Boolean expressions are written above the lines between the inverters. The input A is inverted to $\overline{A}$ (not A). The $\overline{A}$ is then inverted again to form $\overline{\overline{A}}$ (not not A). The double inverted A ($\overline{\overline{A}}$) is equal to the original A, as shown in Fig. 3-13d. In the shaded section below the inverters, a 0 bit is the input. The 0 bit is complemented to a 1. The 1 bit is complemented again back to a 0. After a digital signal goes through two inverters, it is restored to its original form.

An alternative logic symbol for the NOT gate or inverter is shown in Fig. 3-13e. The *invert bubble* may be on either the input or the output side of the triangular symbol. When the invert bubble appears on the input side of the NOT symbol (as in Fig. 3-13e), the designer is usually trying to suggest that this is an *active LOW* input. An active LOW input requires a LOW to activate some function in the logic circuit. The alternative NOT gate symbol is commonly used in manufacturer's logic diagrams.

The laws of Boolean algebra govern the action of the inverter, or NOT gate. The formal Boolean algebra laws for the NOT gate are as follows:

$$\overline{0} = 1 \qquad \overline{1} = 0$$
$$\text{If } A = 1, \text{ then } \overline{A} = 0$$
$$\text{If } A = 0, \text{ then } \overline{A} = 1$$
$$\overline{\overline{A}} = A$$

You can check these general statements against the truth table and diagrams in Fig. 3-13.

SOLVED PROBLEMS

3.11 In Fig. 3-14, what is the output at point (e) if the input at point (a) is a 0 bit?

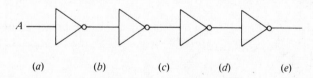

(a) (b) (c) (d) (e)

Fig. 3-14 Inverter problem

Solution:

The output at point (e) is a 0 bit.

3.12 What is the Boolean expression at point (b) in Fig. 3-14?

Solution:

The Boolean expression at point (b) is $\overline{A}$ (not A).

3.13 What is the Boolean expression at point (c) in Fig. 3-14?

Solution:

The Boolean expression at point (c) is $\overline{\overline{A}}$ (not not A). $\overline{\overline{A}}$ equals A according to the laws of Boolean algebra.

3.14   What is the Boolean expression at point (d) in Fig. 3-14?

   Solution:

   The Boolean expression at point (d) is $\bar{\bar{\bar{A}}}$ (not not not A). $\bar{\bar{\bar{A}}}$ equals $\bar{A}$ (not A).

3.15   What is the output at point (d) in Fig. 3-14 if the input at point (a) is a 1 bit?

   Solution:

   The output at point (d) is a 0 bit.

3.16   The NOT gate is said to *invert* its input. List two other words we can use instead of "invert."

   Solution:

   The words *complement* and *negate* also mean to invert.

3.17   The NOT gate can have _____ (one, many) input variable(s).

   Solution:

   The NOT gate can have one input variable.

3-5   COMBINING LOGIC GATES

   Many everyday digital logic problems use several logic gates. The most common *pattern* of gates is shown in Fig. 3-15a. This pattern is called the AND-OR pattern. The outputs of the AND gates (1 and 2) are feeding the inputs of the OR gate (3). You will note that this logic circuit has three inputs (A, B, and C). The output of the entire circuit is labeled Y.

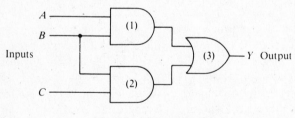

(a) AND-OR logic circuit

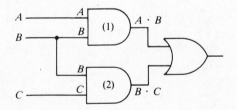

(b) Boolean expressions at the outputs of the AND gates

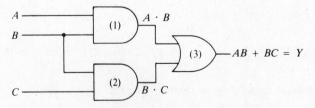

(c) Boolean expression at the output of the OR gate

Fig. 3-15

Let us first determine the Boolean expression that will describe this logic circuit. Begin the examination at gate (1). This is a 2-input AND gate. The output of this gate will be $A \cdot B$ (A AND B). This expression is written at the output of gate (1) in Fig. 3-15b. Gate (2) is also a 2-input AND gate. The output of this gate will be $B \cdot C$ (B AND C). This expression is written at the output of gate (2). Next the outputs of gates (1) and (2) are ORed together by gate (3). Figure 3-15c shows AB being ORed with BC. The resulting Boolean expression is $AB + BC = Y$. The Boolean expression $AB + BC = Y$ is read as (A AND B) OR (B AND C) will equal a 1 at output Y. You will note that the ANDing is done first, and finally the ORing is done.

The next question arises. What is the truth table for the AND-OR logic diagram in Fig. 3-15? Figure 3-16 will help us determine the truth table for the Boolean expression $AB + BC = Y$. The Boolean expression tells us that if *both* variables A AND B are 1, the output will be **1**. Figure 3-16 illustrates that the last two lines of the truth table have 1s in *both* the A and B positions. Therefore an output **1** is placed under the Y column.

$$A \cdot B + B \cdot C = Y$$

Inputs			Output
A	B	C	Y
0	0	0	0
0	0	1	0
0	1	0	0
0	1	1	**1**
1	0	0	0
1	0	1	0
1	1	0	**1**
1	1	1	**1**

Fig. 3-16 Complementing output column of truth table from a Boolean expression

The Boolean expression then goes on to say that another condition will also generate an output of **1**. The expression says that B AND C will also generate a **1** output. Looking at the truth table, it is found that the fifth line from the bottom has 1s in *both* the B AND C positions. The bottom line also has 1s in *both* the B AND C positions. Both of these lines will generate a **1** output. The bottom line already has a **1** under the output column (Y). The fifth line up will also get a **1** in the output column (Y). These are the only combinations that will generate a **1** output. The rest of the combinations are then listed as 0 outputs under column Y.

SOLVED PROBLEMS

3.18   What is the Boolean expression for the AND-OR logic diagram in Fig. 3-17?

Solution:

The Boolean expression for the logic circuit shown in Fig. 3-17 is

$$\overline{A}B + AC = Y$$

The expression is read as (not A AND B) OR (A AND C) equals output Y.

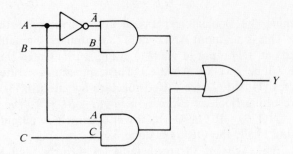

Fig. 3-17 AND-OR logic-circuit problem

3.19 What is the truth table for the logic diagram in Fig. 3-17?

Solution:

Inputs			Output	Inputs			Output
A	B	C	Y	A	B	C	Y
0	0	0	0	1	0	0	0
0	0	1	0	1	0	1	1
0	1	0	1	1	1	0	0
0	1	1	1	1	1	1	1

3.20 What is the Boolean expression for the AND-OR logic diagram in Fig. 3-18?

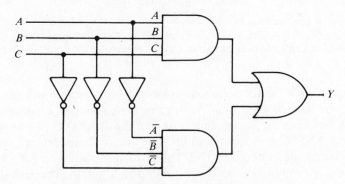

Fig. 3-18 AND-OR logic-circuit problem

Solution:

The Boolean expression for the logic circuit shown in Fig. 3-18 is

$$ABC + \overline{A}\,\overline{B}\,\overline{C} = Y$$

The expression reads as (A AND B AND C) OR (not A AND not B AND not C) equals output Y.

3.21 What is the truth table for the logic diagram in Fig. 3-18?

Solution:

Inputs			Output	Inputs			Output
A	B	C	Y	A	B	C	Y
0	0	0	1	1	0	0	0
0	0	1	0	1	0	1	0
0	1	0	0	1	1	0	0
0	1	1	0	1	1	1	1

3.22 What is the Boolean expression for the AND-OR logic diagram in Fig. 3-19?

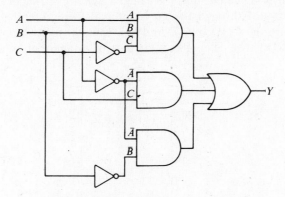

Fig. 3-19 AND-OR logic-circuit problem

Solution:

The Boolean expression for the logic circuit shown in Fig. 3-19 is $AB\overline{C} + \overline{A}C + \overline{A}\,\overline{B} = Y$. The expression reads as (A AND B AND not C) OR (not A AND C) OR (not A AND not B) equals output Y.

3.23 What is the truth table for the logic diagram in Fig. 3-19?

Solution:

Inputs			Output	Inputs			Output
A	B	C	Y	A	B	C	Y
0	0	0	1	1	0	0	0
0	0	1	1	1	0	1	0
0	1	0	0	1	1	0	1
0	1	1	1	1	1	1	0

3-6 USING PRACTICAL LOGIC GATES

Logic functions can be implemented in several ways. In the past, vacuum-tube and relay circuits performed logic functions. Presently tiny *integrated circuits* (ICs) perform as logic gates. These ICs contain the equivalent of miniature resistors, diodes, and transistors.

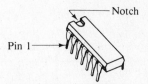

Fig. 3-20 14-pin DIP integrated circuit

A popular type of IC is illustrated in Fig. 3-20. This case style is referred to as a *dual-in-line package* (DIP) by IC manufacturers. This particular IC is called a 14-pin DIP integrated circuit.

Note that immediately *counterclockwise* from the notch on the IC shown in Fig. 3-20 is pin number 1. The pins are numbered counterclockwise from 1 to 14 *when viewed from the top* of the IC. Manufacturers of ICs provide pin diagrams similar to the one in Fig. 3-21 for a 7408 IC. Note that this IC contains four 2-input AND gates; thus it is called a *quadruple 2-input AND gate*. Figure 3-21 shows the IC pins numbered from 1 through 14 in a counterclockwise direction from the notch. The *power connections* to the IC are the GND (pin 7) and V_{CC} (pin 14) pins. All other pins are the inputs and outputs to the four AND gates. The 7408 IC is part of a family of logic devices. It is one of many

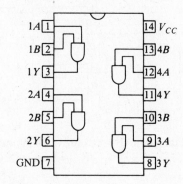

Fig. 3-21 Pin diagram for a 7408 IC

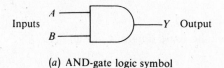

(*a*) AND-gate logic symbol

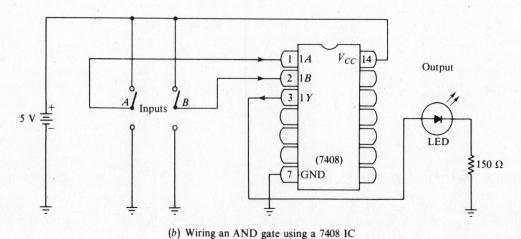

(*b*) Wiring an AND gate using a 7408 IC

Fig. 3-22

devices in the *transistor-transistor logic* (TTL) family of logic circuits. TTL devices are currently among the most popular logic devices.

Given the logic diagram in Fig. 3-22*a*, wire a circuit using a 7408 IC. A wiring diagram for the circuit is shown in Fig. 3-22*b*. A 5-V power supply is used with all TTL devices. The positive (V_{CC}) and negative (GND) power connections are made to pins 14 and 7. Input switches (*A* and *B*) are wired to pins 1 and 2 of the 7408 IC. Note that, if a switch is in the *up position*, a logical 1 (+5 V) is applied to the input of the AND gate. At the right, a light-emitting diode (LED) and 150-ohm (Ω) limiting resistor are connected to ground. If the output at pin 3 is HIGH (+5 V), current will flow through the LED. Lighting the LED indicates a HIGH, or a binary 1, at the output of the AND gate.

The truth table in Fig. 3-23 shows the results of operating the 2-input AND circuit. The LED in Fig. 3-22*b* lights only when *both* input switches (*A* and *B*) are at +5 V.

Inputs		Output	
A	*B*		LED
Voltage	Voltage	Voltage	lights?
GND	GND	GND	no
GND	+5 V	GND	no
+5 V	GND	GND	no
+5 V	+5 V	near +5 V	yes

Fig. 3-23 Truth table for a TTL-type AND gate

Manufacturers of integrated circuits also produce other logic functions. Figure 3-24 illustrates pin diagrams for two basic TTL ICs. Figure 3-24*a* is the pin diagram for a *quadruple 2-input OR gate*. In other words, the 7432 IC contains four 2-input OR gates. It could be wired and tested in a manner similar to the testing of the AND gate shown in Fig. 3-22*b*.

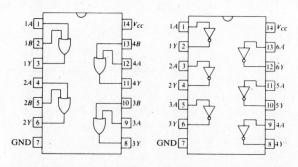

(*a*) Pin diagram for a 7432 IC   (*b*) Pin diagram for a 7404 IC

Fig. 3-24

The 7404 IC shown in Fig. 3-24*b* is also a TTL device. The 7404 IC contains six NOT gates, or inverters. The 7404 is described by the manufacturer as a *hex inverter* IC. Note that each IC has its power connections (V_{CC} and GND). A 5-V dc power supply is always used with TTL logic circuits.

Two variations of DIP ICs are illustrated in Fig. 3-25. The integrated circuit shown in Fig. 3-25*a* has 16 pins with pin 1 identified by using a dot instead of a notch. The IC shown in Fig. 3-25*b* is a 24-pin DIP integrated circuit with pin 1 located immediately counterclockwise (when viewed from the top) from the notch.

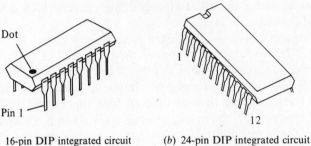

(a) 16-pin DIP integrated circuit   (b) 24-pin DIP integrated circuit

Fig. 3-25

The 7408, 7432, and 7404 ICs you studied in this section were all from the TTL logic family. The newer *complementary metal oxide semiconductor* (CMOS) family of ICs has been gaining popularity because of its low power requirements. Logic gates (AND, OR, and NOT) also are available in DIP IC form in the CMOS family. Typical DIP ICs might be the CMOS 74C08 quad 2-input AND gate, 74C04 hex inverter, or the 74C32 quad 2-input OR gate. The 74CXX series of CMOS gates is *not* directly compatible with the TTL 7400 series of integrated circuits.

SOLVED PROBLEMS

3.24 What logic function is performed by the circuit illustrated in Fig. 3-26?

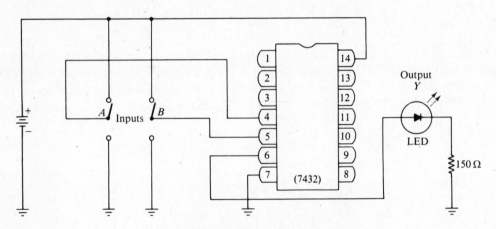

Fig. 3-26 Logic-circuit problem

Solution:

The 7432 IC performs as a 2-input OR gate when wired as shown in Fig. 3-26.

3.25 Write the Boolean expression for the circuit shown in Fig. 3-26.

Solution:

The Boolean expression for the 2-input OR function (Fig. 3-26) is $A + B = Y$.

3.26 What is the *voltage* of the power supply at the left in Fig. 3-26? The 7432 IC is a TTL device.

Solution:

TTL devices use a 5-V dc power supply.

3.27 If both switches A and B in Fig. 3-26 are in the *down position* (toward ground) the output LED will be _____ (lit, not lit).

Solution:

When both inputs are 0, the output of the OR gate will be 0 and the output LED will be not lit.

3.28 In Fig. 3-26, if switch A is up and switch B is down, the output LED will be _____ (lit, not lit).

Solution:

When input A is 1 and input B is 0 (Fig. 3-26), the output of the OR gate will be 1 and the output LED will be lit.

3.29 Pins 7 and 14 of the 7432 IC are _____ (input, output, power) connections.

Solution:

Pins 7 and 14 of the 7432 IC are power connections.

3.30 A _____ (+5 V, GND) voltage at pin 4 of the 7432 IC will cause pin 6 to go to a HIGH logic level.

Solution:

The output (pin 6) goes HIGH anytime as input (such as pin 4) is HIGH (near +5 V).

3.31 The letters TTL stand for _____.

Solution:

The letters TTL stand for the popular transistor-transistor logic family of integrated circuits.

3.32 The _____ (CMOS, TTL) logic family is characterized by its very low power consumption.

Solution:

The CMOS logic family is noted for its very low power consumption.

3.33 Integrated circuits from the TTL and CMOS families _____ (can, cannot) be interchanged in a digital circuit.

Solution:

TTL and CMOS ICs cannot be interchanged in a digital circuit. They may have the same logic function or even have the same pin diagram, but their input and output characteristics are quite different.

Supplementary Problems

3.34 Draw the logic symbol for a 6-input AND gate. Label the inputs A, B, C, D, E, and F. Label the output Y.
Ans. See Fig. 3-27.

3.35 Draw the logic symbol for a 7-input OR gate. Label the inputs A, B, C, D, E, F, and G. Label the output Y.
Ans. See Fig. 3-28.

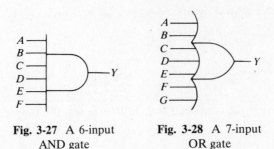

Fig. 3-27 A 6-input
AND gate

Fig. 3-28 A 7-input
OR gate

3.36   Describe the pulse train at output Y of the AND gate shown in Fig. 3-29, if input B is 0.
            Ans.   A 0 will disable the AND gate, and the output will be 0.

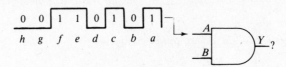

Fig. 3-29 Pulse-train problem

3.37   Describe the pulse train at output Y of the AND gate shown in Fig. 3-29 if input B is 1.
            Ans.   The output waveform will look exactly like the input waveform at input A (Fig. 3-29).

3.38   Describe the pulse train at output Y of the OR gate shown in Fig. 3-30 if input B is 0.
            Ans.   The output waveform will look exactly like the input waveform at input A (Fig. 3-30).

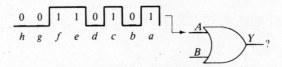

Fig. 3-30 Pulse-train problem

3.39   Describe the pulse train at output Y of the OR gate shown in Fig. 3-30 if input B is 1.
            Ans.   The output will always be 1.

3.40   Write the Boolean expression for the logic circuit shown in Fig. 3-31.
            Ans. $A \cdot \overline{B} + \overline{B} \cdot C = Y$ or $A\overline{B} + \overline{B}C = Y$

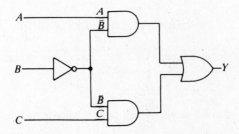

Fig. 3-31 AND-OR logic-circuit problem

3.41 Draw the truth table for the logic circuit shown in Fig. 3-31.
         Ans.

Inputs			Output	Inputs			Output
C	*B*	*A*	*Y*	*C*	*B*	*A*	*Y*
0	0	0	0	1	0	0	1
0	0	1	1	1	0	1	1
0	1	0	0	1	1	0	0
0	1	1	0	1	1	1	0

3.42 Write the Boolean expression for the logic circuit shown in Fig. 3-32.
         Ans. $\overline{A} \cdot B \cdot \overline{C} + \overline{B} \cdot \overline{C} = Y$ or $\overline{A}B\overline{C} + \overline{B}\,\overline{C} = Y$

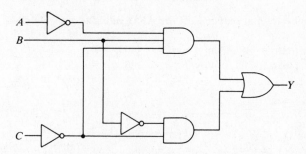

Fig. 3-32 AND-OR logic-circuit problem

3.43 Draw the truth table for the logic circuit shown in Fig. 3-32.
         Ans.

Inputs			Output	Inputs			Output
C	*B*	*A*	*Y*	*C*	*B*	*A*	*Y*
0	0	0	1	1	0	0	0
0	0	1	1	1	0	1	0
0	1	0	1	1	1	0	0
0	1	1	0	1	1	1	0

3.44 Write the Boolean expression for the logic circuit shown in Fig. 3-33.
         Ans. $\overline{A} \cdot \overline{B} \cdot C + \overline{A} \cdot B \cdot \overline{C} + A \cdot \overline{B} \cdot \overline{C} = Y$ or $\overline{A}\,\overline{B}C + \overline{A}B\overline{C} + A\overline{B}\,\overline{C} = Y$

3.45 Draw the truth table for the logic circuit shown in Fig. 3-33.
         Ans.

Inputs			Output	Inputs			Output
C	*B*	*A*	*Y*	*C*	*B*	*A*	*Y*
0	0	0	0	1	0	0	1
0	0	1	1	1	0	1	0
0	1	0	1	1	1	0	0
0	1	1	0	1	1	1	0

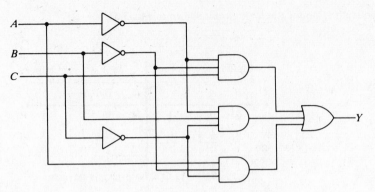

Fig. 3-33 AND-OR logic-circuit problem

3.46 Describe the pulse train at output Y of the AND gate shown in Fig. 3-34.
Ans.

pulse $a = 0$	pulse $c = 0$	pulse $e = 0$	pulse $g = 0$
pulse $b = 1$	pulse $d = 0$	pulse $f = 1$	pulse $h = 1$

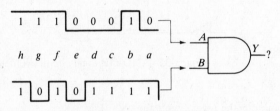

Fig. 3-34 Pulse-train problem

3.47 Describe the pulse train at output Y of the OR gate shown in Fig. 3-35.
Ans.

pulse $a = 0$	pulse $c = 1$	pulse $e = 1$	pulse $g = 1$
pulse $b = 1$	pulse $d = 1$	pulse $f = 1$	pulse $h = 0$

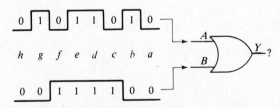

Fig. 3-35 Pulse-train problem

3.48 Write the Boolean expression for the logic circuit shown in Fig. 3-36.
Ans. $A \cdot B \cdot C \cdot D + \overline{A} \cdot \overline{C} = Y$ or $ABCD + \overline{A}\overline{C} = Y$

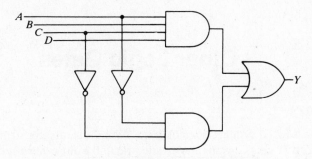

Fig. 3-36 AND-OR logic-circuit problem

3.49 Draw the truth table for the logic circuit shown in Fig. 3-36. Note that the circuit has four input variables. The truth table will have 16 possible combinations.
Ans.

Inputs				Output	Inputs				Output
D	C	B	A	Y	D	C	B	A	Y
0	0	0	0	1	1	0	0	0	1
0	0	0	1	0	1	0	0	1	0
0	0	1	0	1	1	0	1	0	1
0	0	1	1	0	1	0	1	1	0
0	1	0	0	0	1	1	0	0	0
0	1	0	1	0	1	1	0	1	0
0	1	1	0	0	1	1	1	0	0
0	1	1	1	0	1	1	1	1	1

3.50 The part number 74C08 is a quad 2-input AND gate from the _____ (CMOS, TTL) family of ICs.
Ans. The 74C08 is a part number from the CMOS family of ICs. The C in the center of the part number means the part is a CMOS-type IC.

Chapter 4

Other Logic Gates

4-1 INTRODUCTION

The most complex digital systems, such as large computers, are constructed from basic logic gates. The AND, OR, and NOT gates are the most fundamental. Four other useful logic gates can be made from these fundamental devices. The other gates are called the NAND gate, the NOR gate, the exclusive-OR gate, and the exclusive-NOR gate. At the end of this chapter, you will know the logic symbol, truth table, and Boolean expression for each of the *seven logic gates* used in digital systems.

4-2 THE NAND GATE

Consider the logic diagram at the top of Fig. 4-1. An AND gate is connected to an inverter. Inputs A and B are ANDed to form the Boolean expression $A \cdot B$. The $A \cdot B$ is then inverted by the NOT gate. On the right side of the inverter, the overbar is added to the Boolean expression. The Boolean expression for the entire circuit is $\overline{A \cdot B} = Y$. It is said that this is a *not-AND* or NAND circuit.

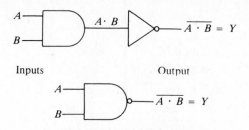

Fig. 4-1 The NAND gate

The standard logic symbol for the NAND gate is shown in the bottom diagram in Fig. 4-1. Note that the NAND symbol is an AND symbol with a small bubble at the output. The bubble is sometimes called an *invert bubble*. The invert bubble provides a simplified method of representing the NOT gate shown in the top diagram in Fig. 4-1.

The truth table describes the exact operation of a logic gate. The *truth table* for the NAND gate is illustrated in the unshaded columns of Fig. 4-2. The AND-gate truth table is also given to show how each output is inverted to give the NAND output. Some students find it useful to think of the NAND gate as an AND gate that puts out a 0 when it is enabled (when both inputs are 1).

Inputs		Output	
B	A	AND	NAND
0	0	0	1
0	1	0	1
1	0	0	1
1	1	1	0

Fig. 4-2 The AND- and NAND-gate truth tables

48

The NAND function has traditionally been the *universal gate* in digital circuits. The NAND gate is widely used in most digital systems.

Consider the NAND gate truth table in Fig. 4-2. The *unique output* from the NAND gate is a LOW when all inputs are HIGH.

SOLVED PROBLEMS

4.1 Write the Boolean expression for a 3-input NAND gate.

Solution:

$$\overline{A \cdot B \cdot C} = Y \text{ or } \overline{ABC} = Y$$

4.2 Draw the logic symbol for a 3-input NAND gate.

Solution:

See Fig. 4-3.

Fig. 4-3 A 3-input NAND gate

4.3 Draw the truth table for a 3-input NAND gate.

Solution:

Inputs			Output	Inputs			Output
C	B	A	Y	C	B	A	Y
0	0	0	1	1	0	0	1
0	0	1	1	1	0	1	1
0	1	0	1	1	1	0	1
0	1	1	1	1	1	1	0

4.4 What would the output pulse train shown in Fig. 4-4 look like if input *B* were 0?

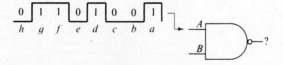

Fig. 4-4 Pulse-train problem

Solution:

The output of the NAND gate shown in Fig. 4-4 would always be 1.

4.5   What would the output pulse train shown in Fig. 4-4 look like if input B were 1?

Solution:

The output would be an inverted copy of the waveform at input A (Fig. 4-4). The output pulses would be as follows:

pulse $a = 0$ pulse $c = 1$ pulse $e = 1$ pulse $g = 0$
pulse $b = 1$ pulse $d = 0$ pulse $f = 0$ pulse $h = 1$

4.6   Draw a logic diagram of how you could connect a 2-input NAND gate to perform as an inverter. Label inverter input as A. Label inverter output as $\overline{A}$.

Solution:

See Fig. 4-5. There are two possibilities.

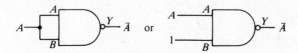

Fig. 4-5   Wiring the NAND gate as an inverter

4-3   THE NOR GATE

Consider the logic diagram in Fig. 4-6. An inverter has been connected to the output of an OR gate. The Boolean expression at the input to the inverter is $A + B$. The inverter then complements the ORed terms, which are shown in the Boolean expression with an overbar. Adding the overbar produces the Boolean expression $\overline{A + B} = Y$. This is a *not-OR* function. The not-OR function can be drawn as a single logic symbol called a *NOR gate*. The standard symbol for the NOR gate is illustrated in the bottom diagram of Fig. 4-6. Note that a small invert bubble has been added to the OR symbol to form the NOR symbol.

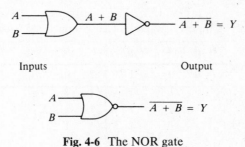

Fig. 4-6   The NOR gate

The truth table in Fig. 4-7 details the operation of the NOR gate. Note that the output column of the NOR gate is the complement (has been inverted) of the shaded OR column. In other words, the

Inputs		Output	
B	A	OR	NOR
0	0	0	1
0	1	1	0
1	0	1	0
1	1	1	0

Fig. 4-7   The OR- and NOR-gate truth tables

NOR gate puts out a 0 where the OR gate would produce a 1. The small invert bubble at the output of the NOR symbol serves as a reminder of the 0 output idea.

Consider the NOR gate truth table in Fig. 4-7. The *unique output* from the NOR gate is a HIGH when all inputs are LOW.

SOLVED PROBLEMS

4.7 Write the Boolean expression for a 3-input NOR gate.

Solution:

$$\overline{A + B + C} = Y$$

4.8 Draw the logic symbol for a 3-input NOR gate.

Solution:

See Fig. 4-8.

Fig. 4-8 A 3-input NOR gate

4.9 What is the truth table for a 3-input NOR gate?

Solution:

Inputs			Output	Inputs			Output
C	B	A	Y	C	B	A	Y
0	0	0	1	1	0	0	0
0	0	1	0	1	0	1	0
0	1	0	0	1	1	0	0
0	1	1	0	1	1	1	0

4.10 What would the output pulse train shown in Fig. 4-9 look like if input *B* were 1?

Fig. 4-9 Pulse-train problem

Solution:

The output of the NOR gate in Fig. 4-9 would always be 0.

4.11   What would the output pulse train shown in Fig. 4-9 look like if input B were 0?

Solution:

The output pulse would be the one shown in Fig. 4-9 but inverted. The pulses would be as follows:

pulse $a = 0$ pulse $c = 1$ pulse $e = 0$ pulse $g = 1$
pulse $b = 1$ pulse $d = 0$ pulse $f = 0$ pulse $h = 0$

4-4 THE EXCLUSIVE-OR GATE

The *exclusive-OR gate* is referred to as the "any but not all" gate. The exclusive-OR term is often shortened to read as *XOR*. A truth table for the XOR function is shown in Fig. 4-10. Careful examination shows that this truth table is similar to the OR truth table except that, when both inputs are 1, the XOR gate generates a 0. The XOR gate is enabled *only when an odd number of 1s appear at the inputs.* Lines 2 and 3 of the truth table have odd numbers of 1s, and therefore the output is enabled with a 1. Lines 1 and 4 of the truth table contain even numbers $(0, 2)$ of 1s, and therefore the XOR gate is disabled and a 0 appears at the output. The XOR gate could be referred to as an odd-bits check circuit.

Inputs		Output
B	A	Y
0	0	0
0	1	1
1	0	1
1	1	0

Fig. 4-10   The exclusive-OR-gate truth table

A Boolean expression for the XOR gate can be developed from the truth table in Fig. 4-10. The expression is $A \cdot \overline{B} + \overline{A} \cdot B = Y$. With this Boolean expression, a logic circuit can be developed by using AND gates, OR gates, and inverters. Such a logic circuit is drawn in Fig. 4-11a. This logic circuit will perform the XOR logic function.

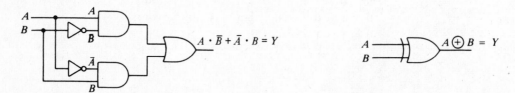

(a) Logic circuit that performs the XOR function   (b) Standard logic symbol for the XOR gate

Fig. 4-11

The standard logic symbol for the XOR gate is shown in Fig. 4-11b. Both logic symbol diagrams in Fig. 4-11 would produce the same truth table (XOR). The Boolean expression at the right in Fig. 4-11b is a *simplified XOR expression*. The $\oplus$ symbol signifies the XOR function in Boolean algebra. It is said that inputs A and B in Fig. 4-11b are exclusively ORed together.

SOLVED PROBLEMS

4.12 Write the Boolean expression (simplified form) for a 3-input XOR gate.

Solution:

$$A \oplus B \oplus C = Y$$

4.13 Draw the logic symbol for a 3-input XOR gate.

Solution:

See Fig. 4-12.

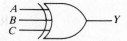

Fig. 4-12 A 3-input XOR gate

4.14 What is the truth table for a 3-input XOR gate? Remember that an odd number of 1s generates a 1 output.

Solution:

Inputs			Output	Inputs			Output
C	B	A	Y	C	B	A	Y
0	0	0	0	1	0	0	1
0	0	1	1	1	0	1	0
0	1	0	1	1	1	0	0
0	1	1	0	1	1	1	1

4.15 The XOR gate might be considered an _____ (even-, odd-) number-of-1s detector.

Solution:

The XOR gate generates a 1 when an *odd* number of 1 bits are present. For this reason it might be considered as an odd-number-of-1s detector.

4.16 What will the pulse train at the output of the XOR gate shown in Fig. 4-13 look like?

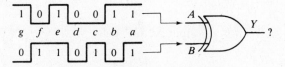

Fig. 4-13 Pulse-train problem

Solution:

The output pulses from the XOR gate shown in Fig. 4-13 will be as follows:

pulse $a = 0$ pulse $c = 1$ pulse $e = 0$ pulse $g = 1$
pulse $b = 1$ pulse $d = 0$ pulse $f = 1$

4-5 THE EXCLUSIVE-NOR GATE

The output of an XOR gate is shown inverted in Fig. 4-14. The output of the inverter on the right side is called the *exclusive-NOR* (XNOR) function. The XOR gate produces the expression $A \oplus B$. When this is inverted, it forms the Boolean expression for the XNOR gate, $\overline{A \oplus B} = Y$. The standard logic symbol for the XNOR gate is shown in the bottom diagram of Fig. 4-14. Note that the symbol is an XOR symbol with an invert bubble attached to the output.

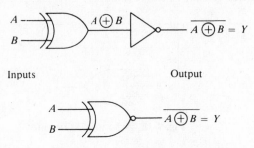

Fig. 4-14 The XNOR gate

The right-hand column of the truth table in Fig. 4-15 details the operation of the XNOR gate. Note that all outputs of the XNOR gate are the complements of the XOR-gate outputs. While the XOR gate is an odd-number-of-1s detector, the XNOR gate detects *even numbers of 1s*. The XNOR gate will produce a 1 output when an *even number of 1s* appear at the inputs.

Inputs		Output	
B	A	XOR	XNOR
0	0	0	1
0	1	1	0
1	0	1	0
1	1	0	1

Fig. 4-15 The XOR- and XNOR-gate truth tables

SOLVED PROBLEMS

4.17 Write the Boolean expression for a 3-input XNOR gate.

Solution:

$\overline{A \oplus B \oplus C} = Y$

4.18 Draw the logic symbol for a 3-input XNOR gate.

Solution:

See Fig. 4-16.

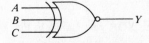

Fig. 4-16 A 3-input XNOR gate

4.19   Construct a truth table for a 3-input XNOR gate. Remember that an even number of 1s generates a 1 output.

Solution:

Inputs			Output	Inputs			Output
C	B	A	Y	C	B	A	Y
0	0	0	1	1	0	0	0
0	0	1	0	1	0	1	1
0	1	0	0	1	1	0	1
0	1	1	1	1	1	1	0

4.20   What will the pulse train at the output of the XNOR gate shown in Fig. 4-17 look like?

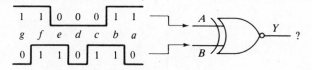

Fig. 4-17 Pulse-train problem

Solution:

The output pulses from the XNOR gate shown in Fig. 4-17 will be as follows:

pulse $a = 0$ pulse $c = 0$ pulse $e = 0$ pulse $g = 0$
pulse $b = 1$ pulse $d = 1$ pulse $f = 1$

4-6 CONVERTING GATES WHEN USING INVERTERS

When using logic gates, the need will arise to convert to another logic function. An easy method of converting is to use inverters placed at the outputs or inputs of gates. It has been shown that an inverter connected at the output of an AND gate produces the NAND function. Also, an inverter placed at the output of an OR gate produces the NOR function. The chart in Fig. 4-18 illustrates these and other conversions.

Placing inverters at all the inputs of a logic gate produces the results illustrated in Fig. 4-19. In the first line the inputs to an AND gate are being inverted (the plus symbol indicates addition in this figure). This produces the NOR function at the output of the AND gate. The second line of Fig. 4-19 shows the inputs of an OR gate being inverted. This produces the NAND function. The first two examples suggest new symbols for the NOR and NAND functions. Figure 4-20 illustrates two logic symbols sometimes used for the NOR and NAND functions. Figure 4-20*a* is an *alternative logic symbol* for a NOR gate. Figure 4-20*b* is an *alternative logic symbol* for a NAND gate. These symbols are encountered in some manufacturers' literature.

The effect of inverting both the inputs and output of a logic gate is shown in Fig. 4-21. Again the plus symbol stands for addition. This technique is probably not used often because of the large number of gates needed. Note that this is the method of converting from the AND to the OR to the AND function. This is also the method of converting from the NAND to the NOR to the NAND function.

Original gate		Add inverter to output		New logic function
	+		=	NAND
	+		=	AND
	+		=	NOR
	+		=	OR

(+) symbol means adding on this chart

Fig. 4-18 Effect of inverting outputs of gates

Add inverters to inputs		Original gate		New logic function
	+		=	NOR
	+		=	NAND
	+		=	OR
	+		=	AND

(+) symbol means adding on this chart

Fig. 4-19 Effect of inverting inputs of gates

$\overline{A + B} = Y$

(a) NOR gate symbol

$\overline{A \cdot B} = Y$

(b) NAND gate symbol

Fig. 4-20 Alternative logic symbols

Add inverters to inputs		Original gate		Add inverter to output		New logic function
	+		+		=	OR
	+		+		=	AND
	+		+		=	NOR
	+		+		=	NAND

(+) symbol means adding on this chart

Fig. 4-21 Effect of inverting both inputs and outputs of gates

SOLVED PROBLEMS

4.21   Given an OR gate and inverters, draw a logic diagram that will perform the 2-input NAND function.

   Solution:
       See Fig. 4-22.

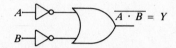

Fig. 4-22   2-input NAND function

4.22   Given an OR gate and inverters, draw a logic diagram that will perform the 3-input AND function.

   Solution:
       See Fig. 4-23.

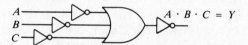

Fig. 4-23   3-input AND function

4.23   Given a NAND gate and inverters, draw a logic diagram that will perform the 2-input OR function.

   Solution:
       See Fig. 4-24.

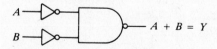

Fig. 4-24   2-input OR function

4.24   Given a NAND gate and inverters, draw a logic diagram that will perform the 3-input AND function.

   Solution:
       See Fig. 4-25.

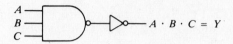

Fig. 4-25 3-input AND function

4.25 Given an AND gate and inverters, draw a logic diagram that will perform the 2-input NOR function.

Solution:

See Fig. 4-26.

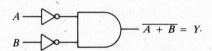

Fig. 4-26 2-input NOR function

4-7 NAND AS A UNIVERSAL GATE

Consider the logic circuit shown in Fig. 4-27a. This is referred to as an *AND-OR pattern of gates*. The AND gates feed into the final OR gate. The Boolean expression for this circuit is shown at the right as $\overline{A} \cdot B + A \cdot B = Y$. In constructing the circuit, you need three different types of gates (AND gates, an OR gate, and an inverter). From a manufacturer's catalog, you would find that three different ICs would be needed to implement the circuit shown in Fig. 4-27a.

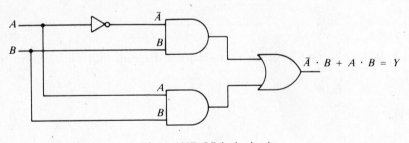

(a) An AND-OR logic circuit

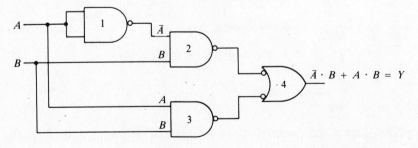

(b) An equivalent NAND logic circuit

Fig. 4-27

It was mentioned earlier that the NAND gate is considered a universal gate. Figure 4-27b shows NAND gates being used to implement the logic $\overline{A} \cdot B + A \cdot B = Y$. This logic is the same as that performed by the AND-OR circuit shown in Fig. 4-27a. Remember that the OR-looking gate (gate 4) with the invert bubbles at the inputs is a NAND gate. The circuit shown in Fig. 4-27b is simpler because all the gates are NAND gates. It is found that only one IC (a quadruple 2-input NAND gate) is needed to implement the NAND logic of Fig. 4-27b. Fewer ICs are needed to implement the NAND logic circuit than the AND-OR pattern of logic gates.

It is customary when converting from AND-OR logic to NAND logic to draw the AND-OR pattern first. This can be done from the Boolean expression. The AND-OR logic diagram would be similar to that in Fig. 4-27a. A NAND gate is then substituted for each inverter, AND gate, and OR gate. The NAND logic pattern will then be similar to the circuit shown in Fig. 4-27b.

A clue to *why* the AND-OR logic can be replaced by NAND logic is shown in Fig. 4-27b. Note the two invert bubbles between the output of gate 2 and the input of gate 4. *Two invert bubbles cancel one another*. That leaves the AND-OR symbols just as in Fig. 4-27a. The *double inversion* also takes place in Fig. 4-27b between gates 3 and 4. This leaves AND gate 3 feeding OR gate 4. NAND gate 1 acts as an inverter when its inputs are tied together as shown in Fig. 4-27b.

SOLVED PROBLEMS

4.26   Redraw the AND-OR circuit shown in Fig. 4-11a by using five 2-input NAND gates. The NAND logic circuit should perform the logic $A \cdot \overline{B} + \overline{A} \cdot B = Y$.

Solution:

   See Fig. 4-28.

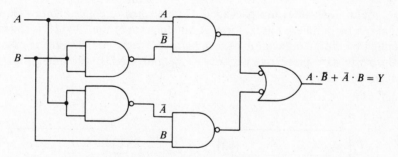

Fig. 4-28   Solution using NAND logic

4.27   Draw a logic diagram for the Boolean expression $\overline{A} \cdot \overline{B} + A \cdot B = Y$. Use inverters, AND gates, and OR gates.

Solution:

   See Fig. 4-29.

4.28   Redraw the logic diagram of Prob. 4.27 by using only five 2-input NAND gates. The circuit should perform the logic $\overline{A} \cdot \overline{B} + A \cdot B = Y$.

Solution:

   See Fig. 4-30.

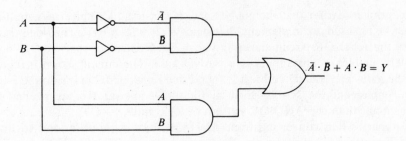

Fig. 4-29 AND-OR logic circuit

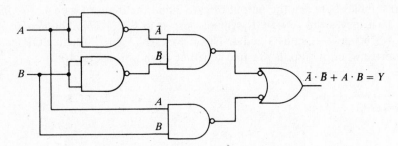

Fig. 4-30 Equivalent NAND logic circuit

4-8 USING PRACTICAL LOGIC GATES

The most useful logic gates are packaged as integrated circuits. Figure 4-31 illustrates two TTL logic gates that can be purchased in IC form. A pin diagram of the 7400 IC is shown in Fig. 4-31a. The 7400 is described by the manufacturer as a *quadruple 2-input NAND gate* IC. Note that the 7400 IC does have the customary power connections (V_{CC} and GND). All other pins are the inputs and outputs from the four 2-input NAND gates.

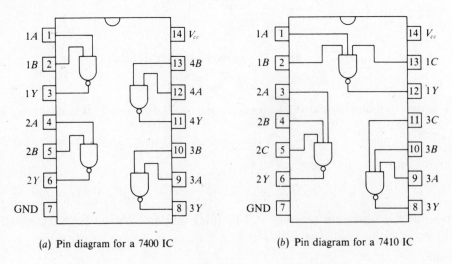

(a) Pin diagram for a 7400 IC   (b) Pin diagram for a 7410 IC

Fig. 4-31

Three 3-input NAND gates are housed in the 7410 TTL IC. The pin diagram for the 7410 IC is shown in Fig. 4-31b. This device is described by the manufacturer as a *triple 3-input NAND gate* IC. NAND gates with more than three inputs also are available.

The 7400 and 7410 ICs were from the common TTL logic family. Manufacturers also produce a variety of NAND, NOR, and XOR gates in CMOS-type ICs. Typical NAND gates might be the CMOS 74C00 quad 2-input NAND gate, 74C30 8-input NAND gate, and 4012 dual 4-input NAND gate DIP ICs. Some CMOS NOR gates in DIP IC form are the 74C02 quad 2-input NOR gate and the 4002 dual 4-input NOR gate. Several exclusive-OR gates are produced in CMOS; examples are the 74C86 quad 2-input XOR gate and the 4030 quad 2-input XOR gate. Note that CMOS ICs come in both a 74C00 series and a 4000 series. It must be remembered that without special interfacing, *TTL and CMOS ICs are not compatible.*

SOLVED PROBLEMS

4.29 Construct the truth table for the circuit shown in Fig. 4-32.

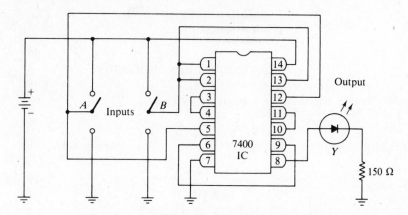

Fig. 4-32 Wiring diagram of a logic-circuit problem

Solution:

Inputs		Output
B	A	Y
0	0	0
0	1	1
1	0	0
1	1	1

4.30 What is the voltage of the power supply at the left in Fig. 4-32? The 7400 IC is a TTL device.

Solution:

A TTL device uses a 5-V dc power supply.

4.31 If both switches (A and B) shown in Fig. 4-32 are in the *up* position (at logical 1), the output LED will be _____ (lit, not lit).

Solution:

When both inputs are 1, the output of the circuit will be 1 and the output LED will be lit.

4.32 The 7400 IC is described by the manufacturer as a quadruple _____ _____ _____.

Solution:

The 7400 IC is a quadruple 2-input NAND gate.

4.33 The circuit shown in Fig. 4-32 could be described as a(n) _____ (AND-OR, NAND) logic circuit.

Solution:

The circuit shown in Fig. 4-32 uses NAND logic.

4.34 The 4012 is a dual 4-input NAND gate IC using the _____ (CMOS, TTL) technology.

Solution:

The 4000 series part numbers designate CMOS digital ICs.

Supplementary Problems

4.35 Write the Boolean expression for a 4-input NAND gate.
Ans. $\overline{A \cdot B \cdot C \cdot D} = Y$ or $\overline{ABCD} = Y$

4.36 Draw the logic symbol for a 4-input NAND gate.   *Ans.* See Fig. 4-33.

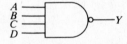

Fig. 4-33 A 4-input NAND gate

4.37 Construct the truth table for a 4-input NAND gate.
Ans.

Inputs				Output	Inputs				Output
D	C	B	A	Y	D	C	B	A	Y
0	0	0	0	1	1	0	0	0	1
0	0	0	1	1	1	0	0	1	1
0	0	1	0	1	1	0	1	0	1
0	0	1	1	1	1	0	1	1	1
0	1	0	0	1	1	1	0	0	1
0	1	0	1	1	1	1	0	1	1
0	1	1	0	1	1	1	1	0	1
0	1	1	1	1	1	1	1	1	0

4.38 What would the output pulse train shown in Fig. 4-34 look like if input *C* were 0?
Ans. The output of the NAND gate would be 1 at all times.

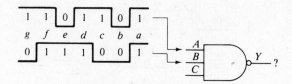

Fig. 4-34 Pulse-train problem

4.39   What would the output pulse train shown in Fig. 4-34 look like if input C were 1?
     Ans.   pulse $a = 0$ pulse $c = 1$ pulse $e = 1$ pulse $g = 1$
               pulse $b = 1$ pulse $d = 0$ pulse $f = 0$

4.40   Write the Boolean expression for a 4-input NOR gate.   *Ans.* $\overline{A + B + C + D} = Y$

4.41   Draw the logic symbol for a 4-input NOR gate.   *Ans.*   See Fig. 4-35.

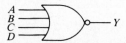

Fig. 4-35 A 4-input NOR gate

4.42   Construct the truth table for a 4-input NOR gate.
     Ans.

Inputs				Output	Inputs				Output
D	C	B	A	Y	D	C	B	A	Y
0	0	0	0	1	1	0	0	0	0
0	0	0	1	0	1	0	0	1	0
0	0	1	0	0	1	0	1	0	0
0	0	1	1	0	1	0	1	1	0
0	1	0	0	0	1	1	0	0	0
0	1	0	1	0	1	1	0	1	0
0	1	1	0	0	1	1	1	0	0
0	1	1	1	0	1	1	1	1	0

4.43   What would the output pulse train shown in Fig. 4-36 look like if input C were 1?
     Ans.   The output of the NOR gate would always be 0.

4.44   What would the output pulse train shown in Fig. 4-36 look like if input C were 0?
     Ans.   pulse $a = 0$ pulse $c = 1$ pulse $e = 0$ pulse $g = 1$
               pulse $b = 1$ pulse $d = 0$ pulse $f = 0$

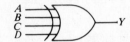

Fig. 4-36 Pulse-train problem

4.45   Write the Boolean expression for a 4-input XOR gate.
            Ans. $A \oplus B \oplus C \oplus D = Y$

4.46   Draw the logic symbol for a 4-input XOR gate.   *Ans.*   See Fig. 4-37.

Fig. 4-37 A 4-input XOR gate

4.47   Construct the truth table for a 4-input XOR gate.
            Ans.

Inputs				Output	Inputs				Output
D	C	B	A	Y	D	C	B	A	Y
0	0	0	0	0	1	0	0	0	1
0	0	0	1	1	1	0	0	1	0
0	0	1	0	1	1	0	1	0	0
0	0	1	1	0	1	0	1	1	1
0	1	0	0	1	1	1	0	0	0
0	1	0	1	0	1	1	0	1	1
0	1	1	0	0	1	1	1	0	1
0	1	1	1	1	1	1	1	1	0

4.48   What would the pulse train at the output of the XOR gate shown in Fig. 4-38 look like?
            Ans.   pulse $a = 0$ pulse $c = 1$ pulse $e = 0$ pulse $g = 0$
                     pulse $b = 1$ ·   pulse $d = 1$ pulse $f = 0$ pulse $h = 1$

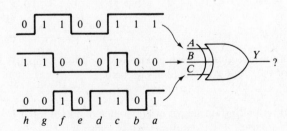

Fig. 4-38 Pulse-train problem

4.49   Write the Boolean expression for a 4-input XNOR gate.
           Ans. $\overline{A \oplus B \oplus C \oplus D} = Y$

4.50   Draw the logic symbol for a 4-input XNOR gate.   *Ans.*   See Fig. 4-39.

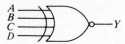

Fig. 4-39 A 4-input XNOR gate

4.51   Construct a truth table for a 4-input XNOR gate.
           Ans.

Inputs				Output	Inputs				Output
D	C	B	A	Y	D	C	B	A	Y
0	0	0	0	1	1	0	0	0	0
0	0	0	1	0	1	0	0	1	1
0	0	1	0	0	1	0	1	0	1
0	0	1	1	1	1	0	1	1	0
0	1	0	0	0	1	1	0	0	1
0	1	0	1	1	1	1	0	1	0
0	1	1	0	1	1	1	1	0	0
0	1	1	1	0	1	1	1	1	1

4.52   What would the pulse train at the output of the XNOR gate shown in Fig. 4-40 look like?
           Ans.   pulse $a = 1$ pulse $c = 0$ pulse $e = 0$ pulse $g = 1$
                    pulse $b = 0$ pulse $d = 1$ pulse $f = 1$ pulse $h = 0$

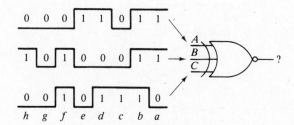

Fig. 4-40 Pulse-train problem

4.53   Given an OR gate and inverters, draw a logic diagram that will perform the 3-input NAND function.
           Ans.   See Fig. 4-41.

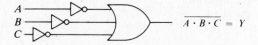

Fig. 4-41 3-input NAND function

4.54   Given a NOR gate and inverters, draw a logic diagram that will perform the 3-input AND function.
     Ans.   See Fig. 4-42.

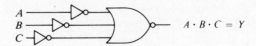

Fig. 4-42   3-input AND function

4.55   Given a NOR gate and inverters, draw a logic diagram that will perform the 5-input OR function.
     Ans.   See Fig. 4-43.

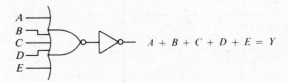

Fig. 4-43   5-input OR function

4.56   Draw a logic diagram for the Boolean expression $\overline{A} \cdot B \cdot C + A \cdot \overline{B} \cdot C + A \cdot B \cdot \overline{C} = Y$. Use inverters, AND gates, and an OR gate.   *Ans.*   See Fig. 4-44.

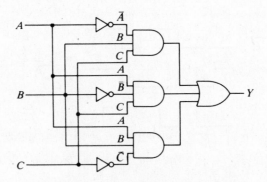

Fig. 4-44   An AND-OR logic circuit

4.57   Redraw the logic diagram for Prob. 4.56 by using three 2-input NAND gates and four 3-input NAND gates.   *Ans.*   See Fig. 4-45.

4.58   Write the Boolean expression for the circuit shown in Fig. 4-46.
     Ans. $A \cdot B + \overline{A} \cdot \overline{B} \cdot C = Y$

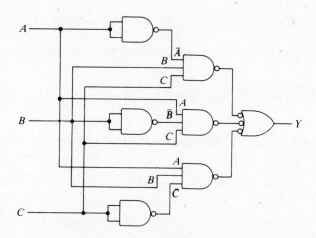

Fig. 4-45 An equivalent NAND logic circuit

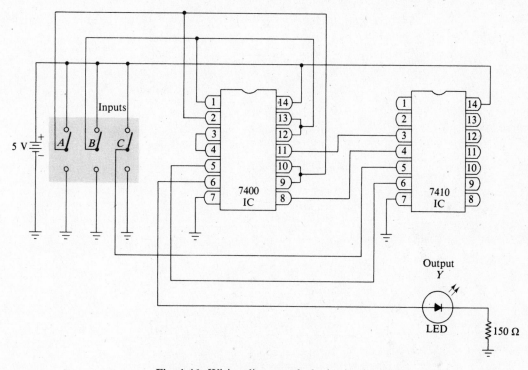

Fig. 4-46 Wiring diagram of a logic-circuit problem

4.59   Construct the truth table for the circuit shown in Fig. 4-46.
     Ans.

Inputs			Output	Inputs			Output
C	*B*	*A*	*Y*	*C*	*B*	*A*	*Y*
0	0	0	0	1	0	0	1
0	0	1	0	1	0	1	0
0	1	0	0	1	1	0	0
0	1	1	1	1	1	1	1

4.60   If switches *A*, *B*, and *C* shown in Fig. 4-46 are in the *up* position (logical 1), the output LED will be _____ (lit, not lit).

     Ans.   When all inputs are 1, the output of the circuit will be 1 according to the truth table and the output LED will be lit.

4.61   The unique output for the _____ logic gate is a LOW when all inputs are HIGH.

     Ans.   NAND

4.62   The unique output for the _____ logic gate is a HIGH when all inputs are LOW.

     Ans.   NOR

4.63   The _____ logic gate generates a HIGH output when an *odd number* of inputs are HIGH.

     Ans.   exclusive-OR or XOR.

Chapter 5

Simplifying Logic Circuits: Mapping

5-1 INTRODUCTION

Consider the Boolean expression $A \cdot \bar{B} + \bar{A} \cdot B + A \cdot B = Y$, a logic diagram for which is in Fig. 5-1a. Note that six gates must be used to implement this logic circuit, which performs the logic detailed in the truth table (Fig. 5-1c). From examination of the truth table, it is determined that a *single 2-input OR gate* will perform the function. It is found that the OR gate shown in Fig. 5-1b will be the simplest method of performing this logic. The logic circuits in Fig. 5-1a and b perform exactly the same logic function. Obviously a designer would choose the simplest, least expensive circuit, shown in Fig. 5-1b. It has been shown that the unsimplified Boolean expression ($A \cdot \bar{B} + \bar{A} \cdot B + A \cdot B = Y$) could be simplified to $A + B = Y$. The simplification was done by simple examination of the truth table and recognizing the OR pattern. Many Boolean expressions can be greatly simplified. Several systematic methods of simplification will be examined in this chapter.

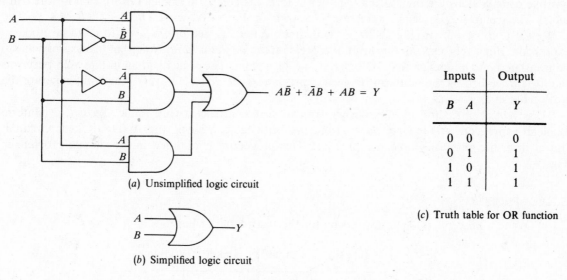

(a) Unsimplified logic circuit

(b) Simplified logic circuit

Inputs		Output
B	A	Y
0	0	0
0	1	1
1	0	1
1	1	1

(c) Truth table for OR function

Fig. 5-1

In this chapter simple logic gates will be used to implement combinational logic. Other techniques also are commonly used for simplifying more complex logic problems. They include the use of *data selectors* (multiplexers), *decoders*, *PLAs* (programmable logic arrays), *ROMs* (read-only memories) and *PROMs* (programmable read-only memories).

5-2 SUM-OF-PRODUCTS BOOLEAN EXPRESSIONS

It is customary when starting a logic-design problem to first construct a truth table. The table details the exact operation of the digital circuit. Consider the truth table in Fig. 5-2a. It contains the three variables C, B, and A. Note that only two combinations of variables will generate a 1 output. These combinations are shown in the shaded second and eighth lines of the truth table. From line 2, we say that "a not C AND a not B AND an A input will generate a 1 output." This is shown near the right side of line 2 as the Boolean expression $\bar{C} \cdot \bar{B} \cdot A$. The other combination of variables that will generate a 1 is shown in line 8 of the truth table. Line 8 reads as "a C AND a B AND an A input will generate a 1 output." The Boolean expression for line 8 is shown at the right as $C \cdot B \cdot A$. These two

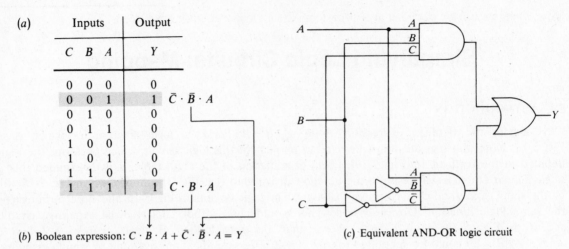

(b) Boolean expression: $C \cdot B \cdot A + \bar{C} \cdot \bar{B} \cdot A = Y$

(c) Equivalent AND-OR logic circuit

Fig. 5-2

possible combinations are then ORed together to form the complete Boolean expression for the truth table. The complete Boolean expression is shown in Fig. 5-2b as $C \cdot B \cdot A + \bar{C} \cdot \bar{B} \cdot A = Y$. The $C \cdot B \cdot A + \bar{C} \cdot \bar{B} \cdot A = Y$ in Fig. 5-2b is sometimes called a *sum-of-products* form of a Boolean expression. Engineers also call this form of expression the *minterm form*. Note that this expression can be translated into a familiar AND-OR pattern of logic gates. The logic diagram in Fig. 5-2c performs the logic described by the minterm Boolean expression $C \cdot B \cdot A + \bar{C} \cdot \bar{B} \cdot A = Y$ and will generate the truth table in Fig. 5-2a.

It is typical procedure in logic-design work to *first* construct a truth table. *Second*, a minterm Boolean expression is then determined from the truth table. *Finally*, the AND-OR logic circuit is drawn from the minterm Boolean expression. This procedure is outlined in the sample problem in Fig. 5-2.

SOLVED PROBLEMS

5.1 Write a *minterm* Boolean expression for the truth table in Fig. 5-3.

Inputs			Output	Inputs			Output
C	B	A	Y	C	B	A	Y
0	0	0	0	1	0	0	0
0	0	1	0	1	0	1	0
0	1	0	0	1	1	0	1
0	1	1	1	1	1	1	0

Fig. 5-3

Solution:

$$\bar{C} \cdot B \cdot A + C \cdot B \cdot \bar{A} = Y \quad \text{or} \quad \bar{C}BA + CB\bar{A} = Y$$

5.2 The Boolean expression developed in Prob. 5.1 was a _____ (maxterm, minterm) expression. This type of expression is also called the _____ (product-of-sums, sum-of-products) form.

Solution:

This type of Boolean expression $(\bar{C} \cdot B \cdot A + C \cdot B \cdot \bar{A} = Y)$ is called the minterm form or the sum-of-products form.

5.3   Diagram a logic circuit that will perform the logic in the truth table in Fig. 5-3.

Solution:

See Fig. 5-4.

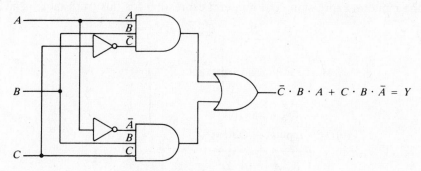

Fig. 5-4 Logic-diagram solution

5.4   Write a sum-of-products Boolean expression for the truth table in Fig. 5-5.

Solution:

$$\overline{C} \cdot \overline{B} \cdot \overline{A} + \overline{C} \cdot B \cdot A + C \cdot \overline{B} \cdot A = Y$$

Inputs			Output	Inputs			Output
C	B	A	Y	C	B	A	Y
0	0	0	1	1	0	0	0
0	0	1	0	1	0	1	1
0	1	0	0	1	1	0	0
0	1	1	1	1	1	1	0

Fig. 5-5

5.5   Diagram a logic circuit that will perform the logic in the truth table in Fig. 5-5.

Solution:

See Fig. 5-6.

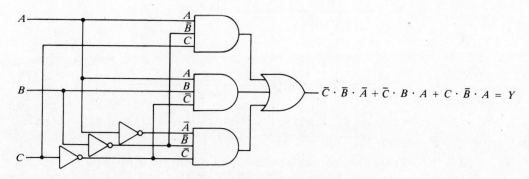

Fig. 5-6 Logic-diagram solution

5-3   PRODUCT-OF-SUMS BOOLEAN EXPRESSIONS

Consider the OR truth table in Fig. 5-7b. The Boolean expression for this truth table can be written in two forms, as was observed in the introductory section. The minterm Boolean expression is developed from the output 1s in the truth table. Each 1 in the output column becomes a term to be ORed in the minterm expression. The minterm expression for this truth table is given in Fig. 5-7c as

$$B \cdot A + B \cdot \overline{A} + \overline{B} \cdot A = Y$$

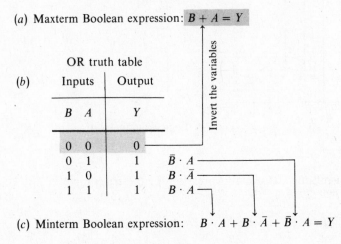

(a) Maxterm Boolean expression: $B + A = Y$

OR truth table

(b)

Inputs		Output
B	A	Y
0	0	0
0	1	1	$\overline{B} \cdot A$
1	0	1	$B \cdot \overline{A}$
1	1	1	$B \cdot A$

Invert the variables

(c) Minterm Boolean expression: $B \cdot A + B \cdot \overline{A} + \overline{B} \cdot A = Y$

Fig. 5-7

The truth table in Fig. 5-7 can also be described by using a *maxterm form* of Boolean expression. This type of expression is developed from the 0s in the output column of the truth table. For each 0 in the output column, an ORed term is developed. Note that the *input variables are inverted and then ORed*. The maxterm Boolean expression for this truth table is given in Fig. 5-7a. The maxterm expression for the OR truth table is shown as $B + A = Y$. This means the same thing as the familiar OR expression $A + B = Y$. For the truth table in Fig. 5-7, the maxterm Boolean expression turns out to be the simplest. Both the minterm and maxterm expressions accurately describe the logic of the truth table in Fig. 5-7.

Consider the truth table in Fig. 5-8a. The minterm expression for this truth table would be rather long. The *maxterm* Boolean expression is developed from the variables in lines 5 and 8. Each of these

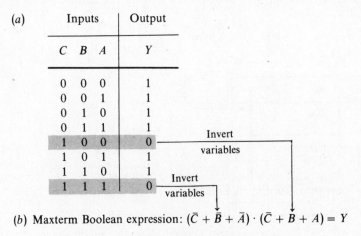

(a)

Inputs			Output
C	B	A	Y
0	0	0	1
0	0	1	1
0	1	0	1
0	1	1	1
1	0	0	0
1	0	1	1
1	1	0	1
1	1	1	0

Invert variables

Invert variables

(b) Maxterm Boolean expression: $(\overline{C} + \overline{B} + \overline{A}) \cdot (\overline{C} + B + A) = Y$

Fig. 5-8 Developing a maxterm expression

lines has a 0 in the output column. The variables are inverted and ORed with parentheses around them. The terms are then ANDed. The complete maxterm Boolean expression is given in Fig. 5-8b. The maxterm expression is also called the *product-of-sums form* of a Boolean expression. The product-of-sums term comes from the arrangement of the sum ($+$) and product ($\cdot$) symbols.

A maxterm Boolean expression would be implemented by using an OR-AND pattern of logic gates illustrated in Fig. 5-9. Note that the outputs of the two OR gates are feeding into an AND gate. The maxterm expression $(\overline{C} + \overline{B} + \overline{A}) \cdot (\overline{C} + B + A) = Y$ is implemented by using the OR-AND pattern of gates in Fig. 5-9.

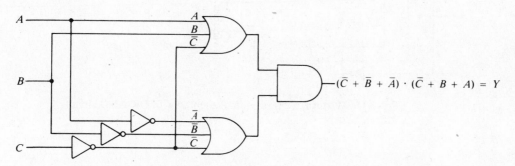

Fig. 5-9 Maxterm expression implemented with OR-AND circuit

SOLVED PROBLEMS

5.6   Write a *maxterm* Boolean expression for the truth table in Fig. 5-10.

Inputs			Output	Inputs			Output
C	B	A	Y	C	B	A	Y
0	0	0	1	1	0	0	1
0	0	1	0	1	0	1	1
0	1	0	1	1	1	0	0
0	1	1	1	1	1	1	1

Fig. 5-10

Solution:

$$(C + B + \overline{A}) \cdot (\overline{C} + \overline{B} + A) = Y$$

5.7   The Boolean expression developed in Prob. 5.6 is a _____ (maxterm, minterm) expression. This type of expression is also called the _____ (product-of-sums, sum-of-products) form.

Solution:

The type of Boolean expression developed in Prob. 5-6 is called the maxterm form or the product-of-sums form.

5.8   Diagram a logic circuit that will perform the logic in the truth table in Fig. 5-10.

   Solution:

   See Fig. 5-11.

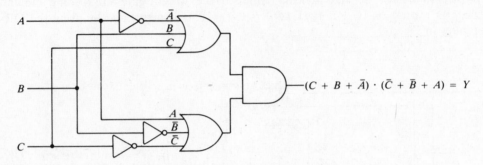

Fig. 5-11 Maxterm expression implemented with OR-AND circuit

5.9   The logic diagram of Prob. 5.8 is called the _____ (AND-OR, OR-AND) pattern of logic gates.

   Solution:

   The pattern of gates shown in Fig. 5-11 is called the OR-AND pattern.

5.10   Write the product-of-sums Boolean expression for the truth table in Fig. 5-12.

Inputs			Output	Inputs			Output
C	B	A	Y	C	B	A	Y
0	0	0	0	1	0	0	1
0	0	1	1	1	0	1	0
0	1	0	1	1	1	0	1
0	1	1	0	1	1	1	1

Fig. 5-12

   Solution:

   $$(C + B + A) \cdot (C + \bar{B} + \bar{A}) \cdot (\bar{C} + B + \bar{A}) = Y$$

5.11   Diagram a logic circuit that will perform the logic in the truth table in Fig. 5-12.

   Solution:

   See Fig. 5-13.

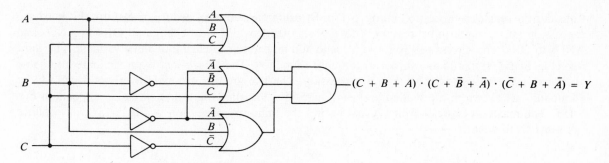

Fig. 5-13 Maxterm expression implemented with OR-AND circuit

5-4 USING DE MORGAN'S THEOREMS

Boolean algebra, the algebra of logic circuits, has many laws or theorems. *De Morgan's theorems* are very useful. They allow for easy transfer back and forth from the minterm to the maxterm form, of Boolean expression. They also allow for elimination of overbars that are over several variables.

De Morgan's theorems can be stated as follows:

$$\text{first theorem } \overline{A + B} = \overline{A} \cdot \overline{B} \qquad \text{second theorem } \overline{A \cdot B} = \overline{A} + \overline{B}$$

The first theorem changes the basic OR situation to an AND situation. A practical example of the first theorem is illustrated in Fig. 5-14a. The NOR gate on the left is equal in function to the AND gate (with inverted inputs) on the right. Note that the conversion is from the basic OR situation to the basic AND situation as shown by the gates in Fig. 5-14(a). This conversion is useful in getting rid of the long overbar on the NOR and can be used in converting from a minterm to a maxterm expression. The AND-looking symbol at the right in Fig. 5-14a would produce a NOR truth table.

$\overline{A + B} = Y$ $\overline{A} \cdot \overline{B} = Y$

(a) NOR functions

$\overline{A \cdot B} = Y$ $\overline{A} + \overline{B} = Y$

(b) NAND functions

Fig. 5-14 Applications of De Morgan's theorems

The second theorem changes the basic AND situation to an OR situation. A practical example of the second theorem is illustrated in Fig. 5-14b. The NAND gate on the left is equal in function to the OR gate (with inverted inputs) on the right. Again the long overbar is eliminated, and conversions can be done from maxterm to minterm forms of Boolean expressions. The OR-looking symbol at the right in Fig. 5-14b would produce a NAND truth table.

The symbols at the right in Fig. 5-14 are the alternate symbols used for the NOR and NAND logic functions. Figure 5-14 illustrates but one use of De Morgan's theorems.

Four steps are needed to transform a basic AND situation to an OR situation (or an OR to an AND situation). The four steps, based on De Morgan's theorems, are as follows:

1. Change all ORs to ANDs and all ANDs to ORs.
2. Complement each individual variable (add overbar to each).
3. Complement the entire function (add overbar to entire function).
4. Eliminate all groups of double overbars.

Consider the maxterm expression in Fig. 5-15a. By using the above procedure, transform this maxterm expression into a minterm expression. The *first step* (Fig. 5-15b) is to change all ORs to ANDs and ANDs to ORs. The *second step* (Fig. 5-15c) is to add an overbar to each individual variable. The *third step* (Fig. 5-15d) is to add an overbar to the entire function. The *fourth step* is to eliminate all double overbars and rewrite the final minterm expression. The five groups of double overbars which will be eliminated are shown in the shaded areas in Fig. 5-15e. The final minterm expression appears in Fig. 5-15f. The maxterm expression in Fig. 5-15a and the minterm expression in Fig. 5-15f will produce the same truth table.

$$\overline{(\overline{A} + \overline{B} + \overline{C}) \cdot (A + B + \overline{C})} = Y$$

(a) Maxterm expression

$$\overline{\overline{A} \cdot \overline{B} \cdot \overline{C} + \overline{A} \cdot \overline{B} \cdot \overline{C}}$$

(d) Third step

$$\overline{\overline{A} \cdot \overline{B} \cdot \overline{C} + A \cdot B \cdot \overline{C}}$$

(b) First step

$$\overline{\overline{A} \cdot \overline{B} \cdot \overline{C} + \overline{A} \cdot \overline{B} \cdot \overline{C}}$$

(e) Fourth step

$$\overline{\overline{A} \cdot \overline{\overline{B}} \cdot \overline{C} + \overline{\overline{A}} \cdot \overline{B} \cdot \overline{\overline{C}}}$$

(c) Second step

$$A \cdot B \cdot C + \overline{A} \cdot \overline{B} \cdot C = Y$$

(f) Minterm expression

Fig. 5-15 From maxterm to minterm expressions using De Morgan's theorems

SOLVED PROBLEMS

5.12 Convert the Boolean expression $\overline{(A + \overline{B} + \overline{C}) \cdot (\overline{A} + B + \overline{C})} = Y$ to its minterm form. Show each step as in Fig. 5-15.

Solution:

Maxterm expression	$\overline{(A + \overline{B} + \overline{C}) \cdot (\overline{A} + B + \overline{C})} = Y$
First step	$\overline{A \cdot \overline{B} \cdot \overline{C} + \overline{A} \cdot B \cdot \overline{C}}$
Second step	$\overline{\overline{A} \cdot \overline{\overline{B}} \cdot \overline{\overline{C}} + \overline{\overline{A}} \cdot \overline{B} \cdot \overline{\overline{C}}}$
Third step	$\overline{\overline{A} \cdot \overline{\overline{B}} \cdot \overline{\overline{C}} + \overline{\overline{A}} \cdot \overline{B} \cdot \overline{\overline{C}}}$
Fourth step	eliminate double overbars
Minterm expression	$\overline{A} \cdot B \cdot C + A \cdot \overline{B} \cdot C = Y$

5.13 Convert the Boolean expression $\overline{C} \cdot \overline{B} \cdot \overline{A} + C \cdot B \cdot \overline{A} = Y$ to its maxterm form. Show each step in the procedure.

Solution:

Minterm expression	$\overline{\overline{C} \cdot \overline{B} \cdot \overline{A} + C \cdot B \cdot \overline{A}} = Y$
First step	$\overline{(\overline{C} + \overline{B} + \overline{A}) \cdot (C + B + \overline{A})}$
Second step	$\overline{(\overline{\overline{C}} + \overline{\overline{B}} + \overline{\overline{A}}) \cdot (\overline{C} + \overline{B} + \overline{\overline{A}})}$
Third step	$\overline{(\overline{\overline{C}} + \overline{\overline{B}} + \overline{\overline{A}}) \cdot (\overline{C} + \overline{B} + \overline{\overline{A}})}$
Fourth step	eliminate double overbars
Maxterm expression	$(C + B + A) \cdot (\overline{C} + \overline{B} + A) = Y$

5.14   Convert the Boolean expression $\overline{A \cdot B} = Y$ to a sum-of-products form.

Solution:

$A + B = Y$

5.15   Convert the Boolean expression $\overline{A + B} = Y$ to a product-of-sums form.

Solution:

$A \cdot B = Y$

5-5 USING NAND LOGIC

All digital systems can be constructed from the fundamental AND, OR, and NOT gates. Because of their low cost and availability, NAND gates are widely used to replace AND, OR, and NOT gates. There are several steps in converting from AND-OR logic to NAND logic:

1. Draw an AND-OR logic circuit.
2. Place a bubble at the output of each AND gate.
3. Place a bubble at each input to the OR gate.
4. Check the logic levels on lines coming from the inputs and going to the outputs.

Consider the minterm Boolean expression in Fig. 5-16a. To implement this expression by using NAND logic, the steps outlined above will be followed. The *first step* (Fig. 5-16b) is to diagram an AND-OR logic circuit. The *second step* is to place a bubble (small circle) at the output of each AND gate. This changes the AND gates to NAND gates. Figure 5-16c shows bubbles added to gates 1 and 2. The *third step* is to place a bubble (small circle) at each input of the OR gate. This will convert the OR gate to a NAND gate. Figure 5-16c shows three bubbles added to the inputs of gate 3. The *fourth step* involves examination of the input and output lines from the AND and OR symbols to see if any of the logic levels have been changed by the addition of bubbles. On examination of the circuit shown in Fig. 5-16c, it is found that the added bubble at point X has changed the input logic level on OR symbol 3. The AND-OR diagram in Fig. 5-16b shows that a HIGH logic level is connected from input E to the OR gate. The HIGH, or 1, activates the OR gate. A HIGH must also arrive at the input of symbol 3 in Fig. 5-16c. This is accomplished by adding the shaded inverter in input line E. In actual practice, a NAND gate is used as the inverter. The double inversion will deliver the HIGH logic level to the OR symbol to activate the OR. The invert bubbles between gates 1 and 3 cancel one another. Likewise, the invert bubbles between gates 2 and 3 cancel. The NAND logic circuit shown in Fig. 5-16c will produce the same truth table as that for the AND-OR circuit.

$(A \cdot B) + (C \cdot D) + E = Y$

(a)

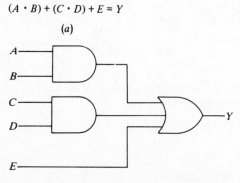

(b) Equivalnet AND-OR logic circuit

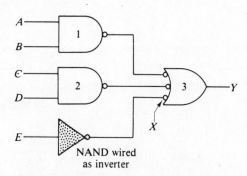

NAND wired
as inverter

(c) Equivalent NAND logic circuit

Fig. 5-16

Using NAND logic does not always simplify a circuit. The example shown in Fig. 5-16 shows that the AND-OR circuit would probably be preferred over the NAND circuit because of the fewer gates used. Most manufacturers of ICs do produce a good variety of all types of gates. The logic designer can usually select the logic that produces the simplest circuitry.

SOLVED PROBLEMS

5.16 Diagram an AND-OR logic circuit for the Boolean expression $A \cdot B + \bar{C} + D \cdot E = Y$.

Solution:

See Fig. 5-17.

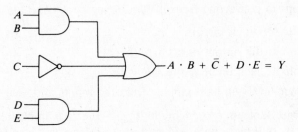

Fig. 5-17 AND-OR logic-circuit solution

5.17 Diagram a NAND logic circuit from the AND-OR circuit in Prob. 5.16. The NAND circuit should perform the logic in the expression $A \cdot B + \bar{C} + D \cdot E = Y$.

Solution:

See Fig. 5-18.

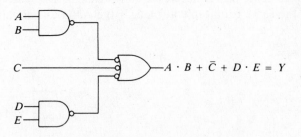

Fig. 5-18 NAND logic-circuit solution

5.18 Diagram an AND-OR logic circuit for the Boolean expression $A + (B \cdot C) + \bar{D} = Y$.

Solution:

See Fig. 5-19.

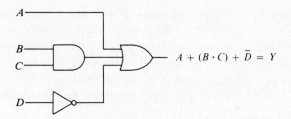

Fig. 5-19 AND-OR logic-circuit solution

5.19 Diagram a NAND logic circuit from the AND-OR circuit in Prob. 5.18. The NAND circuit should perform the logic in the expression $A + (B \cdot C) + \bar{D} = Y$.

Solution:

See Fig. 5-20.

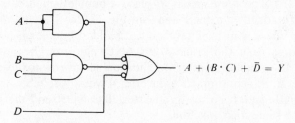

Fig. 5-20 NAND logic-circuit solution

5-6 USING NOR LOGIC

The NAND gate was the "universal gate" used for substituting in an AND-OR logic pattern. When a maxterm Boolean expression forms an OR-AND gate pattern, the NAND gate does not work well. The NOR gate becomes the "universal gate" when substituting in OR-AND logic patterns. The NOR gate is not as widely used as the NAND gate.

Consider the maxterm Boolean expression written in Fig. 5-21a. The expression is drawn as an OR-AND logic diagram in Fig. 5-21b. The OR-AND pattern is redrawn with NOR gates in Fig. 5-21c. Each OR gate and AND gate is replaced by a NOR gate. Gates 1 and 2 in Fig. 5-21c are shown as the

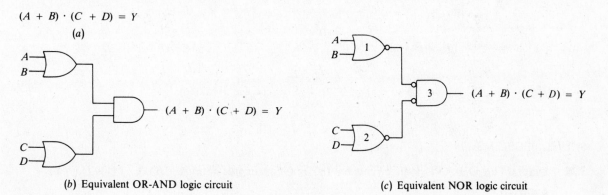

$(A + B) \cdot (C + D) = Y$

(a)

(b) Equivalent OR-AND logic circuit

(c) Equivalent NOR logic circuit

Fig. 5-21

standard NOR symbols. Gate 3 is the alternate NOR symbol. The substitution works because the two invert bubbles between gates 1 and 3 cancel each other. Likewise, the two invert bubbles between gates 2 and 3 cancel. This leaves the two OR symbols (1 and 2) driving an AND symbol (3). This is the pattern used in the original OR-AND logic diagram in Fig. 5-21b.

The procedure for converting from a maxterm Boolean expression to a NOR logic circuit is similar to that used in NAND logic. The steps for converting the NOR logic are as follows:

1. Draw an OR-AND logic circuit.
2. Place a bubble at each input to the AND gate.
3. Place a bubble at the output of each OR gate.
4. Check the logic levels on lines coming from the inputs and going to the output.

Consider the maxterm Boolean expression in Fig. 5-22a. To implement this expression by using NOR logic, the four steps outlined above will be followed. The *first step* (Fig. 5-22b) is to draw an OR-AND logic circuit. The *second step* is to place a bubble (small circle) at each input to the AND gate. This changes it to a NOR gate. The "AND looking" symbol with the three bubbles at the inputs is then a NOR gate (Fig. 5-22c). The *third step* is to place a bubble (small circle) at the output of each OR gate. The bubbles are added to gates 1 and 2 in Fig. 5-22c. The *fourth step* is to examine the input and output lines for changes in logic levels due to added bubbles. The bubble added at point Z in Fig. 5-22c is a change from the original OR-AND pattern. The inverting effect of bubble Z is canceled by adding the shaded inverter 4. The double inversion (inverter 4 and invert bubble Z) cancels in input line E. In actual practice, inverter 4 would probably be a NOR gate. By shorting all inputs together, a NOR gate becomes an inverter. The NOR and the OR-AND circuits pictured in Fig. 5-22 perform the same logic function.

The NOR gate was used as a "universal gate" in the previous example. Using NOR logic may or may not simplify the circuit. In this case the OR-AND circuit might be preferred.

$$(A + B) \cdot (C + D) \cdot E = Y$$

(a)

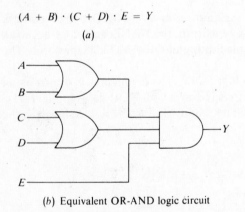

(b) Equivalent OR-AND logic circuit

(c) Equivalent NOR logic circuit

Fig. 5-22

SOLVED PROBLEMS

5.20 Diagram an OR-AND logic circuit for the Boolean expression $(A + B) \cdot \overline{C} \cdot (D + E) = Y$.

Solution:
See Fig. 5-23.

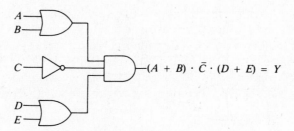

Fig. 5-23   OR-AND logic-circuit solution

5.21   Diagram a NOR logic circuit from the OR-AND circuit in Prob. 5.20. The NOR circuit should perform the logic in the Boolean expression $(A + B) \cdot \bar{C} \cdot (D + E) = Y$.

Solution:

See Fig. 5-24.

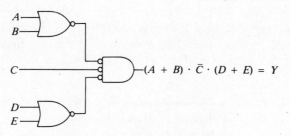

Fig. 5-24   NOR logic-circuit problem

5.22   Diagram an OR-AND logic circuit for the Boolean expression $\bar{A} \cdot (B + C) \cdot D = Y$.

Solution:

See Fig. 5-25.

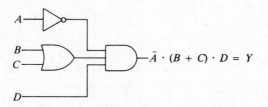

Fig. 5-25   OR-AND logic-circuit solution

5.23   Diagram a NOR logic circuit from the OR-AND circuit in Prob. 5.22. The NOR circuit should perform the logic in the Boolean expression $\bar{A} \cdot (B + C) \cdot D = Y$.

Solution:

See Fig. 5-26.

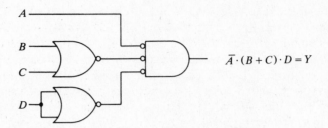

Fig. 5-26 NOR logic-circuit solution

5-7 KARNAUGH MAPS

Boolean algebra is the basis for any simplification of logic circuits. One of the easiest ways to simplify logic circuits is to use the *Karnaugh map* method. This graphic method is based on Boolean theorems. It is only one of several methods used by logic designers to simplify logic circuits. Karnaugh maps are sometimes referred to as *K maps*.

The *first step* in the Karnaugh mapping procedure is to develop a minterm Boolean expression from a truth table. Consider the familiar truth table in Fig. 5-27*a*. Each 1 in the *Y* column of the truth table produces two variables ANDed together. These ANDed groups are then ORed to form a sum-of-products (minterm) type of Boolean expression (Fig. 5-27*b*). This expression will be referred to as the *unsimplified* Boolean expression. The *second step* in the mapping procedure is to plot 1s in the Karnaugh map in Fig. 5-27*c*. Each ANDed set of variables from the minterm expression is placed in

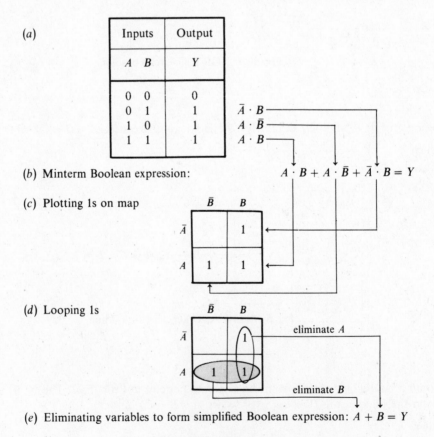

Fig. 5-27 Using a map

the *appropriate* square of the map. The map is just a very special output column of the truth table. The *third step* is to *loop* adjacent groups of two, four, or eight 1s together. Figure 5-27d shows two loops drawn on the map. Each loop contains two 1s. The *fourth step* is to eliminate variables. Consider first the shaded loop in Fig. 5-27d. Note that a B and a $\bar{B}$ (not B) are contained within the shaded loop. *When a variable and its complement are within a loop, that variable is eliminated.* From the shaded loop, the B and $\bar{B}$ terms are eliminated, leaving the A variable (Fig. 5-27e). Next consider the unshaded loop in Fig. 5-27d. It contains an A and a $\bar{A}$ (not A). The A and $\bar{A}$ terms are eliminated, leaving only the B variable (Fig. 5-27e). The *fifth step* is to OR the remaining variables. The final *simplified* Boolean expression is $A + B = Y$ (Fig. 5-27e). The simplified expression is that of a 2-input OR gate.

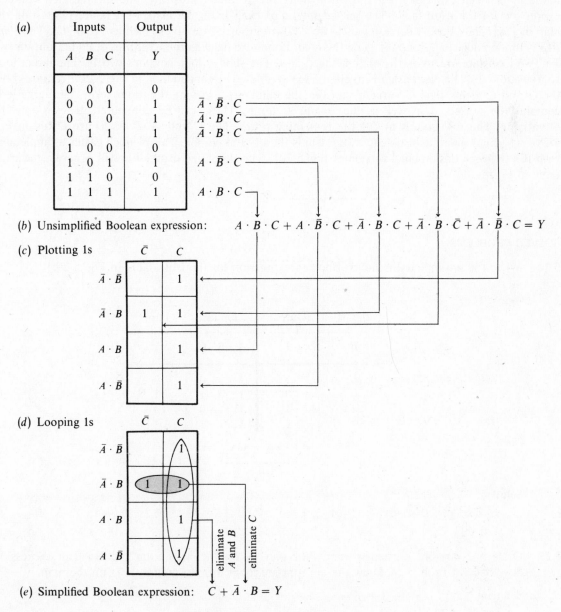

Fig. 5-28 Using a three-variable map

In summary, the steps in simplifying a logic expression using a Karnaugh map are as follows:

1. Write a minterm Boolean expression from the truth table.
2. Plot a 1 on the map for each ANDed group of variables. (The number of 1s in the Y column of the truth table will equal the number of 1s on the map.)
3. Draw loops around adjacent groups of two, four, or eight 1s on the map. (The loops may overlap.)
4. Eliminate the variable(s) that appear(s) with its (their) complement(s) within a loop, and save the variable(s) that is (are) left.
5. Logically OR the groups that remain to form the simplified minterm expression.

Consider the truth table in Fig. 5-28a. The *first step* in using the Karnaugh map is to write the minterm Boolean expression for the truth table. Figure 5-28b illustrates the unsimplified minterm expression for the truth table. The *second step* is plotting 1s on the map. Five 1s are plotted on the map in Fig. 5-28c. Each 1 corresponds to an ANDed group of variables (such as $A \cdot B \cdot C$). The *third step* is to loop adjacent groups of 1s on the map. Loops are placed around groups of eight, four, or two 1s. Two loops are drawn on the map in Fig. 5-28d. The shaded loop contains two 1s. The larger loop contains four 1s. The *fourth step* is to eliminate variables. The shaded loop in Fig. 5-28d contains both the C and $\overline{C}$ terms. The C variable can thus be eliminated, leaving the $\overline{A} \cdot B$ term. The large loop contains the A and $\overline{A}$ as well as the B and $\overline{B}$ terms. These can be eliminated, leaving only the C variable. The *fifth step* is to OR the remaining terms. The C and $\overline{A} \cdot B$ terms are ORed in Fig. 5-28e. The final simplified Boolean expression is then $C + \overline{A} \cdot B = Y$. This is much easier to implement with ICs than the unsimplified version of Fig. 5-28b. The simplified expression will generate the truth table in Fig. 5-28a.

SOLVED PROBLEMS

5.24 Write the unsimplified minterm Boolean expression for the truth table in Fig. 5-29.

Inputs			Output	Inputs			Output
A	B	C	Y	A	B	C	Y
0	0	0	1	1	0	0	0
0	0	1	0	1	0	1	1
0	1	0	1	1	1	0	0
0	1	1	0	1	1	1	1

Fig. 5-29

Solution:
$$\overline{A} \cdot \overline{B} \cdot \overline{C} + \overline{A} \cdot B \cdot \overline{C} + A \cdot \overline{B} \cdot C + A \cdot B \cdot C = Y$$

5.25 Draw a 3-variable Karnaugh map. Plot four 1s on the map from the Boolean expression developed in Prob. 5.24. Draw the appropriate loops around groups of 1s on the map.

Solution:
See Fig. 5-30.

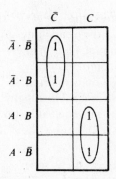

Fig. 5-30 Karnaugh map solution

5.26 Write the simplified Boolean expression based on the Karnaugh map from Prob. 5.25.

Solution:

$$\bar{A} \cdot \bar{C} + A \cdot C = Y$$

5-8 KARNAUGH MAPS WITH FOUR VARIABLES

Consider the truth table with four variables in Fig. 5-31a. The *first step* in simplification by using a Karnaugh map is to write the minterm Boolean expression. The lengthy unsimplified minterm expression appears in Fig. 5-31b. An ANDed group for four variables is written for each 1 in the Y column of the truth table. The *second step* is to plot 1s on the Karnaugh map. Nine 1s are plotted on the map in Fig. 5-31c. Each 1 on the map represents an ANDed group of terms from the unsimplified expression. The *third step* is to loop adjacent groups of 1s. Adjacent groups of eight, four, or two 1s are looped. Larger loops provide more simplification. Two loops have been drawn in Fig. 5-31c. The larger loop contains eight 1s. The *fourth step* is to eliminate variables. The large loop in Fig. 5-31c eliminates the A, B, and C variables. This leaves the D term. The small loop contains two 1s and

(a)

Inputs				Output
A	B	C	D	Y
0	0	0	0	0
0	0	0	1	1
0	0	1	0	0
0	0	1	1	1
0	1	0	0	0
0	1	0	1	1
0	1	1	0	1
0	1	1	1	1
1	0	0	0	0
1	0	0	1	1
1	0	1	0	0
1	0	1	1	1
1	1	0	0	0
1	1	0	1	1
1	1	1	0	0
1	1	1	1	1

(b) Unsimplified minterm expression

$$\bar{A} \cdot \bar{B} \cdot \bar{C} \cdot D + \bar{A} \cdot \bar{B} \cdot C \cdot D + \bar{A} \cdot B \cdot \bar{C} \cdot D + \bar{A} \cdot B \cdot C \cdot \bar{D}$$
$$+ \bar{A} \cdot B \cdot C \cdot D + A \cdot \bar{B} \cdot \bar{C} \cdot D + A \cdot \bar{B} \cdot C \cdot D$$
$$+ A \cdot B \cdot \bar{C} \cdot D + A \cdot B \cdot C \cdot D = Y$$

(c) Plotting and looping 1s on map

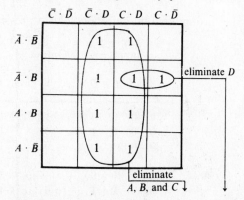

(d) Simplified Boolean expression: $D + \bar{A} \cdot B \cdot C = Y$

Fig. 5-31 Using a four-variable map

eliminates the D variable. That leaves the $\overline{A} \cdot B \cdot C$ term. The *fifth step* is to logically OR the remaining terms. Figure 5-31d shows the remaining groups ORed to form the simplified minterm expression $D + \overline{A} \cdot B \cdot C = Y$. The amount of simplification in this example is obvious when the two Boolean expressions in Fig. 5-31 are compared.

Consider the 3-variable Karnaugh map in Fig. 5-32a. The letters have been omitted from the edges of the map to simplify the illustration. How many loops can be drawn on this map? There are *no adjacent groups of 1s*, and therefore no loops are drawn in Fig. 5-32a. No simplification is possible in the example shown in Fig. 5-32a.

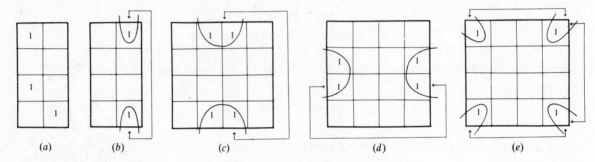

| (a) | (b) | (c) | (d) | (e) |

Fig. 5-32 Some unusual looping variations

The 3-variable Karnaugh map in Fig. 5-32b contains two 1s. Think of the top and bottom edges of the map as being connected as if rolled into a tube. The 1s can then be looped into a group of two, as shown in Fig. 5-32b. One variable can thus be eliminated.

Consider the 4-variable Karnaugh maps in Fig. 5-32c and d. The top and bottom edges of the map are considered connected for looping purposes in Fig. 5-32c. The 1s can then be looped into a group of four 1s, and two terms can be eliminated. In Fig. 5-32d the right edge of the map is considered connected to the left edge. The four 1s are looped into a single loop. Two variables are thus eliminated.

Another looping variation is illustrated in Fig. 5-32e. The corners of the map are considered connected as if the map were wrapped around a ball. The four 1s in the corners of the map are then looped into a single loop. The single loop of four 1s thereby eliminates two variables.

SOLVED PROBLEMS

5.27 Write the unsimplified minterm Boolean expression for the truth table in Fig. 5-33.

Inputs				Output	Inputs				Output
A	B	C	D	Y	A	B	C	D	Y
0	0	0	0	1	1	0	0	0	0
0	0	0	1	0	1	0	0	1	0
0	0	1	0	0	1	0	1	0	1
0	0	1	1	0	1	0	1	1	1
0	1	0	0	1	1	1	0	0	0
0	1	0	1	0	1	1	0	1	0
0	1	1	0	0	1	1	1	0	1
0	1	1	1	0	1	1	1	1	1

Fig. 5-33

Solution:

$$\overline{A}\cdot\overline{B}\cdot\overline{C}\cdot\overline{D}+\overline{A}\cdot B\cdot\overline{C}\cdot\overline{D}+A\cdot\overline{B}\cdot C\cdot\overline{D}+A\cdot\overline{B}\cdot C\cdot D+A\cdot B\cdot C\cdot\overline{D}+A\cdot B\cdot C\cdot D=Y$$

5.28 Draw a 4-variable Karnaugh map. Plot six 1s on the map from the Boolean expression developed in Prob. 5.27. Draw the appropriate loops around groups of 1s on the map.

Solution:

See Fig. 5-34.

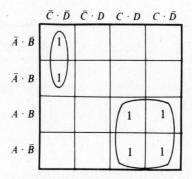

Fig. 5-34 Karnaugh map solution

5.29 Write the simplified Boolean expression based on the Karnaugh map from Prob. 5.28.

Solution:

$$A\cdot C+\overline{A}\cdot\overline{C}\cdot\overline{D}=Y$$

5.30 Write the unsimplified sum-of-products Boolean expression for the truth table in Fig. 5-35.

Inputs				Output	Inputs				Output
A	B	C	D	Y	A	B	C	D	Y
0	0	0	0	1	1	0	0	0	0
0	0	0	1	0	1	0	0	1	0
0	0	1	0	1	1	0	1	0	0
0	0	1	1	0	1	0	1	1	0
0	1	0	0	1	1	1	0	0	0
0	1	0	1	0	1	1	0	1	0
0	1	1	0	1	1	1	1	0	0
0	1	1	1	0	1	1	1	1	1

Fig. 5-35

Solution:

$$\overline{A}\cdot\overline{B}\cdot\overline{C}\cdot\overline{D}+\overline{A}\cdot\overline{B}\cdot C\cdot\overline{D}+\overline{A}\cdot B\cdot\overline{C}\cdot\overline{D}+\overline{A}\cdot B\cdot C\cdot\overline{D}+A\cdot B\cdot C\cdot D=Y$$

5.31 Draw a 4-variable Karnaugh map. Plot five 1s on the map from the Boolean expression developed in Prob. 5.30. Draw the appropriate loops around groups of 1s on the map.

Solution:

See Fig. 5-36.

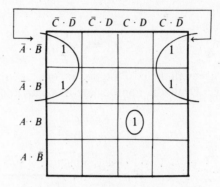

Fig. 5-36 Karnaugh map solution

5.32   Write the simplified Boolean expression based on the Karnaugh map from Prob. 5.31.

Solution:

$$\overline{A} \cdot \overline{D} + A \cdot B \cdot C \cdot D = Y$$

5-9   USING MAPS WITH MAXTERM EXPRESSIONS

A different form of the Karnaugh map is used with maxterm Boolean expressions. The steps for simplifying maxterm expressions are as follows:

1.   Write a maxterm Boolean expression from the truth table. (Note the inverted form in Fig. 5-37a.)

2.   Plot a 1 on the map for each ORed group of variables. The number of 0s in the Y column of the truth table will equal the number of 1s on the map.

3.   Draw loops around adjacent groups of two, four, or eight 1s on the map.

4.   Eliminate the variable(s) that appear(s) with its (their) complement(s) within a loop, and save the variable(s) that is (are) left.

5.   Logically AND the groups that remain to form the simplified maxterm expression.

Consider the truth table in Fig. 5-37a. The *first step* in simplifying a maxterm expression by using a Karnaugh map is to write the expression in unsimplified form. Figure 5-37a illustrates how a maxterm is written for *each 0 in the Y column* of the truth table. The terms of the ORed group are *inverted* from the way they appear in the truth table. The ORed groups are then ANDed to form the unsimplified maxterm Boolean expression in Fig. 5-37b. The *second step* is to plot 1s on the map for each ORed group. The three maxterms in the unsimplified expression are placed as three 1s on the *revised* Karnaugh map (Fig. 5-37c). The *third step* is to loop adjacent groups of eight, four, or two 1s on the map. Two loops have been drawn on the map in Fig. 5-37c. Each loop contains two 1s. The *fourth step* is to eliminate variables. The shaded loop in Fig. 5-37c is shown to eliminate the A variable. This leaves the maxterm $(B + C)$. The partially unshaded loop is shown to eliminate the B variable. This leaves the maxterm $(\overline{A} + C)$. The *fifth step* is the ANDing of the remaining terms. Figure 5-37d shows the two maxterms being ANDed to form the simplified maxterm Boolean expression $(B + C) \cdot (\overline{A} + C) = Y$. Compare this simplified maxterm expression with the simplified minterm expression from Fig. 5-28e. These two expressions were developed from the *same* truth table. The minterm expression $(C + \overline{A} \cdot B = Y)$ is slightly easier to implement by using logic gates.

The maxterm mapping procedure and Karnaugh map are different from those used for minterm expressions. Both techniques should be tried on a truth table to find the less costly logic circuit.

A 4-variable Karnaugh map *for maxterm expressions* is illustrated in Fig. 5-38. Note the special pattern of letters on the left and top edges of the map. *Care must always be used to position all the terms correctly when drawing maps.*

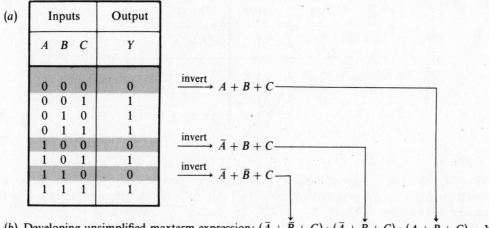

(b) Developing unsimplified maxterm expression: $(\bar{A} + \bar{B} + C) \cdot (\bar{A} + B + C) \cdot (A + B + C) = Y$

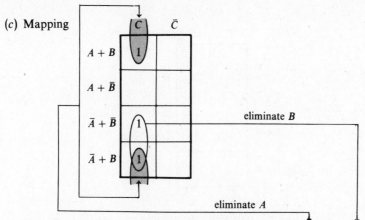

(d) Eliminating variables to yield simplified expression: $(B + C) \cdot (\bar{A} + C) = Y$

Fig. 5-37 Mapping with maxterm expressions

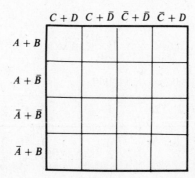

Fig. 5-38 A four-variable maxterm Karnaugh map

SOLVED PROBLEMS

5.33 Write the unsimplified maxterm Boolean expression for the truth table in Fig. 5-39. (Be sure to note the inverted form.)

Solution:

$$(A + B + \bar{C}) \cdot (\bar{A} + B + \bar{C}) \cdot (\bar{A} + \bar{B} + C) \cdot (\bar{A} + \bar{B} + \bar{C}) = Y$$

Inputs			Output	Inputs			Output
A	B	C	Y	A	B	C	Y
0	0	0	1	1	0	0	1
0	0	1	0	1	0	1	0
0	1	0	1	1	1	0	0
0	1	1	1	1	1	1	0

Fig. 5-39

5.34 Draw a 3-variable Karnaugh map for maxterm expressions. Plot four 1s on the map for the maxterm Boolean expression developed in Prob. 5.33. Draw the appropriate loops around groups of 1s on the map.

Solution:

See Fig. 5-40.

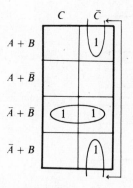

Fig. 5-40 Maxterm map solution

5.35 Write the simplified Boolean expression based on the Karnaugh map from Prob. 5.34.

Solution:

$(\overline{A} + \overline{B}) \cdot (B + \overline{C}) = Y$

5.36 Write the unsimplified product-of-sums Boolean expression for the truth table in Fig. 5-41.

Inputs				Output	Inputs				Output
A	B	C	D	Y	A	B	C	D	Y
0	0	0	0	0	1	0	0	0	0
0	0	0	1	0	1	0	0	1	1
0	0	1	0	0	1	0	1	0	0
0	0	1	1	1	1	0	1	1	1
0	1	0	0	1	1	1	0	0	1
0	1	0	1	1	1	1	0	1	1
0	1	1	0	1	1	1	1	0	1
0	1	1	1	1	1	1	1	1	1

Fig. 5-41

Solution:

$(A + B + C + D) \cdot (A + B + C + \overline{D}) \cdot (A + B + \overline{C} + D) \cdot (\overline{A} + B + C + D) \cdot (\overline{A} + B + \overline{C} + D) = Y$

5.37   Draw a 4-variable product-of-sums Karnaugh map. Plot five 1s on the map for the Boolean expression developed in Prob. 5.36. Draw the appropriate loops around groups of 1s on the map.

Solution:

See Fig. 5-42.

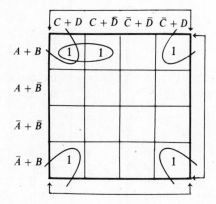

Fig. 5-42   Maxterm Karnaugh map solution

5.38   Write the simplified product-of-sums Boolean expression based on the Karnaugh map from Prob. 5.37.

Solution:

$$(A + B + C) \cdot (B + D) = Y$$

5-10   DON'T CARES ON KARNAUGH MAPS

Consider the table for BCD (8421) numbers given in Fig. 5-43. Note that the binary numbers 0000 to 1001 on the table are used to specify decimal numbers from 0 to 9. For convenience, the table is completed in the shaded section, which shows other possible combinations of the variables D, C, B, and A. These six combinations (1010, 1011, 1100, 1101, 1110, and 1111) are not used by the BCD code. These combinations are called *don't cares* when plotted on a Karnaugh map. The don't cares may have some effect on simplifying any logic diagram that might be constructed.

Suppose a problem specifying that a warning light would come ON when the BCD count reached 1001 (decimal 9); see the truth table in Fig. 5-44. See 1 is placed in the output column (Y) of the truth table after the input 1001. The Boolean expression for this table (above the shaded section) is $D \cdot \overline{C} \cdot \overline{B} \cdot A = Y$. This is shown to the right of the table. The "not used" combinations in the shaded section of the truth table might have some effect on this problem. A Karnaugh map is drawn in Fig. 5-45b. The 1 for the $D \cdot \overline{C} \cdot \overline{B} \cdot A$ term is plotted on the map. The six *don't cares* (X's from the truth table) are plotted as X's on the map. An X on the map means that square can be either a 1 or a 0. A loop is drawn around adjacent 1s. The X's on the map can be considered 1s, so the single loop is drawn around the 1 and three X's. Remember that only groups of two, four, or eight adjacent 1s and X's are looped together. The loop contains four squares, which will eliminate two variables. The B and C variables are eliminated, leaving the simplified Boolean expression $D \cdot A = Y$ in Fig. 5-45c.

As was said earlier, unused combinations from a truth table are called don't cares. They are shown as X's on a Karnaugh map. Including don't cares (X's) in loops on a map helps to further simplify Boolean expressions.

BCD (8421) number				Decimal equivalent
D	C	B	A	
8s	4s	2s	1s	
0	0	0	0	0
0	0	0	1	1
0	0	1	0	2
0	0	1	1	3
0	1	0	0	4
0	1	0	1	5
0	1	1	0	6
0	1	1	1	7
1	0	0	0	8
1	0	0	1	9
1	0	1	0	not used
1	0	1	1	not used
1	1	0	0	not used
1	1	0	1	not used
1	1	1	0	not used
1	1	1	1	not used

Fig. 5-43 Table of BCD numbers

Inputs				Output
D	C	B	A	Y
8s	4s	2s	1s	
0	0	0	0	0
0	0	0	1	0
0	0	1	0	0
0	0	1	1	0
0	1	0	0	0
0	1	0	1	0
0	1	1	0	0
0	1	1	1	0
1	0	0	0	0
1	0	0	1	1	$D \cdot \bar{C} \cdot \bar{B} \cdot A$
1	0	1	0	X
1	0	1	1	X
1	1	0	0	X
1	1	0	1	X
1	1	1	0	X
1	1	1	1	X

Fig. 5-44

$$D \cdot \bar{C} \cdot \bar{B} \cdot A = Y$$

(*a*) Unsimplified Boolean expression

$$D \cdot A = Y$$

(*c*) Simplified Boolean expression

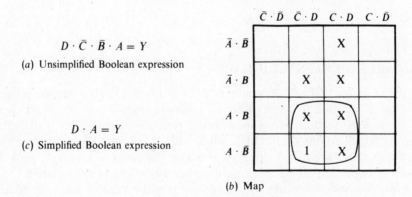

(*b*) Map

Fig. 5-45 Using a map

SOLVED PROBLEMS

5.39 Write the unsimplified minterm Boolean expression for the BCD truth table in Fig. 5-46.

Solution:

$$D \cdot \bar{C} \cdot \bar{B} \cdot \bar{A} + D \cdot \bar{C} \cdot \bar{B} \cdot A = Y$$

Inputs				Output	Inputs				Output
D	C	B	A	Y	D	C	B	A	Y
8s	4s	2s	1s		8s	4s	2s	1s	
0	0	0	0	0	0	1	0	1	0
0	0	0	1	0	0	1	1	0	0
0	0	1	0	0	0	1	1	1	0
0	0	1	1	0	1	0	0	0	1
0	1	0	0	0	1	0	0	1	1

Fig. 5-46

5.40 Draw a 4-variable minterm Karnaugh map. Plot two 1s and six X's (for the don't cares) on the map based on the truth table in Fig. 5-46. Draw the appropriate loops around groups of 1s and X's on the map.

Solution:

See Fig. 5-47.

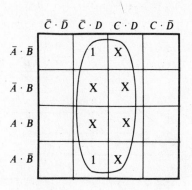

Fig. 5-47 Karnaugh map solution

5.41 Write the simplified Boolean expression based on the Karnaugh map from Prob. 5.40.

Solution:

$D = Y$

5-11 KARNAUGH MAPS WITH FIVE VARIABLES

A three-dimensional Karnaugh map can be used to solve logic problems with five variables. The map used to simplify 5-variable minterm Boolean expressions is shown in Fig. 5-48c. Notice that both the top (E) plane and bottom ($\bar{E}$) plane are duplicates of the 4-variable minterm map used in Sec. 5-8. The procedure for simplifying a minterm logic expression using a 5-variable Karnaugh map is like those used previously.

Consider the truth table with five variables in Fig. 5-48a. The *first step* in simplification is to write the minterm Boolean expression. The lengthy unsimplified minterm Boolean expression appears in Fig. 5-48b. An ANDed group of five variables is written for each 1 in the Y column of the truth table. The *second step* is to plot 1s on the 5-variable map. Seven 1s are plotted on the map in Fig. 5-48c.

(a)

	Inputs				Output		Inputs				Output
A	B	C	D	E	Y	A	B	C	D	E	Y
0	0	0	0	0	0	1	0	0	0	0	1
0	0	0	0	1	0	1	0	0	0	1	1
0	0	0	1	0	1	1	0	0	1	0	0
0	0	0	1	1	1	1	0	0	1	1	0
0	0	1	0	0	0	1	0	1	0	0	0
0	0	1	0	1	0	1	0	1	0	1	0
0	0	1	1	0	0	1	0	1	1	0	1
0	0	1	1	1	0	1	0	1	1	1	0
0	1	0	0	0	0	1	1	0	0	0	0
0	1	0	0	1	0	1	1	0	0	1	0
0	1	0	1	0	1	1	1	0	1	0	0
0	1	0	1	1	1	1	1	0	1	1	0
0	1	1	0	0	0	1	1	1	0	0	0
0	1	1	0	1	0	1	1	1	0	1	0
0	1	1	1	0	0	1	1	1	1	0	0
0	1	1	1	1	0	1	1	1	1	1	0

(b) Unsimplified minterm expression

$$\bar{A}\cdot\bar{B}\cdot\bar{C}\cdot D\cdot\bar{E} + \bar{A}\cdot\bar{B}\cdot\bar{C}\cdot D\cdot E + \bar{A}\cdot B\cdot\bar{C}\cdot D\cdot\bar{E} + \bar{A}\cdot B\cdot\bar{C}\cdot D\cdot E + A\cdot\bar{B}\cdot\bar{C}\cdot\bar{D}\cdot\bar{E}$$
$$+ A\cdot\bar{B}\cdot\bar{C}\cdot\bar{D}\cdot E + A\cdot\bar{B}\cdot C\cdot D\cdot\bar{E} = Y$$

(c) Plotting and looping 1s on map

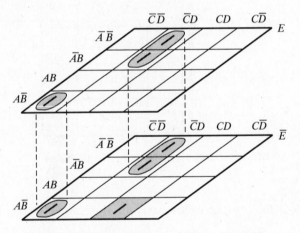

Fig. 5-48 Karnaugh map solution—5 variable

(d) Simplified minterm expression

$$A\cdot\bar{B}\cdot\bar{C}\cdot\bar{D} + \bar{A}\cdot\bar{C}\cdot D + A\cdot\bar{B}\cdot C\cdot D\cdot\bar{E} = Y$$

Each 1 on the map represents an ANDed group of terms from the unsimplified minterm expression. The *third step* is to loop adjacent groups of 1s. Adjacent groups of eight, four, or two 1s are looped. Two loops have been drawn in Fig. 5-48c. The larger loop contains four 1s and forms a cylinder between the top and bottom planes of the map. The smaller loop contains two 1s and forms the cylinder at the lower left in Fig. 5-48c. The single 1 near the bottom of the map does not have any 1s adjacent to it on either the E or $\bar{E}$ plane. The *fourth step* is to eliminate variables. The large loop (cylinder) in Fig. 5-48c eliminates the B and E variables leaving the term $\bar{A}\cdot\bar{C}\cdot D$. The smaller loop (cylinder) contains two 1s and eliminates the E term leaving the term $A\cdot\bar{B}\cdot\bar{C}\cdot\bar{D}$. The single 1 near the bottom is not looped and allows no simplification. The *fifth step* is to logically OR the remaining terms. Figure 5-48d shows the remaining groups ORed, yielding the simplified minterm expression of $A \cdot \overline{B} \cdot \overline{C} \cdot \overline{D} + \overline{A} \cdot \overline{C} \cdot D + A \cdot \overline{B} \cdot C \cdot D \cdot \overline{E} = Y$. The amount of simplification in this example is obvious when the two Boolean expressions in Fig. 5-48 are compared.

SOLVED PROBLEMS

5.42 Write the unsimplified minterm Boolean expression for the truth table in Fig. 5-49.

Inputs					Output	Inputs					Output
A	B	C	D	E	Y	A	B	C	D	E	Y
0	0	0	0	0	0	1	0	0	0	0	0
0	0	0	0	1	0	1	0	0	0	1	1
0	0	0	1	0	1	1	0	0	1	0	0
0	0	0	1	1	1	1	0	0	1	1	0
0	0	1	0	0	0	1	0	1	0	0	0
0	0	1	0	1	0	1	0	1	0	1	0
0	0	1	1	0	1	1	0	1	1	0	0
0	0	1	1	1	1	1	0	1	1	1	0
0	1	0	0	0	0	1	1	0	0	0	0
0	1	0	0	1	0	1	1	0	0	1	1
0	1	0	1	0	1	1	1	0	1	0	0
0	1	0	1	1	1	1	1	0	1	1	0
0	1	1	0	0	0	1	1	1	0	0	0
0	1	1	0	1	0	1	1	1	0	1	0
0	1	1	1	0	1	1	1	1	1	0	0
0	1	1	1	1	1	1	1	1	1	1	0

Fig. 5-49

Solution:

$$\overline{A} \cdot \overline{B} \cdot \overline{C} \cdot D \cdot \overline{E} + \overline{A} \cdot \overline{B} \cdot \overline{C} \cdot D \cdot E + \overline{A} \cdot \overline{B} \cdot C \cdot D \cdot \overline{E} + \overline{A} \cdot \overline{B} \cdot C \cdot D \cdot E +$$
$$\overline{A} \cdot B \cdot \overline{C} \cdot D \cdot \overline{E} + \overline{A} \cdot B \cdot \overline{C} \cdot D \cdot E + \overline{A} \cdot B \cdot C \cdot D \cdot \overline{E} + \overline{A} \cdot B \cdot C \cdot D \cdot E +$$
$$A \cdot \overline{B} \cdot \overline{C} \cdot \overline{D} \cdot E + A \cdot B \cdot \overline{C} \cdot \overline{D} \cdot E = Y$$

5.43 Draw a 5-variable Karnaugh map. Plot ten 1s on the map from the Boolean expression developed in Prob. 5.42. Draw the appropriate loops around groups of 1s on the map.

Solution:

See Fig. 5-50.

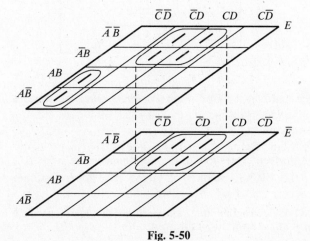

Fig. 5-50

5.44   Write the simplified Boolean expression based on the Karnaugh map from Prob. 5.43.

Solution:

$$A \cdot \overline{C} \cdot \overline{D} \cdot E + \overline{A} \cdot D = Y$$

Supplementary Problems

5.45   Write a minterm Boolean expression for the truth table in Fig. 5-51.
Ans. $\overline{A} \cdot \overline{B} \cdot C + \overline{A} \cdot B \cdot \overline{C} + A \cdot \overline{B} \cdot \overline{C} + A \cdot B \cdot C = Y$

Inputs			Output	Inputs			Output
A	B	C	Y	A	B	C	Y
0	0	0	0	1	0	0	1
0	0	1	1	1	0	1	0
0	1	0	1	1	1	0	0
0	1	1	0	1	1	1	1

Fig. 5-51

5.46   Draw an AND-OR logic diagram that will perform the logic specified by the Boolean expression developed in Prob. 5.45.   *Ans.*   See Fig. 5-52.

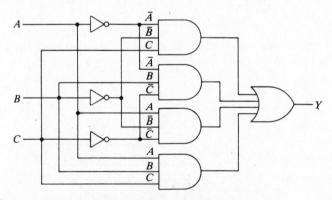

Fig. 5-52   AND-OR logic circuit

5.47   Write the maxterm Boolean expression for the truth table in Fig. 5-51.
Ans. $(A + B + C) \cdot (A + \overline{B} + \overline{C}) \cdot (\overline{A} + B + \overline{C}) \cdot (\overline{A} + \overline{B} + C) = Y$

5.48   Draw an OR-AND logic diagram that will perform the logic specified by the Boolean expression developed in Prob. 5.47.   *Ans.*   See Fig. 5-53.

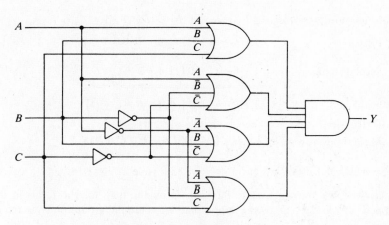

Fig. 5-53 OR-AND logic circuit

5.49 Use De Morgan's theorem to convert the Boolean expression

$$\overline{(A+\bar{B}+C+D)\cdot(A+\bar{B}+\bar{C}+D)} = Y$$

to its minterm form. Show each step as in Fig. 5-15.

Ans. Maxterm expression $\overline{(A+\bar{B}+C+D)\cdot(A+\bar{B}+\bar{C}+D)} = Y$

First step $\overline{\overline{A\cdot\bar{B}\cdot C\cdot D} + \overline{A\cdot\bar{B}\cdot\bar{C}\cdot D}}$

Second step $\overline{\overline{\bar{A}\cdot\bar{\bar{B}}\cdot\bar{C}\cdot\bar{D}} + \overline{\bar{A}\cdot\bar{\bar{B}}\cdot\bar{\bar{C}}\cdot\bar{D}}}$

Third step $\overline{\bar{A}\cdot\bar{\bar{B}}\cdot\bar{C}\cdot\bar{D}} + \overline{\bar{A}\cdot\bar{\bar{B}}\cdot\bar{\bar{C}}\cdot\bar{D}}$

Fourth step   eliminate double overbars

Minterm expression $\bar{A}\cdot B\cdot\bar{C}\cdot\bar{D} + \bar{A}\cdot B\cdot C\cdot\bar{D} = Y$

5.50 Draw a 4-variable minterm Karnaugh map. Plot two 1s on the map for the terms in the minterm expression developed in Prob. 5.49. Draw the appropriate loops around groups of 1s on the map.
Ans. See Fig. 5-54.

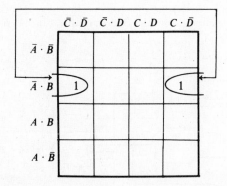

Fig. 5-54 Completed minterm Karnaugh map

5.51 Write the simplified minterm Boolean expression based on the Karnaugh map from Prob. 5.50.
Ans. $\bar{A}\cdot B\cdot\bar{D} = Y$

5.52 Use De Morgan's theorem to convert the Boolean expression $\bar{A}\cdot\bar{B}\cdot\bar{C}\cdot\bar{D} + \bar{A}\cdot B\cdot\bar{C}\cdot\bar{D} = Y$ to its maxterm form. Show each step as in Fig. 5-15.

Ans.   Minterm expression $\overline{\overline{A} \cdot \overline{B} \cdot \overline{C} \cdot D + \overline{A} \cdot B \cdot \overline{C} \cdot \overline{D}} = Y$

First step $\overline{(\overline{A} + \overline{B} + \overline{C} + \overline{D}) \cdot (\overline{A} + B + \overline{C} + \overline{D})}$

Second step $\overline{\left(\overline{\overline{A}} + \overline{\overline{B}} + \overline{\overline{C}} + \overline{\overline{D}}\right) \cdot \left(\overline{\overline{A}} + \overline{B} + \overline{\overline{C}} + \overline{\overline{D}}\right)}$

Third step $\left(\overline{\overline{A}} + \overline{\overline{B}} + \overline{\overline{C}} + \overline{\overline{D}}\right) \cdot \left(\overline{\overline{A}} + \overline{B} + \overline{\overline{C}} + \overline{\overline{D}}\right)$

Fourth step eliminate double overbars

Maxterm expression $(A + B + C + D) \cdot (A + \overline{B} + C + D) = Y$

5.53   Draw a 4-variable Karnaugh map. Plot two 1s on the map for the terms in the maxterm expression developed in Prob. 5.52. Draw the appropriate loops around groups of 1s on the map.
Ans.   See Fig. 5-55.

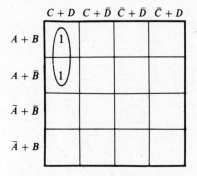

Fig. 5-55 Completed maxterm map

5.54   Write the simplified maxterm Boolean expression based on the Karnaugh map from Prob. 5.53.
Ans. $A + C + D = Y$

5.55   Draw an AND-OR logic circuit from the Boolean expression $A \cdot B + \overline{C} \cdot D + \overline{E} + \overline{F} = Y$.
Ans.   See Fig. 5-56.

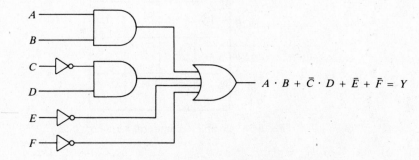

Fig. 5-56 AND-OR logic circuit

5.56   Draw a NAND logic circuit for the AND-OR circuit in Prob. 5.55. The NAND logic circuit should perform the logic in the expression $A \cdot B + \overline{C} \cdot D + \overline{E} + \overline{F} = Y$.   *Ans.*   See Fig. 5-57.

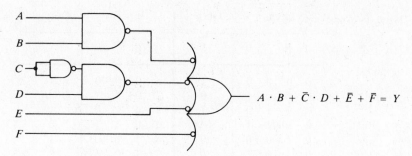

Fig. 5-57 NAND logic circuit

5.57   Draw an OR-AND logic circuit for the Boolean expression $\overline{A} \cdot (\overline{B} + C) \cdot \overline{D} \cdot E = Y$.
   Ans.   See Fig. 5-58.

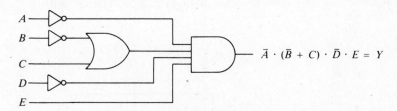

Fig. 5-58 OR-AND logic circuit

5-58   Draw a NOR logic circuit for the OR-AND circuit in Prob. 5.57. The NOR circuit should perform the logic in the expression $\overline{A} \cdot (\overline{B} + C) \cdot \overline{D} \cdot E = Y$.   *Ans.*   See Fig. 5-59.

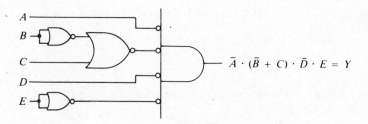

Fig. 5-59 NOR logic circuit

5.59   NOR logic can be easily substituted in an _____ (AND-OR, OR-AND) circuit.
   Ans.   NOR logic can be substituted for OR-AND circuits.

5.60   Write the unsimplified sum-of-products Boolean expression for the truth table in Fig. 5-60.
   Ans. $\overline{A} \cdot \overline{B} \cdot \overline{C} \cdot \overline{D} + \overline{A} \cdot \overline{B} \cdot C \cdot \overline{D} + \overline{A} \cdot B \cdot C \cdot \overline{D} + A \cdot \overline{B} \cdot \overline{C} \cdot \overline{D} + A \cdot \overline{B} \cdot C \cdot \overline{D} + A \cdot B \cdot \overline{C} \cdot \overline{D} +$
   $A \cdot B \cdot C \cdot \overline{D} = Y$

Inputs				Output	Inputs				Output
A	B	C	D	Y	A	B	C	D	Y
0	0	0	0	1	1	0	0	0	1
0	0	0	1	0	1	0	0	1	0
0	0	1	0	1	1	0	1	0	1
0	0	1	1	0	1	0	1	1	0
0	1	0	0	0	1	1	0	0	1
0	1	0	1	0	1	1	0	1	0
0	1	1	0	1	1	1	1	0	1
0	1	1	1	0	1	1	1	1	0

Fig. 5-60

5.61   Draw a 4-variable minterm Karnaugh map. Plot seven 1s on the map from the Boolean expression developed in Prob. 5.60. Draw the appropriate loops around groups of 1s on the map.
Ans.   See Fig. 5-61.

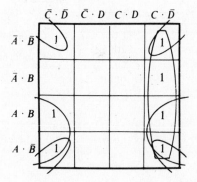

Fig. 5-61 Completed minterm map

5.62   Write the simplified minterm Boolean expression based on the Karnaugh map from Prob. 5.61.
Ans. $C \cdot \overline{D} + A \cdot \overline{D} + \overline{B} \cdot \overline{D} = Y$

5.63   Write the unsimplified product-of-sums Boolean expression for the truth table in Fig. 5-60.
Ans. $(A + B + C + \overline{D}) \cdot (A + B + \overline{C} + \overline{D}) \cdot (A + \overline{B} + C + D) \cdot (A + \overline{B} + C + \overline{D}) \cdot (A + \overline{B} + \overline{C} + \overline{D})$
$\cdot (\overline{A} + B + C + \overline{D}) \cdot (\overline{A} + B + \overline{C} + \overline{D}) \cdot (\overline{A} + \overline{B} + C + \overline{D}) \cdot (\overline{A} + \overline{B} + \overline{C} + \overline{D}) = Y$

5.64   Draw a 4-variable maxterm Karnaugh map. Plot nine 1s on the map from the Boolean expression developed in Prob. 5.63. Draw the appropriate loops around groups of 1s on the map.
Ans.   See Fig. 5-62.

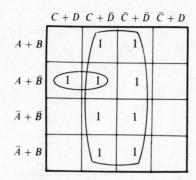

Fig. 5-62 Completed maxterm map

5.65 Write the simplified maxterm Boolean expression based on the Karnaugh map from Prob. 5.64.
Ans. $(A + \bar{B} + C) \cdot \bar{D} = Y$

5.66 The simplified _____ (maxterm, minterm) form of Boolean expression is the easiest circuit to implement for the truth table in Fig. 5-60.
Ans. The *maxterm* expression $(A + \bar{B} + C) \cdot \bar{D} = Y$ appears to be simpler to implement with logic gates than the minterm expression $C \cdot \bar{D} + A \cdot \bar{D} + \bar{B} \cdot \bar{D} = Y$.

5.67 Design a logic circuit that will respond with a 1 when even numbers (decimals 0, 2, 4, 6, 8) appear at the inputs. Figure 5-63 is the BCD (8421) truth table you will use in this problem. *Write* the unsimplified *minterm* Boolean expression for the truth table.
Ans. $\bar{D} \cdot \bar{C} \cdot \bar{B} \cdot \bar{A} + \bar{D} \cdot \bar{C} \cdot B \cdot \bar{A} + \bar{D} \cdot C \cdot \bar{B} \cdot \bar{A} + \bar{D} \cdot C \cdot B \cdot \bar{A} + D \cdot \bar{C} \cdot \bar{B} \cdot \bar{A} = Y$. The expression represents the 1s in the Y column of the truth table. Six other groups of don't cares (X's) might also be considered and will be plotted on the map.

Inputs				Output	Inputs				Output
D	C	B	A		D	C	B	A	
				Y					Y
8s	4s	2s	1s		8s	4s	2s	1s	
0	0	0	0	1	1	0	0	0	1
0	0	0	1	0	1	0	0	1	0
0	0	1	0	1	1	0	1	0	X
0	0	1	1	0	1	0	1	1	X
0	1	0	0	1	1	1	0	0	X
0	1	0	1	0	1	1	0	1	X
0	1	1	0	1	1	1	1	0	X
0	1	1	1	0	1	1	1	1	X

Fig. 5-63 Truth table with don't cares

5.68 Draw a 4-variable minterm Karnaugh map. Plot five 1s and six X's (for the don't cares) on the map based on the truth table in Fig. 5-63. Draw the appropriate loops around groups of 1s and X's on the map.
Ans. See Fig. 5-64.

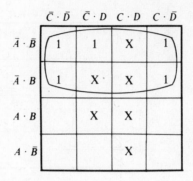

Fig. 5-64 Completed minterm maps using don't cares

5.69 Write the simplified Boolean expression based on the Karnaugh map from Prob. 5.68.
Ans. $\overline{A} = Y$

5.70 Write the simplified Boolean expression based on the Karnaugh map from Prob. 5.68 *without using the don't cares* for simplification.
Ans. $\overline{A} \cdot \overline{D} + \overline{A} \cdot \overline{B} \cdot \overline{C} = Y$. The use of the don't cares greatly aids simplification in this problem because using them reduces the expression to $\overline{A} = Y$.

5.71 In this chapter, individual logic gates were used to simplify combinational logic problems. List several more complex ICs used for logic circuit simplification.
Ans. Several ICs used to simplify combinational logic problems are data selectors (multiplexers), decoders, PLAs, ROMs, and PROMs.

5.72 Write the unsimplified minterm Boolean expression for the truth table in Fig. 5-65.
Ans. $\overline{A} \cdot \overline{B} \cdot C \cdot \overline{D} \cdot E + \overline{A} \cdot \overline{B} \cdot C \cdot D \cdot E + \overline{A} \cdot B \cdot C \cdot \overline{D} \cdot E + \overline{A} \cdot B \cdot C \cdot D \cdot E +$
$A \cdot B \cdot \overline{C} \cdot D \cdot \overline{E} + A \cdot B \cdot \overline{C} \cdot D \cdot E + A \cdot B \cdot C \cdot D \cdot \overline{E} + A \cdot B \cdot C \cdot D \cdot E = Y$

Inputs					Output	Inputs					Output
A	B	C	D	E	Y	A	B	C	D	E	Y
0	0	0	0	0	0	1	0	0	0	0	0
0	0	0	0	1	0	1	0	0	0	1	0
0	0	0	1	0	0	1	0	0	1	0	0
0	0	0	1	1	0	1	0	0	1	1	0
0	0	1	0	0	0	1	0	1	0	0	0
0	0	1	0	1	1	1	0	1	0	1	0
0	0	1	1	0	0	1	0	1	1	0	0
0	0	1	1	1	1	1	0	1	1	1	0
0	1	0	0	0	0	1	1	0	0	0	0
0	1	0	0	1	0	1	1	0	0	1	0
0	1	0	1	0	0	1	1	0	1	0	1
0	1	0	1	1	0	1	1	0	1	1	1
0	1	1	0	0	0	1	1	1	0	0	0
0	1	1	0	1	1	1	1	1	0	1	0
0	1	1	1	0	0	1	1	1	1	0	1
0	1	1	1	1	1	1	1	1	1	1	1

Fig. 5-65

5.73 Draw a 5-variable minterm Karnaugh map. Plot eight 1s on the map from the Boolean expression developed in Prob. 5.72. Draw the appropriate loops around groups of 1s on the map.
Ans. See Fig. 5-66.

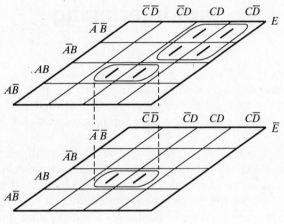

Fig. 5-66

5.74 Write the simplified minterm Boolean expression based on the Karnaugh map from Prob. 5.73.
Ans. $A \cdot B \cdot D + \overline{A} \cdot C \cdot E = Y$

Chapter 6

TTL and CMOS ICs: Characteristics and Interfacing

6-1 INTRODUCTION

The growing popularity of digital circuits is due in part to the availability of inexpensive integrated circuits (ICs). Manufacturers have developed many *families* of digital ICs—groups that can be used together in building a digital system. The ICs in a family are said to be compatible, and they can be easily connected.

Digital ICs can be categorized as either *bipolar* or *unipolar*. Bipolar digital ICs are fabricated from parts comparable to discrete bipolar transistors, diodes, and resistors. The TTL (transistor-transistor logic) family is the most popular of the ICs using the bipolar technology. Unipolar digital ICs are fabricated from parts comparable to insulated-gate field-effect transistors (IGFETs). The CMOS (complementary-metal-oxide semiconductor) family is a widely used group of ICs based on the metal-oxide semiconductor (MOS) technology.

Integrated circuits are sometimes grouped by manufacturers as to their circuit complexity. Circuit complexity of ICs is defined as follows:

1. *SSI* (*small-scale integration*):
   Number of gates: less than 12
   Typical digital devices: gates and flip-flops

2. *MSI* (*medium-scale integration*):
   Number of gates: 12 to 99
   Typical digital devices: adders, counters, decoders, encoders, multiplexers, and
                               demultiplexers, registers

3. *LSI* (*large-scale integration*):
   Number of gates: 100 to 9999
   Typical digital devices: digital clocks, smaller memory chips, calculators

4. *VLSI* (*very-large-scale integration*):
   Number of gates: 10,000 to 99,999
   Typical digital devices: microprocessors, larger memory chips, advanced calculators

5. *ULSI* (*ultra-large-scale integration*):
   Number of gates: over 100,000
   Typical digital devices: advanced microprocessors

Many digital IC families are available to the digital circuit designer, and some of them are listed below:

1. *Bipolar families*:
   RTL   resistor-transistor logic
   DTL   diode-transistor logic
   TTL   transistor-transistor logic
         (types: standard TTL, low-power TTL, high-speed TTL, Schottky TTL,
         advanced low-power Schottky TTL, advanced Schottky TTL)
   ECL   emitter-coupled logic
         (also called CML, current-mode logic)
   HTL   high-threshold logic
         (also called HNIL, high-noise-immunity logic)
   IIL   integrated-injection logic

2. *MOS families*:
   PMOS   *P*-channel metal-oxide semiconductor
   NMOS   *N*-channel metal-oxide semiconductor
   CMOS   complementary metal-oxide semiconductor

The TTL and CMOS technologies are commonly used to fabricate SSI and MSI integrated circuits. Those circuits include such functional devices as logic gates, flip-flops, encoders and decoders, multiplexers, latches, and registers. MOS devices (PMOS, NMOS, and CMOS) dominate in the fabrication of LSI and VLSI devices. NMOS is especially popular for use in microprocessors and memories. CMOS is popular for use in very low power applications such as calculators, wrist watches, and battery-powered computers.

6-2 DIGITAL IC TERMS

Several terms that appear in IC manufacturers' literature assist the technician in using and comparing logic families. Some of the more important terms and characteristics of digital ICs will be outlined.

How is a logical 0 (LOW) or a logical 1 (HIGH) defined? Figure 6-1*a* shows an inverter (such as the 7404) from the TTL IC family. The manufacturers specify that, for proper operation, a *LOW input*

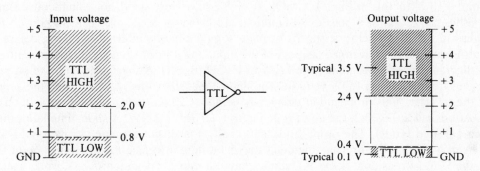

(*a*) TTL input and output voltage levels

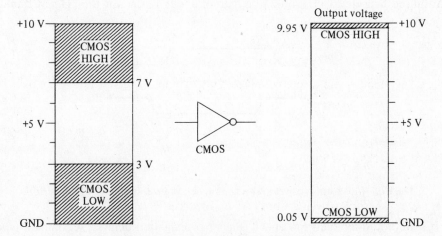

(*b*) CMOS input and output voltage levels

Fig. 6-1 Defining logical HIGH and LOW

must range from GND to 0.8 V. Likewise, a *HIGH input* must range from 2.0 V to 5.0 V. The unshaded section in Fig. 6-1(*a*) between 0.8 V and 2.0 V on the input side is the forbidden region. An input voltage of 0.5 V would then be a LOW input, whereas an input of 2.6 V would be a HIGH input. An input of 1.5 V would yield unpredictable results and is considered a forbidden input. The forbidden region might also be called the uncertain or undefined region.

The expected outputs are shown on the right side of the TTL inverter shown in Fig. 6-1*a*. A *LOW output* would typically be 0.1 V but could be as high as 0.4 V. A *HIGH output* would typically be 3.5 V but could be as low as 2.4 V. The *HIGH* output depends on the resistance value of the load at the output. The greater the load current, the lower the HIGH output voltage. The unshaded portion on the output voltage side in Fig. 6-1*a*, is the forbidden region.

Observe the difference in the definition of a HIGH from input to output in Fig. 6-1*a*. The input HIGH is defined as greater than 2.0 V, whereas the output HIGH is greater than 2.4 V. The reason for this difference is to provide for *noise immunity*—the digital circuit's insensitivity to undesired electrical signals. The input LOW is less than 0.8 V and the output LOW is 0.4 V or less. Again the margin between those figures is to assure rejection of unwanted noise entering the digital system.

The voltage ranges defining HIGH and LOW are different for each logic family. For comparison, the input and output voltages for a typical CMOS inverter are given in Fig. 6-1*b*. In this example the manufacturer specifies that the HIGH *output* will be nearly the full supply voltage (over $+9.95$ V). A LOW *output* will be within 0.05 V of ground (GND) potential. Manufacturers also specify that a CMOS IC will consider *any input voltage from $+7$ V to $+10$ V as a HIGH*. Figure 6-1*b* also notes that a CMOS IC will consider any input voltage from GND to $+3$ V as a LOW.

CMOS ICs have a wide swing of output voltages approaching both rails of the power supply (GND and $+10$ V in this example). CMOS ICs also have very good noise immunity. Both these characteristics, along with low power consumption, are listed as *advantages* of CMOS over TTL ICs.

Because of the high operating speeds of many digital circuits, internal switching delays become important. Figure 6-2 is a waveform diagram of the input and output from an inverter circuit. At point *a* on the diagram, the input is going from LOW to HIGH (0 to 1). A short time later the output of the inverter goes from HIGH to LOW (1 to 0). The delay time, shown as t_{PLH}, is called the *propagation delay* of the inverter. This propagation delay may be about 20 nanoseconds (ns) for a standard TTL inverter. At point *b* in Fig. 6-2, the input is going from HIGH to LOW. A short time later, the output goes from LOW to HIGH. The propagation delay (t_{PHL}) is shown as about 15 ns for this standard TTL inverter. Note that the propagation delay may be different for the L-to-H transition of the input than for the H-to-L transition. Some IC families have shorter propagation delays, which makes them more adaptable for high-speed operation. Propagation delays range from an average low of about 1.5 ns for the Advanced Schottky TTL family to a high of about 125 ns for the HTL IC family.

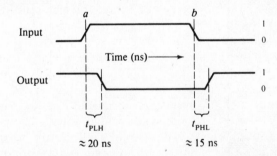

Fig. 6-2 Waveforms showing propagation delays for a standard TTL inverter

CMOS ICs are noted for their low speed (higher propagation delays). A common type of CMOS IC may have a propagation delay of from 25 to 100 ns, depending on the device. However, a newer subfamily of *high-speed CMOS* ICs has reduced the propagation delays. For instance, the 74HCO4 CMOS inverter has a propagation delay of only 8 ns. These high-speed CMOS ICs make this family much more suitable for higher-speed applications.

Integrated circuits are grouped into *families* because they are compatible. Figure 6-3*a* shows the left TTL inverter driving the right load inverter. In this case *conventional current* flows from the load device to the driver gate and to ground as illustrated in Fig. 6-3*a*. It is said that the driver inverter is *sinking the current* (sinks current to ground). This sinking current may be as high as 1.6 mA (milliamperes) from a single TTL load. Note the direction of a sinking current.

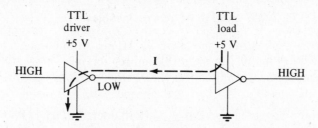

(*a*) Sinking drive current to ground

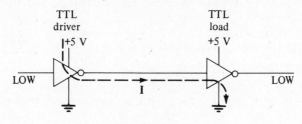

(*b*) Source drive current

Fig. 6-3

When the output of the TTL driver goes HIGH, the situation in Fig. 6-3*b* is created. In this case *conventional current* flows from the driver to the load device as illustrated. It is said that the driver inverter is *sourcing the current*. This sourcing current is quite low when it is driving a single load (perhaps only 40 μA, microamperes).

The current driving capabilities of logic gates vary from family to family. As a general rule, TTL ICs can sink more current than they can source. For instance, a standard TTL gate used for driving a load can sink up to 16 mA, whereas a low-power Schottky TTL gate can sink only a maximum of 8 mA.

CMOS output drive currents are nearly the same when sinking or sourcing current. A typical CMOS gate might have a drive capability of about 0.5 mA. The high-speed CMOS series of ICs (such as the 74HCO2) feature sinking or sourcing drive currents of 4 mA.

It is common in logic circuits to have one gate drive several others. The limitation of *how many* gates can be driven by a single output is called the *fan-out* of a logic circuit. The typical fan-out for TTL logic circuits is 10. This means that a single TTL output can drive up to 10 TTL inputs. The CMOS logic family has a fan-out of 50.

One of the many advantages of ICs over other circuits is their low power dissipation. Some IC families, however, have much lower power dissipation than others. The power consumption might average about 10 milliwatts (mW) per gate in the standard TTL family, whereas it might be as low as 1 mW per gate in the low-power TTL family. The CMOS family is noted for its extremely low power consumption and is widely used in battery-operated portable products.

SOLVED PROBLEMS

6.1   Refer to Fig. 6-1*a*. A 2.2-V input to the TTL inverter is a logical _____ (0, 1) input.

Solution:

A 2.2-V input to a TTL inverter is a logical 1 input, because it is in the HIGH range.

6.2   Refer to Fig. 6-1a. A 2.2-V output from the TTL inverter is a logical _____ output.

Solution:

A 2.2-V output from a TTL inverter is defined as a forbidden output caused by a faulty IC or too heavy a load at the output.

6.3   What are *typical* TTL LOW and HIGH *output* voltages?

Solution:

The typical LOW output voltage from a TTL IC is 0.1 V. The typical HIGH output voltage from a TTL IC is about 3.5 V, but the voltage varies widely with loading.

6.4   A 0.7-V input would be considered a _____ (forbidden, HIGH, LOW) input to a TTL device.

Solution:

See Fig. 6-1a. A 0.7-V input would be considered a LOW input to a TTL IC.

6.5   The time it takes for the output of a digital logic gate to change states after the input changes is called _____ _____.

Solution:

Propagation delay is the time it takes for the output to change after the input has changed logic states. See Fig. 6-2.

6.6   Propagation delays in modern digital ICs are measured in _____ (milli, micro, nano) seconds.

Solution:

Propagation delays in modern digital ICs are measured in nanoseconds. A nanosecond (ns) is 10^{-9} s.

6.7   The number of parallel loads that can be driven by a single digital IC output is a characteristic called _____.

Solution:

Fan-out is the number of parallel loads that can be driven by a single digital IC output.

6.8   The _____ (CMOS, TTL) digital IC family is noted for its very low power consumption.

Solution:

The CMOS digital IC family is noted for its very low power consumption.

6.9   Refer to Fig. 6-1b. An 8.5-V input to the CMOS inverter is a logical _____ (0, 1) input.

Solution:

8.5-V input to a CMOS inverter is a logical 1 input because it is in the HIGH range shown in Fig. 6-1b.

6.10   Refer to Fig. 6-1b. What are the typical CMOS LOW and HIGH output voltages?

Solution:

The typical output CMOS voltages are very near the rails of the power supply. A typical LOW might be 0 V (GND), and a HIGH might be +10 V.

6.11   The _____ (CMOS, TTL) logic family is noted for its very good noise immunity.

    Solution:

        The CMOS family is noted for its good noise immunity.

6.12   Refer to Fig. 6-4*a*. The NAND gate is said to be _____ (sinking, sourcing) current in the logic circuit shown.

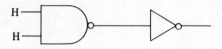

(*a*) NAND gate driving an inverter input

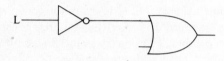

(*b*) Inverter driving an OR input

Fig. 6-4

    Solution:

        The output of the NAND gate shown in Fig. 6-4*a* is LOW. The NAND gate is therefore said to be sinking the drive current.

6.13   Refer to Fig. 6-4*b*. The inverter is said to be _____ (sinking, sourcing) current in the logic circuit shown.

    Solution:

        The output of the inverter shown in Fig. 6-4*b* is HIGH. The inverter is therefore said to be sourcing the drive current.

6-3   TTL INTEGRATED CIRCUITS

    The famous 7400 series of TTL logic circuits was introduced by Texas Instruments in 1964. The TTL family of ICs is still one of the more widely used for constructing logic circuits. TTL ICs are manufactured in a wide variety of SSI and MSI integrated circuits.

    Over the years, improvements in TTL logic circuits have been made, which has led to *subfamilies* of transistor-transistor logic ICs. The following six TTL subfamilies are currently available from National Semiconductor Corporation:

    1. Standard TTL logic
                Typical IC marking: 7404 (function: hex inverter)
    2. Low-power TTL logic
                Typical IC marking: 74L04 (hex inverter)
    3. Low-power Schottky TTL logic
                Typical IC marking: 74LS04 (hex inverter)
    4. Schottky TTL logic
                Typical IC marking: 74S04 (hex inverter)

5. Advanced low-power Schottky TTL logic
   Typical IC marking: 74ALS04 (hex inverter)
6. Advanced Schottky TTL logic
   Typical IC marking: 74AS04 (hex inverter)

The code letters L, LS, S, ALS, and AS are used in the *middle of the 7400 series number* to designate the subfamily. This can be observed above where typical IC markings for the various TTL subfamilies are listed. Notice that no special code letter is used in the middle of a standard TTL logic IC. The subfamilies with the code letter S contain a Schottky barrier diode to increase switching speed. Several companies also use the code letter F (as in 74F04) for a *fast* advanced Schottky TTL IC.

It should be noted that the voltage characteristics of all the TTL subfamilies are the same. Their power and speed characteristics are different, and under some conditions, substituting one subfamily for another might cause trouble. For instance, a technician would *not* want to replace a very fast 74AS04 inverter IC with a much slower 74L04 IC from the low-power TTL logic subfamily.

The internal details of a standard TTL NAND gate are shown in Fig. 6-5. National Semiconductor Corporation's description follows. TTL logic was the first saturating logic integrated circuit family introduced; it set the standard for all future families. It offers a combination of speed, power consumption, output source, and sink capabilities suitable for most applications, and it offers the greatest variety of logic functions. The basic gate (see Fig. 6-5) features a multiple-emitter input configuration for fast switching speeds and active pull-up output to provide a low driving source impedance which also improves noise margin and device speed. Typical device power dissipation is 10 mW per gate, and the typical propagation delay is 10 ns when driving a 15 pF per $400 - \Omega$ load.

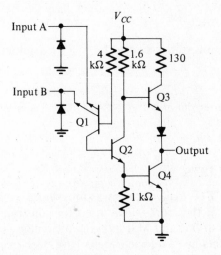

Fig. 6-5 Schematic diagram of a standard TTL NAND gate (*Courtesy of National Semiconductor Corporation*)

Digital logic designers must consider two important factors when selecting a logic family. They are *speed* and *power consumption*. In Fig. 6-6*a* the TTL subfamilies are ranked from best to worst (fastest to slowest) by speed, or low propagation delay. Note that the advanced Schottky subfamily is the fastest. In Fig. 6-6*b* the TTL subfamilies are ranked by power consumption. Note that the low-power TTL is best in power consumption. Both the low-power Schottky and the advanced low-power Schottky are excellent compromise subfamilies with both low power consumption and high speed. Currently, both the low-power Schottky and the advanced low-power Schottky are very popular.

Speed	TTL subfamily
Fastest ↓ Slowest	Advanced Schottky Schottky Advanced low-power Schottky Low-power Schottky Standard TTL Low power

(a) TTL subfamilies ranked by speed

Power consumption	TTL subfamily
Low ↓ High	Low power Advanced low-power Schottky Low-power Schottky Advanced Schottky Standard TTL Schottky

(b) TTL subfamilies ranked by power consumption

Fig. 6-6

Devices in the 7400 TTL series are referred to as commercial grade ICs; they operate over a temperature range of 0 to 70°C. The 5400 TTL series have the same logic functions, but they will operate over a greater temperature range (-55 to 125°C). The 5400 TTL series is sometimes called the military series of TTL logic circuits. ICs in the 5400 series are more expensive.

The NAND gate output shown in Fig. 6-5 is connected between two transistors (Q3 and Q4). This is called a *totem pole* output. For the output to sink current (LOW output), transistor Q4 must be "turned on" or saturated. For a HIGH output as shown in Fig. 6-5, transistor Q3 must be saturated, which will allow the NAND gate to become a source of drive current. Most TTL logic gates have the totem pole type of output.

Some TTL circuits have an *open-collector* output in which transistor Q3 (see Fig. 6-5) is missing. A *pull-up resistor* is used with open-collector outputs. Pull-up resistors are connected from the output to the $+5$-V rail of the power supply outside the logic gate.

A third type of TTL output used on some devices is the *three-state output*. It has three possible outputs (HIGH, LOW, or high impedance). The three-state output will be explored in connection with the three-state buffer.

As a general rule, *outputs of TTL devices cannot be connected together*. That is true of gates with totem pole outputs. TTL outputs can be connected together with no damage if they are of the open-collector or three-state type.

Markings on TTL ICs vary with the manufacturer. Figure 6-7a shows a typical marking on a TTL digital IC. Pin 1 is identified with a dot, a notch, or a colored band across the end of the IC. The manufacturer's logo is shown at the upper left in Fig. 6-7a. In this example the manufacturer is National Semiconductor Corporation. The part number is DM7408N. The core number (generic number) is 7408, which means this is a TTL quad 2-input AND gate IC.

The part number (DM7408N) is further decoded in Fig. 6-7b. The prefix (DM in this example) is a manufacturer's code. The core number of 7408 is divided. The 74 portion means that this is a commercial grade TTL IC from the 7400 series. The 08 identifies the IC by function (quad 2-input AND gate in this example). The N suffix is the manufacturer's code for a dual-in-line package IC.

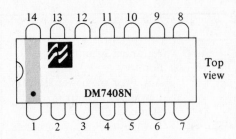

(a) Markings on a typical TTL IC

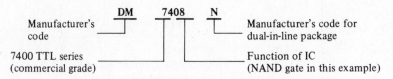

(b) Decoding a typical TTL IC part number

(c) Markings on another TTL IC

Fig. 6-7

Consider the IC shown in Fig. 6-7c. The logo represents Texas Instruments, the manufacturer. The SN portion of the part number is a prefix used by Texas Instruments. The 74 specifies this to be a commercial grade TTL IC. The LS means that this is a low-power Schottky TTL digital IC. The 04 specifies the function of the IC (hex inverter in this example). The trailing N specifies a DIP IC.

Another characteristic of TTL inputs should be understood. Unconnected inputs to a TTL gate are said to be "floating HIGH." In other words, any TTL input left disconnected (floating) will be assumed to be at a logical 1.

SOLVED PROBLEMS

6.14   List six TTL subfamilies.

Solution:

The six TTL subfamilies currently available are standard TTL, low-power, low-power Schottky, Schottky, advanced low-power Schottky, and advanced Schottky.

6.15   The _____ (speed, voltage) characteristics of all the TTL subfamilies are the same.

Solution:

The voltage characteristics of all the TTL subfamilies are the same. They are shown in Fig. 6-1a.

6.16   The TTL logic family was first developed in the _____ (1960s, 1970s).

Solution:

The first TTL logic family was developed in 1964.

6.17   When a designer selects a logic family, what two very important characteristics must be considered?

Solution:

Designers must consider the speed and power consumption characteristics of various logic families in any design.

6.18   Which TTL subfamily is the fastest?

Solution:

Refer to Fig. 6-6a. The advanced Schottky TTL family provides the lowest propagation delays and therefore the best high-speed characteristics.

6.19   Refer to Fig. 6-5. This standard TTL 2-input NAND gate uses _____ (open-collector, totem pole) outputs.

Solution:

The 2-input TTL NAND gate in Fig. 6-5 uses a totem pole output configuration.

6.20   Which *two* TTL subfamilies consume the least power?

Solution:

Refer to Fig. 6-6b. The low-power and advanced low-power Schottky TTL subfamilies are the best for low-power consumption.

6.21   TTL ICs with totem pole outputs _____ (may, may not) have their outputs connected together.

Solution:

TTL totem pole outputs may not have their outputs connected together.

6.22   The _____ (5400, 7400) series of TTL logic devices will operate over a wider temperature range, is more expensive, and is referred to as military grade.

Solution:

The 5400 series of TTL logic devices will operate over a wider temperature range, is more expensive, and is referred to as military grade.

6.23   TTL open-collector outputs require a _____ resistor connected from the output to the +5-V rail of the power supply.

Solution:

TTL open-collector outputs require pull-up resistors.

6.24   Refer to Fig. 6-8. Interpret the markings on this TTL DIP IC.

Fig. 6-8 TTL IC

Solution:

The logo and the DM prefix indicate that National Semiconductor is the manufacturer of this IC. The N suffix indicates that this is a dual-in-line package IC. The 74ALS76 is the generic section of the part number. The 74 means this is a commercial grade 7400 series digital TTL IC. The 76 specifies the function, which is a dual JK flip-flop. The ALS identifies this IC as part of the advanced low-power Schottky TTL subfamily.

6.25   An unconnected TTL input floats at a _____ (HIGH, LOW) logic level.

Solution:

HIGH

6-4   CMOS INTEGRATED CIRCUITS

The first *complementary metal-oxide semiconductor* (*CMOS*) family of ICs was introduced in 1968 by RCA. Since then, it has become very popular. CMOS ICs are growing in popularity because of their extremely low power consumption, high noise immunity, and their ability to operate from an inexpensive nonregulated power supply. Other advantages of CMOS ICs over TTLs are low noise generation and a great variety of available functions. Some analog functions available in CMOS ICs have no equivalents in TTLs.

The schematic diagram of a CMOS inverter is shown in Fig. 6-9a. It is fabricated by using both *N*-channel and *P*-channel MOSFETS (metal-oxide semiconductor field-effect transistors). The bottom transistor (Q1) in Fig. 6-9a is the *N*-channel enhancement-mode MOSFET. The top transistor (Q2) is the *P*-channel enhancement-mode MOSFET. Note that the gate (G), source (S), and drain (D) connections of each FET are labeled.

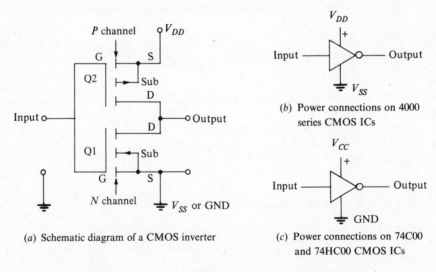

(a) Schematic diagram of a CMOS inverter

(b) Power connections on 4000 series CMOS ICs

(c) Power connections on 74C00 and 74HC00 CMOS ICs

Fig. 6-9

When the input to the CMOS inverter in Fig. 6-9a goes LOW (GND), the negative voltage causes the *P*-channel FET (Q2) to conduct. However, the *N*-channel FET (Q1) is not conducting. This connects the output terminal to the positive (V_{DD}) of the power supply through the low-resistance *P* channel of Q2. The CMOS circuit shown in Fig. 6-9a produced a HIGH (positive) output with a LOW input. This is the proper action for an inverter.

When the input to the CMOS inverter shown in Fig. 6-9a goes HIGH (V_{DD}), the positive voltage causes the N-channel FET (Q1) to conduct. However, the P-channel FET (Q2) is not conducting. This connects the output terminal through the N channel to ground (V_{SS}) of the power supply. In this example a HIGH input generates a LOW output.

The general arrangement of transistors and the operation of the CMOS output shown in Fig. 6-9a are comparable to the TTL totem pole outputs diagrammed in Fig. 6-5. In each case only one of the two output transistors is conducting at a time. The CMOS arrangement is simpler and the currents used to switch the CMOS are extremely small compared to those of the bipolar TTL counterpart.

A logic symbol for the CMOS inverter is shown in Fig. 6-9b. Note especially the labeling on the power supply connections. The V_{DD} and V_{SS} (GND) labels are used with the older 4000 series and many LSI CMOS ICs. The newer 74C00 and 74HC00 families of CMOS digital logic ICs use the V_{CC} and GND labels shown in Fig. 6-9c. This labeling is similar to that of the power connections on TTL ICs.

Manufacturers produce at least three common families of SSI/MSI CMOS integrated circuits. They include the older 4000 series, the 74C00 series, and the newer 74HC00 series.

The CMOS 4000 series has a wide variety of circuit functions. The 4000 series has been improved, and most ICs in this family are now *buffered* and referred to as the 4000B series. Some of the circuit functions available in the 4000 series are logic gates, flip-flops, registers, latches, adders, buffers, bilateral switches, counters, decoders, multiplexers/demultiplexers, and multivibrators (astable and monostable).

A typical 4000 series IC is sketched in Fig. 6-10a. The manufacturer is RCA. Pin 1 is located immediately counterclockwise from the notch. The part number (CD4024BE) is decoded in Fig. 6-10b. The prefix CD is RCA's code for CMOS digital. The suffix E is RCA's code for a plastic dual-in-line package. The generic 4024B is the core number. The 40 identifies this as part of the 4000 series of CMOS ICs. The 24 identifies the function of the IC as a 7-stage binary counter. The B stands for series B or buffered CMOS.

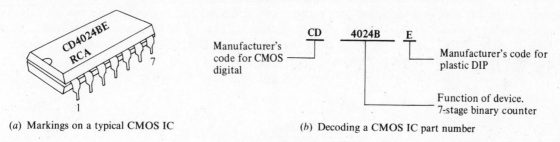

(*a*) Markings on a typical CMOS IC   (*b*) Decoding a CMOS IC part number

Fig. 6-10

The 4000 series of CMOS ICs features a wide voltage supply range from 3 to 15 V. The ICs also have high noise immunity and very low power consumption (10 nW is typical). Many 4000 series devices can drive two low-power TTLs or one low-power Schottky TTL IC.

The 4000 series suffers in the area of speed. Propagation delays may range from 20 to 300 ns depending on the device, temperature, and supply voltage. Static electricity can also be a problem with CMOS ICs. Unfortunately, the power consumption of CMOS devices does increase somewhat as the frequency of operation increases.

The 74C00 series of CMOS digital ICs features functions and pin outs compatible with the industry standard 7400 TTL series. This helps designers already familiar with the 7400 series. The 74C00 family has the same characteristics as the 4000 series.

A typical 74C00 series IC is shown in Fig. 6-11. The logo indicates that the manufacturer is National Semiconductor. Pin 1 is located by using a dot, color band, or notch. The IC has both 4000 series and 74C00 series part numbers. The 74C00 series part number is MM74C192N. The prefix MM

Fig. 6-11 A typical 74C00 CMOS IC

is the manufacturer's code for MOS monolithic. The suffix N is National Semiconductor's code for a plastic DIP IC. 74C192 is the generic part number. The 74C indicates that the IC is part of the 74C00 series of CMOS ICs. The 192 defines the function of the IC, which is a synchronous 4-bit up/down decade counter. This IC can also substitute in the 4000 series family. CD40192BCD is the 4000 series part number.

The 74HC00 series of *high-speed* CMOS digital ICs is an improved version of the 4000 and 74C00 series. The propagation delays have been improved to attain bipolar (74LS) speed. A typical propagation delay of a 74HC00 series gate might be from 8 to 12 ns. The normal CMOS advantages have been retained but with improved output drive capabilities of up to 4 mA for good fan-out. Some 74HC00 series ICs have a fan-out of 10 LS-TTL loads. The 74HC00 series reproduces the most popular 7400 and 4000 series functions. A 2- to 6-V power supply operating range was chosen for the 74C00 series. A subfamily called the 74HCT00 series is used for interfacing from TTL to the 74HC00 series.

Typical markings on a 74HC00 series high-speed CMOS IC are reproduced in Fig. 6-12. Pin 1 is located next to the dot. The manufacturer is National Semiconductor Corporation. Two part numbers appear on the IC; each has the same core number of 74HC32N. The prefix MM is used by National Semiconductor to mean MOS monolithic, and the prefix MC is used by Motorola. The N suffix means a DIP IC. The 74HC means the IC is from the high-speed CMOS family. The 32 describes the function of the IC (quad 2-input OR gate).

Fig. 6-12 Typical markings on a 74HC00 series high-speed CMOS IC

The CMOS technology may be most suitable for large-scale and very-large-scale integrations instead of SSI/MSI ICs. Because of simple internal circuitry and low power consumption, many elements can be squeezed onto a very tiny area of the silicon chip. Some LSI and VLSI ICs that are available in CMOS are microprocessors, memory devices (RAMs, PROMs), microcontrollers, clocks, modems, filters, coders-decoders, and tone generators for telecommunications, analog-to-digital (A/D) and digital-to-analog (D/A) converters, LCD display decoders/drivers, UARTS for serial data transmission, and calculator chips.

Manufacturers suggest that, when CMOS ICs are being worked with, *damage from static discharge and transient voltages can be prevented* by:

1. Storing CMOS ICs in special conductive foam
2. Using battery-powered soldering irons when working on CMOS chips or grounding the tips of ac-operated irons
3. Turning power off when removing CMOS ICs or changing connections on a breadboard
4. Ensuring that input signals do not exceed power supply voltages
5. Turning input signals off before turning circuit power off
6. Connecting *all unused input leads* to either positive or GND of the power supply (only unused CMOS *outputs* may be left unconnected)

SOLVED PROBLEMS

6.26 List three SSI/MSI families of CMOS ICs.

Solution:

Three popular SSI/MSI families of CMOS ICs are the 4000, 74C00, and the 74HC00 series.

6.27 The _____ (CMOS, TTL) family of digital ICs was first introduced in 1964.

Solution:

The TTL family was first introduced in 1964. RCA came out with CMOS in 1968.

6.28 Refer to Fig. 6-9a. If the input is at GND potential, which MOSFET transistor is turned on (conducting)?

Solution:

If the inverter input (Fig. 6-9a) is negative (GND), then the *P*-channel transistor (Q2) will conduct (be turned on). When the input is LOW, the output from the inverter will be HIGH.

6.29 Decode the markings on the IC depicted in Fig. 6-13. To interpret all the markings, a manufacturer's logic manual or *IC Master* would probably be required.

Fig. 6-13   Dual-in-line package IC

Solution:

The manufacturer is National Semiconductor (logo); the core number is 4001B. CD is the manufacturer's code for the CMOS 4000 series of ICs. The suffix N is the manufacturer's code for a plastic DIP. The suffix C is for a temperature range of −40 to 85°C. The 40 indicates the 4000 series of CMOS ICs. The 01 indicates the function of the IC (quad 2-input NOR gate in this example), and B stands for a buffered CMOS IC. A manufacturer's CMOS logic data manual or general manual such as the *IC Master* is needed to find some of this information.

6.30 List several advantages of CMOS ICs over TTL devices.

Solution:

The advantages of CMOS ICs over TTLs are lower power consumption, better noise immunity, lower noise generation, and the ability to operate on an inexpensive, nonregulated power supply.

6.31 List some disadvantages of CMOS ICs compared with TTLs.

Solution:

The disadvantages of CMOS ICs compared with TTLs are poorer speed characteristics, unwanted sensitivity to static discharges and transient voltages, and lower output current drive capabilities.

6.32 Unused CMOS inputs _____ (may, may not) be left disconnected.

Solution:

Unused CMOS inputs may not be left disconnected.

6.33 When _____ (CMOS, TTL) chips are being worked on, battery-operated soldering irons are recommended to protect the circuitry.

Solution:

When CMOS chips are being worked on, battery-operated soldering irons are used to protect the circuits from possible static discharges or transient voltages.

6.34 The _____ (CMOS, TTL) family has better noise immunity.

Solution:

The CMOS family has better noise immunity than TTL ICs.

6.35 The _____ (4000, 74HC00) series of CMOS ICs has lower propagation delays.

Solution:

The 74HC00 series of CMOS ICs has lower propagation delays and can therefore be used at high frequencies.

6-5 INTERFACING TTL AND CMOS ICs

Interfacing is the method of connecting two electronic devices such as logic gates. Manufacturers guarantee that, *within a family* of logic circuits, one gate will drive another. As an example, the two TTL gates shown in Fig. 6-14*a* are simply connected together with no extra parts required and no problems. A second example of two CMOS gates interfaced is illustrated in Fig. 6-14*b*. In both examples the manufacturer has taken great care to make sure the devices would interface easily and properly.

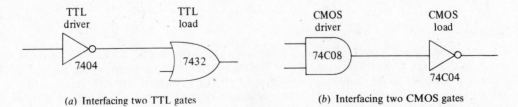

(*a*) Interfacing two TTL gates   (*b*) Interfacing two CMOS gates

Fig. 6-14

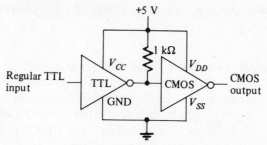

(a) Standard TTL-to-CMOS interfacing using pull-up resistor

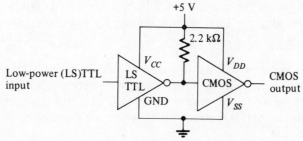

(b) Low-power Schottky TTL-to-CMOS interfacing using a pull-up resistor

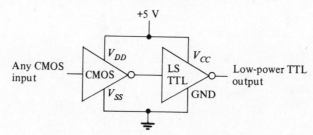

(c) CMOS-to-low-power Schottky TTL interfacing

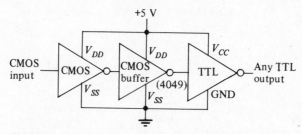

(d) CMOS-to-standard TTL interfacing using a CMOS buffer IC

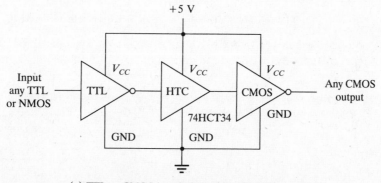

(e) TTL-to-CMOS interfacing using a 74HCT00 series IC

Fig. 6-15 Interfacing TTL and CMOS when both devices operate on a common +5-V supply (*Roger L. Tokheim, Digital Electronics, 3d ed., McGraw-Hill, New York, 1990*)

119

What about interfacing families of ICs such as TTL and CMOS? CMOS and TTL logic levels (voltages) are defined differently. Refer to Fig. 6-1 for details on the definition of LOW and HIGH logic levels of both TTL and CMOS ICs. Because of the differences in voltage levels, CMOS and TTL ICs usually cannot simply be connected directly together as within a family. Current requirements for CMOS and TTL ICs also are different. Therefore, TTL and CMOS ICs usually cannot be connected directly. Special simple *interface techniques* will be outlined.

Interfacing a CMOS with a TTL IC is quite easy if both devices operate on a common +5-V power supply. Figure 6-15 shows five examples of TTL-to-CMOS and CMOS-to-TTL interfacing.

Figure 6-15*a* shows the use of a 1-kΩ *pull-up resistor* for interfacing standard TTL with CMOS ICs. Figure 6-15*b* shows the use of a 2.2-kΩ pull-up resistor for interfacing low-power TTL ICs with CMOS ICs.

CMOS-to-TTL interfacing is even easier. Figure 6-15*c* shows both CMOS and low-power TTL ICs sharing the same +5-V power supply. A direct connection between a CMOS output and *any one* low-power TTL input can be made. Note that the CMOS gate can drive only one low-power TTL input. The exception would be 74HC00 series CMOS, which can drive up to 10 low-power TTL inputs.

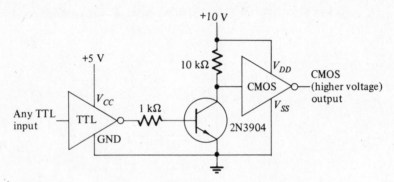

(a) TTL-to-CMOS interfacing using a transistor

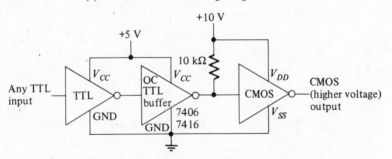

(b) TTL-to-CMOS interfacing using an TTL open-collector buffer IC

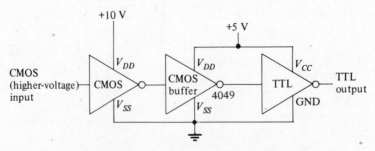

(c) CMOS-to-TTL interfacing using a CMOS buffer IC

Fig. 6-16 Interfacing TTL and CMOS devices when each device uses a different power supply voltage (*Roger L. Tokheim*, Digital Electronics, *3d ed., McGraw-Hill, New York, 1990*)

For more driving power, Fig. 6-15*d* shows the use of a special 4049 CMOS buffer between the CMOS and TTL units. The CMOS buffer can drive up to two standard TTL inputs. A noninverting buffer which is similar to the unit shown in Fig. 6-15*d* is the 4050 CMOS IC.

The problem of voltage incompatibility from TTL (or NMOS) to CMOS can be solved using a pull-up resistor as in Fig. 6-15*a*. A second method of solving this interface problem is shown in Fig. 6-15*e*. The 74HCT00 series of CMOS ICs is designed as an interface element between TTL (or NMOS) and CMOS. A 74HCT34 noninverting IC is used as the interface element between TTL and CMOS ICs in Fig. 6-15*e*.

The 74HCT00 series of CMOS ICs is used for interfacing between LSI NMOS devices and CMOS. The NMOS output characteristics are almost the same as for low-power Schottky TTL ICs.

Interfacing a CMOS device with a TTL device requires some additional components when each operates on a *different voltage power supply*. Figure 6-16 shows three examples of TTL-to-CMOS and CMOS-to-TTL interfacing. Figure 6-16*a* shows the TTL inverter driving a general-purpose NPN transistor. The transistor and associated resistors translate the lower-voltage TTL outputs to the high-voltage inputs needed to operate the CMOS inverter. The CMOS output has a voltage swing from GND to +10 V.

Figure 6-16*b* shows an *open-collector* TTL buffer and a 10-kΩ pull-up resistor being used to translate from the lower TTL to the higher CMOS voltages. The 7406 and 7416 TTL ICs are two

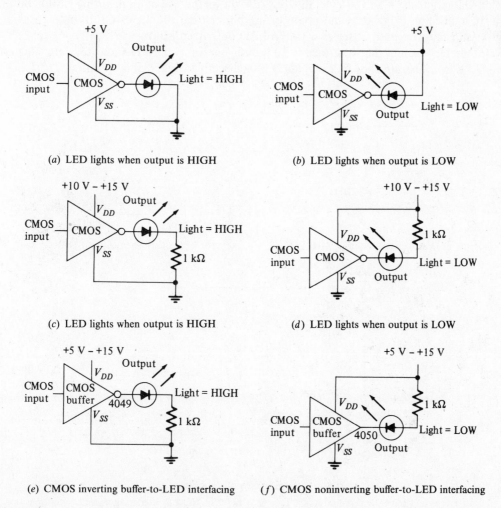

(*a*) LED lights when output is HIGH   (*b*) LED lights when output is LOW

(*c*) LED lights when output is HIGH   (*d*) LED lights when output is LOW

(*e*) CMOS inverting buffer-to-LED interfacing   (*f*) CMOS noninverting buffer-to-LED interfacing

Fig. 6-17 CMOS-to-LED interfacing (*Roger L. Tokheim*, Digital Electronics, *3d ed., McGraw-Hill, New York, 1990*)

inverting, open-collector buffers. The 7407 and 7417 TTL ICs are similar noninverting, open-collector buffers which can also be used in the circuit in Fig. 6-16b.

Interfacing a higher-voltage CMOS inverter to a lower-voltage TTL inverter is illustrated in Fig. 6-16c. The 4049 buffer is used between the higher-voltage CMOS inverter and the lower-voltage TTL IC. Note that the CMOS buffer is powered by the lower-voltage (+5-V) power supply shown in Fig. 6-16c.

Digital circuits can also drive devices other than logic gates. Interfacing CMOS devices with simple LED indicator lamps is easy. Figure 6-17 shows six examples of CMOS ICs driving LED indicators. Figure 6-17a and b shows the CMOS supply voltage at +5 V. At this low voltage, no limiting resistors are needed in series with the LEDs. In Fig. 6-17a, when the output of the CMOS inverter goes HIGH, the LED output indicator lights. The opposite is true in Fig. 6-17b; when the CMOS output goes LOW, the LED indicator lights.

Figure 6-17c and d shows the CMOS ICs being operated on a higher supply voltage (+10 to +15 V). Because of the higher voltage, a 1-kΩ limiting resistor is placed in series with the LED output indicator lights. When the output of the CMOS inverter shown in Fig. 6-17c goes HIGH, the LED output indicator lights. In Fig. 6-17d, however, the LED indicator is shown activated by a LOW at the CMOS output.

Figure 6-17e and f shows CMOS buffers being used to drive LED indicators. The circuits may operate on voltages from +5 to +15 V. Figure 6-17e shows the use of an inverting CMOS buffer (like the 4049 IC), and Fig. 6-17f uses the noninverting buffer (like the 4050 IC). In both cases, a 1-kΩ limiting resistor must be used in series with the LED output indicator.

Several simple circuits that interface a TTL with one or two LED indicators are illustrated in Fig. 6-18. TTL inverters are shown driving the LEDs directly in Fig. 6-18a, b, and c. The LED shown

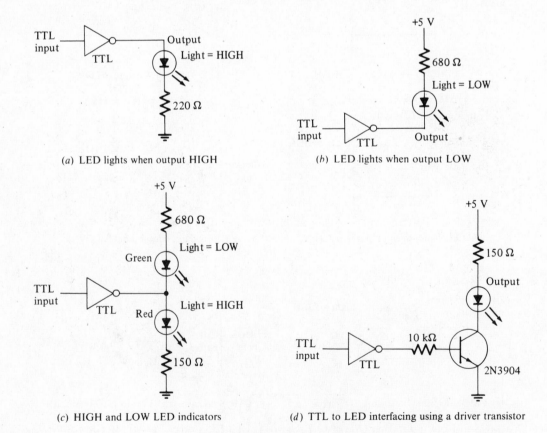

(a) LED lights when output HIGH

(b) LED lights when output LOW

(c) HIGH and LOW LED indicators

(d) TTL to LED interfacing using a driver transistor

Fig. 6-18 TTL-to-LED interfacing

in Fig. 6-18a lights when the inverter's output is HIGH, but the LED shown in Fig. 6-18b lights when the inverter's output is LOW. These ideas are combined to form the circuit shown in Fig. 6-18c. When the red LED lights, the output of the inverter is HIGH, but when the output of the inverter goes LOW, the green LED will light.

The circuit shown in Fig. 6-18c has an additional feature. If the output of the inverter were between HIGH and LOW (in the undefined region), both LEDs would light. This circuit therefore becomes a simple logic indicator for checking logic levels at the outputs of logic circuits. Figure 6-18d shows the use of a driver transistor to turn the LED on and off. When the output of the TTL inverter goes LOW, the transistor is turned off and the LED does not light. When the inverter output goes HIGH, the transistor conducts and causes the LED to light. This circuit reduces the output drive current from the TTL inverter.

SOLVED PROBLEMS

6.36   What is interfacing?

Solution:

Interfacing is the method used to interconnect two separate electronic devices in such a way that their output and input voltages and currents are compatible.

6.37   Show the interfacing of two TTL gates (an OR gate driving an AND gate).

Solution:

See Fig. 6-19. Note that, within a family of logic ICs, a direct connection can usually be made between the output of one gate and the input of the next.

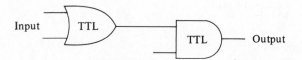

Fig. 6-19 Solution to Prob. 6.37

6.38   Show the interfacing of a CMOS NAND gate driving a low-power Schottky TTL OR gate. Use a +5-V power supply.

Solution:

See Fig. 6-20. By using Fig. 6-15c as a guide, it was determined that the output of the CMOS gate can drive one LS-TTL load.

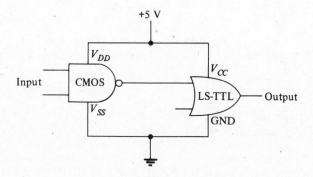

Fig. 6-20 Solution to Prob. 6.38

6.39 Show the interfacing of a standard TTL OR gate driving a CMOS inverter. Use a +5-V power supply.

Solution:

   See Fig. 6-21. By using Fig. 6-15a as a guide, it was determined that a 1-kΩ pull-up resistor was needed to help pull the TTL output to a HIGH that was positive enough to have the CMOS input accept it as a logical 1.

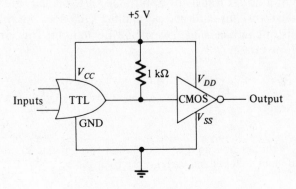

Fig. 6-21 Solution to Prob. 6.39

6.40 Show the interfacing of a standard TTL AND gate (using +5-V supply) driving a CMOS inverter (using a +10-V supply).

Solution:

   An interface circuit using a driver transistor is shown in Fig. 6-22. An open-collector TTL buffer and pull-up resistor could also be used as in the circuit shown in Fig. 6-16b.

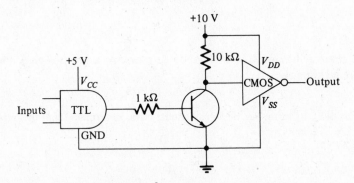

Fig. 6-22 One solution to Prob. 6.40

6.41 Show a TTL NAND gate driving an LED output indicator so the LED goes on when the output of the NAND gate goes HIGH.

Solution:

   See Fig. 6-23. When the output of the NAND gate goes HIGH, the LED is forward-biased, current flows, and the LED lights.

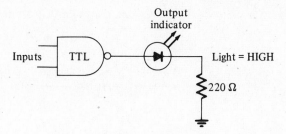

Fig. 6-23 Solution to Prob. 6.41

6.42   Refer to Fig. 6-18*c*. If the output of the inverter drops near ground potential, the _____ (green, red) LED lights to indicate a _____ (HIGH, LOW) logic level.

   Solution:
      When the output of the inverter shown in Fig. 6-18*c* is near GND or LOW, the green LED lights.

6.43   Show the interfacing for a CMOS NAND gate directly driving an LED so the indicator lights when the gate output is HIGH. Use a +10-V power supply.

   Solution:
      See Fig. 6-24.

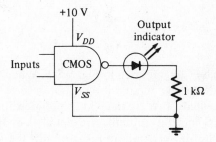

Fig. 6-24 Solution to Prob. 6.43

6-6 INTERFACING TTL AND CMOS WITH SWITCHES

   A common method of entering information into a digital system is by way of switches (or keyboards). This section details several methods of interfacing a switch to either TTL or CMOS ICs.
   Consider the simple switch interface circuit drawn in Fig. 6-25*a*. When the switch is open (not pressed), the input to the TTL inverter is connected directly to the positive of the power supply through the 10-kΩ pull-up resistor; the switch input is HIGH in Fig. 6-25*a* when the switch is open. Pressing the normally open switch in Fig. 6-25*a* grounds the TTL input, driving it LOW. The circuit in Fig. 6-25*a* might be called an *active-LOW* switch interface because the TTL input goes LOW when the switch is activated.

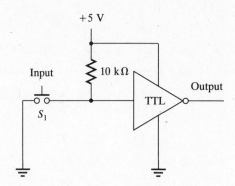

(a) Simple active-LOW switch interface

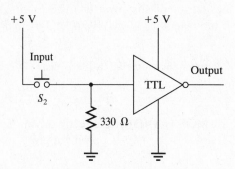

(b) Simple active-HIGH switch interface

Fig. 6-25 Switch-to-TTL interfaces

An *active-HIGH* input switch is diagrammed in Fig. 6-25b. When the switch is activated (pressed), the +5 V is connected directly to the input of the TTL inverter. When the switch is released (opened), the inverter input is pulled LOW by the 330-Ω pull-down resistor.

Two simple switch-to-CMOS interface circuits are detailed in Fig. 6-26. An active-LOW input switch is drawn in Fig. 6-26a. The 100-kΩ pull-up resistor pulls the voltage to +5 V when the input switch is open. The CMOS inverter input goes LOW when the normally open switch is closed in Fig. 6-26a.

An active-HIGH input switch is shown in Fig. 6-26b. The CMOS inverter input is LOW (connected through the pull-down resistor) when the switch is open. When the switch is closed (pressed) in Fig. 6-26b, the input of the inverter is driven HIGH.

Consider the circuit in Fig. 6-27a. Each press and release of the input switch should cause the counter to increment by 1. Unfortunately the counter increases by 1, 2, 3, or sometimes more. This problem is caused by *switch bounce*. When a mechanical switch closes or opens, the contacts do not make or break the circuit cleanly, generating several short voltage spikes. This means that several (instead of one) pulses are fed into the clock (CLK) input of the counter IC on each switch closure.

The counting circuit in Fig. 6-27a needs extra circuitry to eliminate the switch bounce problem. *Switching debouncing* circuitry has been added to the counter circuit in Fig. 6-27b. The TTL decade (0 to 9) counter IC will now count (increment by only 1) on each HIGH-to-LOW cycle of the input switch. The cross-wired NAND gates in the debouncing circuit are sometimes referred to as a *latch* or *RS flip-flop*. Flip-flops are covered in greater detail in Chap. 9.

Two other general-purpose switch debouncing circuits are diagrammed in Fig. 6-28. The debouncing circuit in Fig. 6-28a will drive any 4000 series, 74C00 series, or 74HC00 series CMOS or TTL ICs. Another debouncing circuit is drawn in Fig. 6-28b. This circuit uses *open-collector* 7403 TTL ICs in

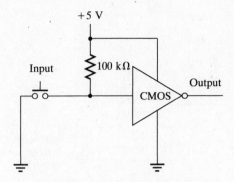

(a) Simple active-LOW switch interface

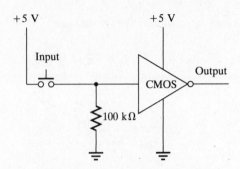

(b) Simple active-HIGH switch interface

Fig. 6-26 Switch-to-CMOS interfaces

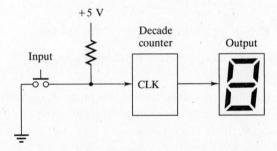

(a) Switch interfacing with decimal counter causes problems

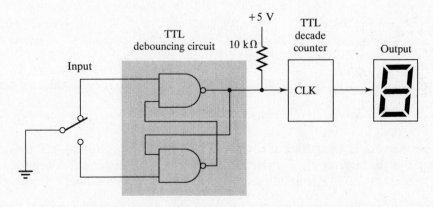

(b) Added switch debouncing circuit makes counter work properly

Fig. 6-27

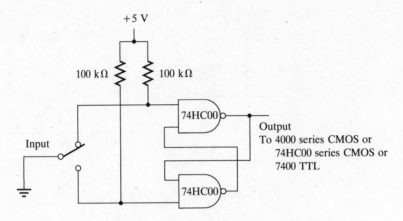

(a) Using a 74HC00 CMOS NAND gate

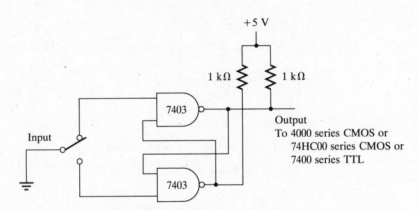

(b) Using a 7403 open-collector TTL gate

Fig. 6-28 General-purpose switch debouncing circuits

the latch with the required pull-up resistors at the outputs of each NAND gate. The switch debouncing circuit in Fig. 6-28b will drive 4000 series, 74C00 series, or 74HC00 series CMOS or TTL ICs.

SOLVED PROBLEMS

6.44   Refer to Fig. 6-25a. Component S_1 is considered an active-_____ (HIGH, LOW) input switch because closing the switch causes the input of the inverter to go _____ (HIGH, LOW).

Solution:

   In Fig. 6.25a, S_1 is an active-LOW input switch because closing S_1 causes the input of the inverter to go LOW.

6.45   Refer to Fig. 6-25b. The 330-Ω resistor is called a pull-_____ (down, up) resistor as it holds the input of the inverter _____ (HIGH, LOW) when the input switch is open (not pressed).

Solution:

   In Fig. 6-25b the resistor is called a pull-down resistor as it holds the input of the inverter LOW when the input switch is open.

6.46   Refer to Fig. 6-27(a). The counter IC does not accurately count the number of times the input switch is pressed because of a problem called switch _____ (bounce, hysteresis).

Solution:

In Fig. 6-27(a), the counter does not accurately count the number of times the input switch is pressed because of a problem called switch bounce.

6.47   Switch debouncing circuits are typically _____ (latches, multiplexers).

Solution:

Switch debouncing circuits are typically latches.

6.48   Refer to Fig. 6-28b. The 7403 TTL NAND gates have _____ (open-collector, totem pole) outputs which require pull-up resistors at the gate outputs.

Solution:

The 7403 TTL NAND gates have open-collector outputs which require pull-up resistors at the gate outputs.

6-7   INTERFACING TTL / CMOS WITH SIMPLE OUTPUT DEVICES

The task of many digital systems is to control simple output devices that may have very different voltage and current characteristics. This section explores simple interface techniques with logic elements driving buzzers, relays, electric motors, and solenoids.

Most logic families do not have the current capabilities to drive output devices directly. Using a logic element to turn on a transistor is a common interface technique. Consider the circuit in Fig. 6-29. This circuit uses the NPN transistor as a switch. When the output of the inverter goes LOW, the voltage between the base (B) and emitter (E) of the bipolar transistor is near 0. This turns the transistor off (very high resistance between E and C terminals), and the buzzer does not sound. When the output of the inverter goes HIGH, the positive voltage on the base (B) of the transistor turns on the transistor (resistance between E and C terminals becomes very low), allowing current to flow through to the buzzer (buzzer sounds). The diode protects against transient voltages (voltage spikes that may be produced within the buzzer). Notice that the interface circuit will work with either TTL or CMOS logic elements.

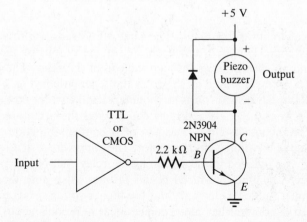

Fig. 6-29 TTL or CMOS interfaced with buzzer using a transistor driver

A relay is an excellent means of isolating a logic element from a high-voltage or high-current circuit. Figure 6-30 illustrates how a logic element could be used with a relay to control an electric motor or solenoid.

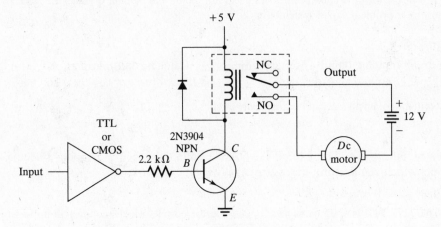

(a) Interfacing TTL or CMOS with an electric motor

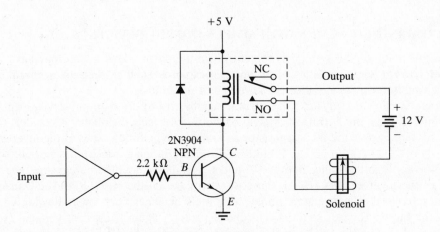

(b) Interfacing TTL or CMOS with a solenoid

Fig. 6-30 Interfacing using a relay

Consider the interface circuit in Fig. 6-30a. The same NPN transistor driver employed previously is used to snap the relay contacts closed and open. When the output of the inverter is LOW, the transistor is turned off and no current flows through the coil of the relay. The spring-loaded normally closed (NC) relay contacts are held closed as shown in the schematic in Fig. 6-30a. When the output of the inverter goes HIGH, the transistor turns on (conducts) and current flows through the coil of the relay. The magnetic force from the energized relay coil attracts the armature (moving part of relay), and the normally open (NO) contacts close. The NO relay contacts function as a simple mechanical switch which turns on the higher-voltage electric motor. The *clamp diode* across the relay coil prevents voltage spikes which might be induced in the system by the relay coil. Notice in Fig. 6-30a that either TTL or CMOS logic circuits may be interfaced in this fashion. Also note the excellent isolation (no electric connection) between the logic elements and the higher-voltage/current motor circuit.

A solenoid is an electrical device that can produce linear motion. The circuit in Fig. 6-30b shows how the output of a TTL or CMOS logic gate can be used to control the higher currents and voltages in the solenoid circuit. Once again the driver transistor is turned on and off by the output of the logic gate. The transistor controls the current through the relay coil. The magnetic force from the relay coil snaps the NO contacts closed when energized. Closing the NO relay contacts completes the higher-voltage circuit energizing the solenoid coil. The solenoid coil causes the core of the solenoid to produce a linear motion.

SOLVED PROBLEMS

6.49   Refer to Fig. 6-29. The buzzer will sound only when the output of the inverter goes _____ (HIGH, LOW) and the transistor _____ (conducts, does not conduct) current.

Solution:

The buzzer in Fig. 6-29 will sound when the output of the inverter goes HIGH and the transistor conducts current.

6.50   Refer to Fig. 6-29. If the output of the inverter goes LOW, the transistor _____ (will, will not) conduct current and the buzzer _____ (is silent, sounds).

Solution:

If the output of the inverter in Fig. 6-29 goes LOW, the transistor will not conduct and the buzzer is silent.

6.51   What is the function of the relay in the circuits in Fig. 6-30?

Solution:

The relay serves to isolate the logic circuitry from the higher-voltage and higher-current motor/solenoid circuits in Fig. 6-30.

6.52   Refer to Fig. 6-30a. The electric motor operates when the output of the logic element (inverter) goes _____ (HIGH, LOW).

Solution:

The motor in Fig. 6-30a operates when the output of the inverter goes HIGH.

6.53   Refer to Fig. 6-30b. What is the purpose of the diode placed in parallel with the relay coil?

Solution:

The diode eliminates harmful voltage spikes that may be generated by the relay coil. It is sometimes called a clamp diode.

6.54   Refer to Fig. 6-29. The transistor acts most like an _____ (amplifier, switch) in this circuit.

Solution:

The transistor acts like a switch in this circuit.

6-8   D/A AND A/D CONVERSION

Digital systems must often be *interfaced* with analog equipment. To review, a *digital signal* is one that has only two discrete voltage levels. An *analog signal* is one that varies *continuously* from a minimum to a maximum voltage or current. Figure 6-31 illustrates a typical situation in which the digital processing unit or system has analog inputs and outputs. The input on the left is a continuous voltage ranging from 0 to 5 V. The special encoder, called an *analog-to-digital converter* (A/D

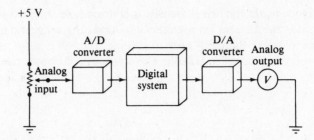

Fig. 6-31 Using A/D and D/A converters in an electronic system

converter), translates the analog input into digital information. On the output side of the digital system shown in Fig. 6-31, a special decoder translates from digital information to an analog voltage. This decoder is called a *digital-to-analog converter* (D/A converter).

The task of a D/A converter is to transform a digital input into an analog output. Figure 6-32*a* illustrates the function of the D/A converter. A binary number is entered at the inputs on the left

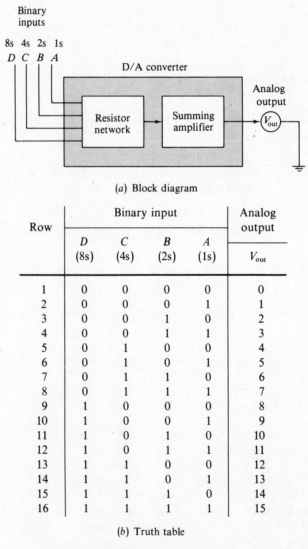

(*a*) Block diagram

Row	Binary input				Analog output
	D (8s)	C (4s)	B (2s)	A (1s)	V_{out}
1	0	0	0	0	0
2	0	0	0	1	1
3	0	0	1	0	2
4	0	0	1	1	3
5	0	1	0	0	4
6	0	1	0	1	5
7	0	1	1	0	6
8	0	1	1	1	7
9	1	0	0	0	8
10	1	0	0	1	9
11	1	0	1	0	10
12	1	0	1	1	11
13	1	1	0	0	12
14	1	1	0	1	13
15	1	1	1	0	14
16	1	1	1	1	15

(*b*) Truth table

Fig. 6-32 D/A converter

with a corresponding output voltage at the right. As with other tasks in electronics, it is well to define exactly the inputs and expected outputs from the system. The truth table in Fig. 6-32*b* details one set of possible inputs and outputs for the D/A converter.

Consider the truth table in Fig. 6-32*b* for the D/A converter. If each of the inputs is LOW, the output voltage (V_{out}) is 0 V as defined in row 1 of the table. Row 2 shows just the 1s input (A) being activated by a HIGH. With the input as LLLH (0001), the output from the D/A converter is 1 V. Row 3 shows only input B activated (0010). This produces a 2-V output. Row 5 shows only input C activated (0100). This yields a 4-V output. Row 9 shows only input D (1000) activated; this produces an 8-V output from the D/A converter. Note that the inputs (D, C, B, A) are *weighted* so that a HIGH at input D generates an 8-V output and a HIGH at input A produces only a 1-V output. The relative weighting of each input is given as 8 for input D, 4 for input C, 2 for input B, and 1 for input A in Fig. 6-32*a*.

A simple D/A converter consists of two functional parts. Figure 6-32*a* shows a block diagram of a D/A converter. The converter is divided into a *resistor network* and a *summing amplifier*. The resistor network weights the 1s, 2s, 4s, and 8s inputs properly, and the summing amplifier *scales* the output voltage according to the truth table. An *op amp*, or *operational amplifier*, is commonly used as the summing amplifier.

A few of the important specifications of commercial D/A converters are resolution, linearity, settling time, power dissipation, type of input (binary, complemented binary, and sign and magnitude), technology (TTL, CMOS, or ECL), and special features. One manual lists more than a hundred different D/A converter ICs having resolutions from 4 to 18 bits.

Consider the simplified block diagram of a commercial A/D converter reproduced in Fig. 6-33*a*. This is the *ADC0804 8-bit microprocessor-compatible A/D converter*. The control lines to the

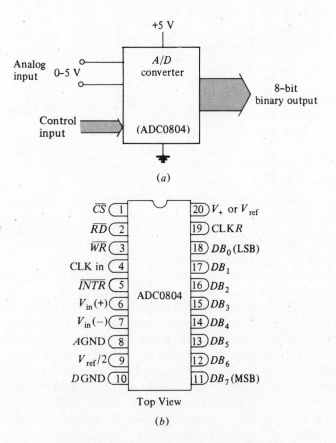

(a)

(b)

Fig. 6-33 ADC0804 8-bit A/D converter IC

ADC0804 A/D converter direct the A/D converter to first *sample and digitize* the analog voltage at the input. Second, the control lines direct the A/D converter to generate an 8-bit binary output. The 8-bit output will be directly proportional to the analog input voltage. If the input voltage is 5 V, the binary output will be 11111111. But if the input voltage were 0 V, the binary output would read 00000000.

A pin diagram of the ADC0804 A/D converter IC is shown in Fig. 6-33b. The ADC0804 IC is a CMOS 8-bit successive approximation A/D converter which is designed to operate with the 8080A microprocessor without extra interfacing. The ADC0804 IC's conversion time is under 100 μs, and all inputs and outputs are TTL compatible. It operates on a 5-V power supply, and it can handle a full range 0- to 5-V analog input between pins 6 and 7. The ADC0804 IC has an on-chip clock generator which needs only an external resistor and capacitor (see Fig. 6-34).

A simple lab setup using the ADC0804 A/D converter is shown in Fig. 6-34. The analog input voltage is developed across the wiper and ground of the 10-kΩ potentiometer. The resolution of the A/D converter is $\frac{1}{255}$ ($2^8 - 1$) of the full-scale analog voltage (5 V in this example). For each increase of 0.02 V ($\frac{1}{255} \times 5$ V = 0.02 V), the binary output increments by 1. Therefore, if the analog input equals 0.1 V, the binary output will be 00000101 (0.1 V/0.02 V = 5, and decimal 5 = 00000101 in binary).

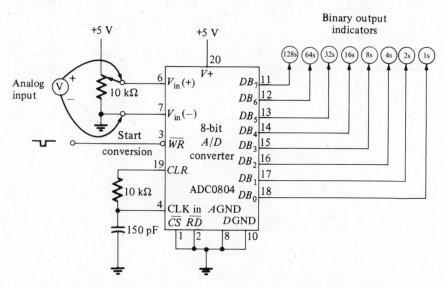

Fig. 6-34 Wiring a test circuit with the ADC0804 8-bit A/D converter IC

The H-to-L transition of the clock pulse at the $\overline{WR}$ input to the ADC0804 IC shown in Fig. 6-34 starts the conversion process. The binary output appears about 100 μs later at the indicators on the right. This A/D converter can make more than 5000 conversions per second. The outputs are three-state buffered, so they can be connected directly to the data bus of a microprocessor-based system. The ADC0804 A/D converter has an *interrupt output* ($\overline{INTR}$, see pin 5, Fig. 6-33b) which signals the microprocessor system when the current analog-to-digital conversion is finished. Interrupts are needed in microprocessor systems when interfacing very "slow" asynchronous devices such as an A/D converter to "very fast" synchronous devices such as a microprocessor.

Important specifications of commercial A/D converters are resolution, linearity, conversion time, power dissipation, type of output (binary, decimal, complemented binary, sign and magnitude, parallel, serial), and special features. One manual lists hundreds of different A/D converter ICs with resolutions between 8 and 20 bits. A/D converters with decimal outputs (like digital voltmeter ICs) are available with resolutions of $3\frac{1}{2}$ and $4\frac{1}{2}$ digits.

SOLVED PROBLEMS

6.55 Explain the fundamental difference between A/D and D/A converters.

Solution:

An A/D converter changes an analog voltage proportionately into a digital output (usually binary). A D/A converter transforms a digital input (usually binary) proportionately into an analog output voltage.

6.56 A simple D/A converter consists of two functional parts, a _____ network and a _____ amplifier.

Solution:

A simple D/A converter consists of two functional parts, a resistor network and a summing amplifier.

6.57 Refer to Fig. 6-32*b*. List the output voltage (V_{out}) for each input combination shown in Fig. 6-35.

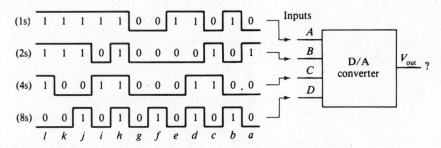

Fig. 6-35 D/A converter pulse-train problem

Solution:

The analog outputs (V_{out}) from the D/A converter in Fig. 6-35 are as follows:

pulse $a = 2$ V   pulse $d = 13$ V   pulse $g = 0$ V   pulse $j = 11$ V
pulse $b = 9$ V   pulse $e = 1$ V   pulse $h = 15$ V   pulse $k = 3$ V
pulse $c = 6$ V   pulse $f = 8$ V   pulse $i = 5$ V   pulse $l = 7$ V

6.58 Refer to Fig. 6-32*a*. The summing amplifier in a D/A converter is commonly a(n) _____ (multiplexer, op amp).

Solution:

An op amp (operational amplifier) is typically used as the summing amplifier in a D/A converter as in Fig. 6-32*a*.

6.59 The ADC0804 IC is an A/D converter with a _____ (parallel, serial) output.

Solution:

The ADC0804 IC is an A/D converter with parallel three-state outputs that may be connected directly to a microprocessor data bus.

6.60 The ADC0804 IC has a resolution of _____ (4, 8, 12) bits.

Solution:

The ADC0804 A/D converter has a resolution of 8 bits, or 1 of 255 ($2^8 - 1 = 255$).

6.61 Refer to Fig. 6-34. If the input voltage is 2 V, the binary output from the A/D converter should be binary _____.

Solution:

The calculation is as follows:

$$\frac{2 \text{ V}}{0.02 \text{ V}} = 100 \qquad \text{decimal } 100 = 01100100 \text{ in binary}$$

The binary output from the A/D converter shown in Fig. 6-34 is 01100100 when the input voltage is 2 V.

6.62 The ADC0804 IC _____ (is, is not) compatible with microprocessor-based systems.

Solution:

The ADC0804 A/D converter IC is compatible with microprocessor-based systems. It has buffered three-state outputs, microprocessor-compatible control inputs, and an interrupt output.

6.63 The ADC0804 A/D converter has a conversion time of about 100 _____ (micro, nano) seconds.

Solution:

The ADC0804 IC has a conversion time of less than 100 μs (microseconds).

6.64 The ADC0804 A/D converter operates _____ (at the same speed as, slower than) a microprocessor.

Solution:

A/D converters operate slower than microprocessors and therefore use an interrupt to signal the system when they are ready to send valid data.

Supplementary Problems

6.65 A group of compatible digital ICs that can be connected directly together to form a digital system is said to form a _____.   *Ans.*   family

6.66 Digital ICs from the (a) and (b) families are the most popular.
Ans. (a) TTL   (b) CMOS

6.67 An IC containing from 12 to 99 equivalent gates is defined as an _____ (LSI, MSI, SSI).   *Ans.*   MSI

6.68 Refer to Fig. 6-1a. A 2.1-V input to the TTL inverter is a logical _____ (0, 1) input.   *Ans.*   1

6.69 Refer to Fig. 6-1a. A 2.1-V output from the TTL inverter is a logical _____ output.
Ans.   A 2.1-V output from the TTL inverter is defined as a forbidden output caused by a defective IC or too heavy a load at the output.

6.70 What is the propagation delay of a digital IC?
Ans. The time it takes for the output to change after the input has changed logic states. The propagation delay for modern digital ICs may range from about 1.5 ns to about 125 ns.

6.71 What is the fan-out of a digital IC?
Ans. The number of parallel loads that can be driven by a single digital IC output.

6.72 The CMOS digital IC family is noted for its _____ (high, low) power consumption.   *Ans.* low

6.73 Refer to Fig. 6-1*b*. A 1-V input to the CMOS inverter is considered a logical _____ (0, 1) input.
Ans. 0 or LOW

6.74 Which TTL subfamily is best for low-power consumption?   *Ans.* low-power TTL (see Fig. 6-6*b*)

6.75 List three types of TTL outputs.   *Ans.* totem pole, open-collector, three-state

6.76 TTL logic devices of the _____ (5400, 7400) series are less expensive and are considered commercial grade ICs.   *Ans.* 7400

6.77 Refer to Fig. 6-36. The manufacturer of the IC shown is _____.
Ans. National Semiconductor Corporation (see logo)

Fig. 6-36 Dual-in-line package IC

6.78 Refer to Fig. 6-36. A _____ (CMOS, TTL) integrated circuit is shown.   *Ans.* TTL

6.79 Refer to Fig. 6-36. What is the function of the IC shown?   *Ans.* quad 2-input NAND gate (7400 IC)

6.80 The letters CMOS stand for _____.   *Ans.* complementary metal-oxide semiconductor

6.81 The _____ (CMOS, TTL) families are generally better suited for use in portable battery-operated equipment.   *Ans.* CMOS

6.82 The _____ (4000, 74HC00) series of CMOS ICs are better suited for high-speed operation.
Ans. 74HC00

6.83 The _____ (4000, 7400) series of ICs might use a 10-V dc power supply.   *Ans.* 4000

6.84 The _____ (CMOS, TTL) families are generally better suited for use when the power source is unregulated, such as a battery source.   *Ans.* CMOS

6.85 ICs of the _____ (CMOS, TTL) family are especially sensitive to static discharges and transient voltages.   *Ans.* CMOS

6.86 If an IC has the marking 74C08, it is a _____ (CMOS, TTL) device.   *Ans.* CMOS

6.87 Refer to Fig. 6-18c. When the output of the TTL inverter goes to about 3 V, the _____ (green, red) LED lights. That indicates a _____ (HIGH, LOW) logic level.   *Ans.*   red, HIGH

6.88 Refer to Fig. 6-18d. When the output of the TTL inverter goes LOW, the transistor _____ (is, is not) conducting and the LED _____ (does not light, lights).   *Ans.*   is not, does not light

6.89 Refer to Fig. 6-15d. The buffer is used for interfacing the CMOS and standard TTL gate because it has _____ (fewer, more) output current drive capabilities than the standard CMOS inverter.
Ans.   more

6.90 In Fig. 6-16b and c, special _____ (buffers, transistors) are used between TTL and CMOS gates to aid in interfacing.   *Ans.*   buffers

6.91 Refer to Fig. 6-25b. Component S_2 is considered an active-_____ (HIGH, LOW) input switch because closing the switch causes the input of the inverter to go _____ (HIGH, LOW).   *Ans.*   HIGH, HIGH

6.92 Refer to Fig. 6-27b. The NAND gates forming the switch debouncing circuit are wired like a latch or _____.   *Ans.*   RS flip-flop

6.93 Refer to Fig. 6-28b. The 7403 TTL NAND gates have open-collector outputs which require _____ (pull-down, pull-up) resistors at the gate outputs.   *Ans.*   pull-up

6.94 When a mechanical switch closes and opens, the contacts do not make and break the circuit cleanly, generating unwanted voltage spikes. This is called switch _____.   *Ans.*   bounce

6.95 Refer to Fig. 6-29. When the output of the inverter goes HIGH, the transistor _____ (blocks current, conducts current) and the buzzer _____ (is silent, sounds).   *Ans.*   conducts current, sounds

6.96 Refer to Fig. 6-30a. The _____ (diode, relay) isolates the logic circuitry from the higher-voltage electric motor circuit.   *Ans.*   relay

6.97 Refer to Fig. 6-30a. When the output of the inverter goes LOW, the transistor _____ (blocks current, conducts current), the _____ (NC, NO) contacts of the relay close, and the electric motor _____ (does not operate, operates).   *Ans.*   blocks current, NC, does not operate

6.98 A special decoder that interfaces a digital system and an analog output is called a(n) _____ _____.
Ans.   D/A converter

6.99 A special encoder that interfaces an analog input and a digital system is called a(n) _____ _____.
Ans.   A/D converter

6.100 Refer to Fig. 6-32b. A 6-V output from the D/A converter could be generated only by a _____ binary input.   *Ans.*   0110

6.101 The abbreviation op amp stands for _____ _____.   *Ans.*   operational amplifier

6.102 A digital voltmeter is one application of a(n) _____ _____.   *Ans.*   A/D converter

6.103 The resolution of an A/D converter can be given as the number of   (a)   or as the percent   (b) .
Ans.   (a) bits   (b) resolution

6.104 The ADC0804 A/D converter has an 8-bit _____ (BCD, binary) output.   *Ans.*   binary

6.105 Refer to Fig. 6-34. If the input voltage is 3 V, the binary output should be _____.
Ans.   10010110 (3 V/0.02 V = 150 = 10010110 in binary)

6.106 Refer to Fig. 6-34. The 10-kΩ resistor and the 150-pF capacitor are associated with the _____ (clock, power supply) of the ADC0804 A/D converter IC.   *Ans.*   clock

6.107 The ADC0804 IC uses the _____ conversion A-to-D conversion technique.
Ans.   successive approximation

Chapter 7

Code Conversion

7-1 INTRODUCTION

One application of logic gates in digital systems is as *code converters*. Common codes used are binary, BCD (8421), octal, hexadecimal, and, of course, decimal. Much of the "mystery" surrounding computers and other digital systems stems from the unfamiliar language of digital circuits. Digital devices can process only 1 and 0 bits. However, it is difficult for humans to understand long strings of 1s and 0s. For that reason, code converters are necessary to convert from the language of people to the language of the machine.

Consider the simple block diagram of a hand-held calculator in Fig. 7-1. The input device on the left is the common keyboard. Between the keyboard and the central processing unit (CPU) of the calculator is an *encoder*. This encoder translates the decimal number pressed on the keyboard into a binary code such as the BCD (8421) code. The CPU performs its operation in binary and puts out a binary code. The decoder translates the binary code from the CPU to a special code which lights the correct segments on the seven-segment display. The decoder thereby translates from binary to decimal. The encoder and decoder in this system are electronic code translators. The encoder can be thought of as translating from the language of people to the language of the machine. The decoder does the opposite; it translates from machine language to human language.

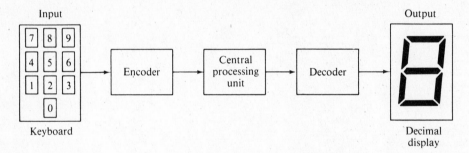

Fig. 7-1 Basic block diagram of a calculator

7-2 ENCODING

The job of the encoder in the calculator is to translate a decimal input to a BCD (8421) number. A logic diagram, in simplified form, for a decimal-to-BCD encoder is shown in Fig. 7-2. The encoder

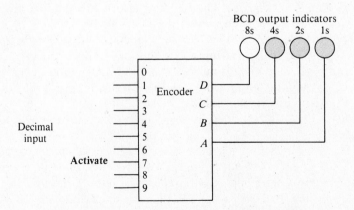

Fig. 7-2 Logic symbol for a decimal-to-BCD encoder

140

has ten inputs on the left and four outputs on the right. The encoder may have *one active input*, which in turn *produces a unique output*. Decimal input 7 is shown being activated in Fig. 7-2. This results in the BCD output of 0111, as shown on the BCD output indicators at the right.

The block diagram for a commercial decimal-to-BCD encoder is shown in Fig. 7-3*a*. The most unusual features are the small bubbles at the inputs and outputs. The bubbles at the inputs mean that the inputs are activated by logical 0s, or LOWs. The bubbles at the outputs mean that the outputs are normally HIGHs, or at logical 1s, but when activated, they go LOW, or to logical 0s. Four inverters have been added to the circuit to invert the output back to its more usual form. Another unusual feature of the encoder is that there is no zero input. A decimal 0 input will mean a 1111 output (at *D*, *C*, *B*, and *A*), which is true when all inputs (1–9) are not connected to anything. When the inputs are not connected, they are said to be *floating*. The 74147 encoder is a TTL device, which means that unconnected inputs will float HIGH.

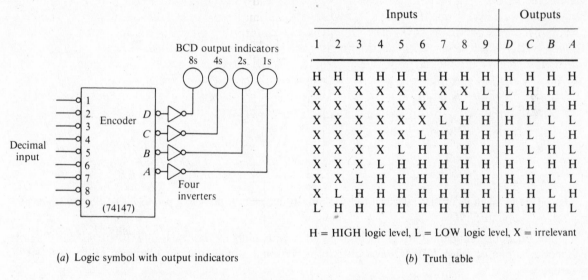

	Inputs									Outputs			
1	2	3	4	5	6	7	8	9	*D*	*C*	*B*	*A*	
H	H	H	H	H	H	H	H	H	H	H	H	H
X	X	X	X	X	X	X	X	L	L	H	H	L
X	X	X	X	X	X	X	L	H	L	H	H	H
X	X	X	X	X	X	L	H	H	H	L	L	L
X	X	X	X	X	L	H	H	H	H	L	L	H
X	X	X	X	L	H	H	H	H	H	L	H	L
X	X	X	L	H	H	H	H	H	H	L	H	H
X	X	L	H	H	H	H	H	H	H	H	L	L
X	L	H	H	H	H	H	H	H	H	H	L	H
L	H	H	H	H	H	H	H	H	H	H	H	L

H = HIGH logic level, L = LOW logic level, X = irrelevant

(*a*) Logic symbol with output indicators   (*b*) Truth table

Fig. 7-3 Commercial TTL 74147 decimal-to-BCD priority encoder

The encoder diagrammed in Fig. 7-3 is called a 10-line–to–4-line priority encoder by the manufacturer. This TTL device is referred to as a 74147 encoder. A truth table for the 74147 encoder is in Fig. 7-3*b*. The first line on the truth table is for *no inputs*. With all inputs floating HIGH, the outputs are HIGH. This is interpreted as a 0000 on the BCD output indicators in Fig. 7-3*a*. The second line of the truth table in Fig. 7-3*b* shows decimal input 9 being activated by a LOW, or 0. This produces an LHHL at outputs *D, C, B, A*. The LHHL is inverted by the four inverters, and the BCD indicators read 1001, which is the BCD indication for a decimal 9.

The second line of the truth table in Fig. 7-3*b* shows inputs 1 through 8 marked with Xs. An X in the table means *irrelevant*. An irrelevant input can be either HIGH or LOW. This encoder has a *priority feature*, which activates the highest number that has a LOW input. If LOWs were simultaneously placed on inputs 9 and 5, the output would be 1001 for the decimal 9. The encoder activates the output of the highest-order input number only.

The logic diagram for the 74147 encoder, as furnished by Texas Instruments, Inc., is shown in Fig. 7-4. All 30 gates inside the single 74147 TTL IC are shown. First try activating the decimal 9 input (LOW at input 9). This 0 input is inverted by inverter 1, and a 1 is applied to NOR gates 2 and 3. NOR gates 2 and 3 are thus activated, putting out LOWs. NOR gates 4 and 5 are deactivated by the presence of 0s at their inputs from the deactivated AND gates 7 through 18. These AND gates (7 through 18) are deactivated by the 0s at their bottom inputs produced by NOR gate 6. The AND gates (7 through 18) make sure that the higher decimal input has priority over small numbers.

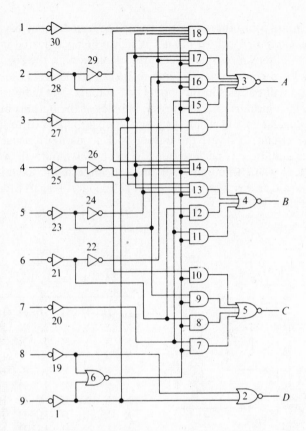

Fig. 7-4 Logic diagram of 74147 decimal-to-BCD priority encoder (*Reprinted by permission of Texas Instruments*)

An encoder using the CMOS technology also is available. The 74HC147 10- to 4-Line Priority Encoder is one of many DIP ICs available from National Semiconductor Corporation in its 74HC00 series.

SOLVED PROBLEMS

7.1   The 74147 encoder translates from the _____ (decimal, Gray) code into the _____ (BCD, octal) code.

Solution:

The 74147 translates from decimals into the BCD code.

7.2   At a given time, an encoder may have _____ (one, many) active input(s) which produce(s) a unique output.

Solution:

By definition an encoder will have only one input activated at any given time. If several inputs appear to be activated by having LOWs, the highest decimal number will be encoded on a unit such as the 74147 encoder.

7.3   In Fig. 7-3*a*, if input 3 is activated with a _____ (HIGH, LOW), the BCD output indicators will read _____ (four bits).

Solution:

A LOW at input 3 will produce a 0011 at the output indicators.

7.4 If both inputs 4 and 5 are activated with LOWs, the output indicators shown in Fig. 7-3a will read _____ (four bits).

Solution:

The 74147 encoder gives priority to input 5, producing a 0101 output on the BCD indicators.

7.5 Refer to Fig. 7-4. A logical _____ (0, 1) is needed to activate input 1.

Solution:

A logical 0 is needed to activate any input on the 74147 encoder.

7.6 Assume that only input 1 is activated in the circuit of Fig. 7-4. Output _____ (A, B, C, D) will be LOW because AND gate 18 is _____ (activated, disabled).

Solution:

Output A will be LOW because the AND gate is activated by all 1s at its inputs.

7.7 List the outputs at the BCD indicators for each of the eight input pulses shown in Fig. 7-5. (Remember the priority feature, which activates the highest number that has a LOW input.)

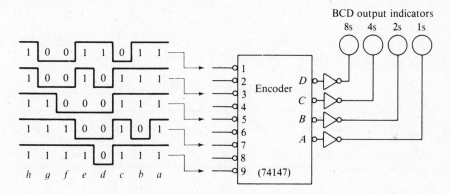

Fig. 7-5 Encoder pulse-train problem

Solution:

The indicators will read the following BCD (8421) outputs:

pulse a = 0000	pulse c = 0001	pulse e = 0111	pulse g = 0011
pulse b = 0111	pulse d = 1001	pulse f = 0101	pulse h = 0000

7-3 DECODING: BCD TO DECIMAL

A *decoder* may be thought of as the opposite of an encoder. To reverse the process described in Sec. 7-2 would produce a decoder that translated from the BCD code to decimals. A block diagram of such a decoder is in Fig. 7-6. The BCD (8421) code forms the input on the left of the decoder. The 10 output lines are shown on the right. Only one output line will be activated at any one time. Indicators (LEDs or lamps) have been attached to the output lines to help show which output is activated. Inputs B and C (B = 2s place, C = 4s place) are activated in Fig. 7-6. This causes the decimal 6 output to be activated, as shown by indicator 6 being lit. If no inputs are activated, the zero output indicator should light. A BCD 0011 input would activate the 3 output indicator.

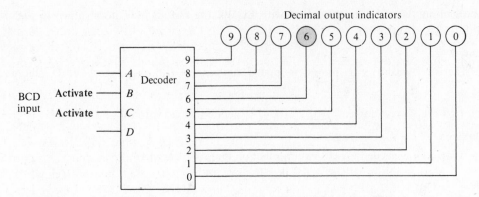

Fig. 7-6 Logic symbol for a BCD-to-decimal decoder

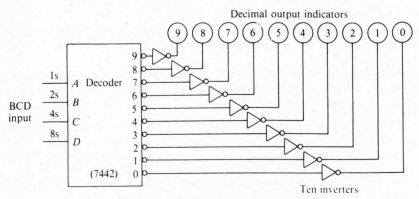

(a) Logic symbol with output indicators

Line	No.	BCD Inputs				Decimal output									
		D	C	B	A	0	1	2	3	4	5	6	7	8	9
Line 1	0	L	L	L	L	L	H	H	H	H	H	H	H	H	H
Line 2	1	L	L	L	H	H	L	H	H	H	H	H	H	H	H
Line 3	2	L	L	H	L	H	H	L	H	H	H	H	H	H	H
Line 4	3	L	L	H	H	H	H	H	L	H	H	H	H	H	H
Line 5	4	L	H	L	L	H	H	H	H	L	H	H	H	H	H
Line 6	5	L	H	L	H	H	H	H	H	H	L	H	H	H	H
Line 7	6	L	H	H	L	H	H	H	H	H	H	L	H	H	H
Line 8	7	L	H	H	H	H	H	H	H	H	H	H	L	H	H
Line 9	8	H	L	L	L	H	H	H	H	H	H	H	H	L	H
Line 10	9	H	L	L	H	H	H	H	H	H	H	H	H	H	L
Line 11		H	L	H	L	H	H	H	H	H	H	H	H	H	H
Line 12		H	L	H	H	H	H	H	H	H	H	H	H	H	H
Line 13		H	H	L	L	H	H	H	H	H	H	H	H	H	H
Line 14		H	H	L	H	H	H	H	H	H	H	H	H	H	H
Line 15		H	H	H	L	H	H	H	H	H	H	H	H	H	H
Line 16		H	H	H	H	H	H	H	H	H	H	H	H	H	H

(Lines 11–16 labeled "Invalid")

H = HIGH   L = LOW

(b) Truth table

Fig. 7-7 Commercial 7442 BCD-to-decimal decoder/driver

A commercial BCD-to-decimal decoder is shown in Fig. 7-7a. This TTL device is given the number 7442 by the manufacturer. The four BCD inputs on the left of the logic symbol are labeled D, C, B, and A. The D input is the 8s input, and the A input is the 1s input. A logical 1, or HIGH, will activate an input. On the right in Fig. 7-7a are 10 outputs from the decoder. The small bubbles attached to the logic symbol indicate that the outputs are *active LOW* outputs. They normally float HIGH except when activated. For convenience, 10 inverters were added to the circuit to drive the decimal-indicator lights. An active output will then be inverted to a logical 1 at the output indicators.

The truth table for the 7442 decoder is in Fig. 7-7b. The first line (representing a decimal 0) shows all inputs LOW (L). With an input of LLLL (0000), the decimal 0 output is activated to a LOW (L) state. The bottom inverter complements this output to a HIGH, which lights decimal output indicator 0. None of the other indicators are lit. Likewise, the fifth line (representing a decimal 4) shows the BCD input as LHLL (0100). Output 4 is activated to a LOW. The LOW is inverted in Fig. 7-7a, thereby lighting output decimal indicator 4. This decoder then has active HIGH inputs and active LOW outputs.

Consider line 11 in Fig. 7-7b. The input is HLHL (1010), and it would normally stand for a decimal 10. Since the BCD code does not contain that number, this is an *invalid* input and no output lamps will light (no outputs are activated). Note that the last six lines of the truth table show invalid inputs with no outputs activated.

The logic diagram for the 7442 BCD-to-decimal decoder is shown in Fig. 7-8. The BCD inputs are on the left, and the decimal outputs are on the right. The labeling at the inputs is somewhat different from that used before.

The A_3 input is the most significant bit (MSB), or the 8s input. The A_0 input is the least significant bit (LSB), or the 1s input. The outputs are labeled with decimal numbers. The decoder's active LOW outputs are shown with bars over the decimal outputs ($\overline{9}$, $\overline{8}$, and so forth).

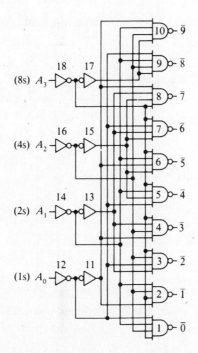

Fig. 7-8 Logic diagram of 7442 BCD-to-decimal decoder

Assume a BCD input of LLLL (0000) on the decoder shown in Fig. 7-8. Through careful tracing from the four inputs through inverters 12, 14, 16, and 18, four logical 1s are seen to be applied to NAND gate 1, which activates the gate and thus puts out a logical 0. All other NAND gates are disabled by 0s at some of their inputs. In like manner, each combination of inputs could be verified by analysis of the logic diagram in Fig. 7-8 for the 7442 decoder. All 18 gates shown in Fig. 7-8 are contained inside the single IC referred to as the 7442 decoder. As is customary, the power connections (V_{CC} and GND) to the IC are not shown in the logic diagram.

BCD-to-decimal decoders also are available in CMOS form from several manufacturers. Several representative CMOS ICs are the 4028, 74C42 and 74HC42 BCD-to-decimal decoders.

SOLVED PROBLEMS

7.8   Refer to Fig. 7-7. When inputs A, B, and C are activated by _____ (0s, 1s), output _____ (decimal number) will be active.

Solution:

When inputs A, B, and C are activated by logical 1s, output 7 will be active.

7.9   Refer to Fig. 7-7. If inputs are at HHHH (1111), which output will be activated?

Solution:

Input HHHH (1111) is an invalid BCD input according to the truth table, and therefore no outputs will be activated.

7.10 Refer to Fig. 7-7. Decimal output indicator _____ (decimal number) will be lit when the input is LHLH (0101).

Solution:

A 0101 input activates output 5, and the inverter lights the output 5 indicator.

7.11 Refer to Fig. 7-8. Gate number _____ (decimal number) is activated when the input is HLLL (1000) to this logic circuit.

Solution:

NAND gate 9 is activated, producing a LOW output at $\overline{8}$ with an HLLL (1000) input.

7.12 List the *active output* for each of the input pulses shown in Fig. 7-9.

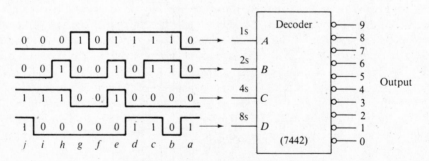

Fig. 7-9 Decoder pulse-train problem

Solution:

The active output (output = LOW) for each of the inputs shown in Fig. 7-9 is as follows:

pulse $a = 8$ pulse $d = 9$ pulse $g = 1$ pulse $j =$ (no active output)
pulse $b = 3$ pulse $e = 7$ pulse $h = 6$
pulse $c =$ (no active output)   pulse $f = 0$ pulse $i = 4$

7-4 DECODING: BCD–TO–SEVEN-SEGMENT CODE

A common task for a digital circuit is to decode from machine language to decimal numbers. A common output device used to display decimal numbers is the *seven-segment display*, shown in Fig. 7-10a. The seven segments are labeled with standard letters from *a* through *g*. The first 10 displays, representing decimal digits 0 through 9, are shown on the left side of Fig. 7-10b. For instance, if segments *b* and *c* of the seven-segment display light, a decimal 1 appears. If segments *a*, *b*, and *c* light, a decimal 7 appears, and so forth.

Seven-segment displays are manufactured by using several technologies. Each of the seven segments may be a thin filament which glows. This type of display is called an *incandescent* display, and it is similar to a regular lamp. Another type of display is the *gas-discharge tube*, which operates at

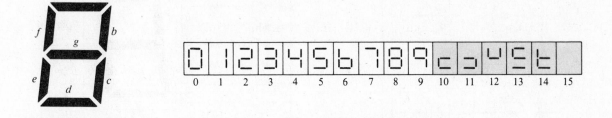

(a) Segment identification   (b) Decimal number with typical display

Fig. 7-10 Seven-segment display

high voltages. This unit gives off an orange glow. A *fluorescent tube* display gives off a greenish glow when lit and operates at low voltages. The newer *liquid-crystal display* (LCD) creates black numbers on a silvery background. LCD displays are extremely popular on hand-held calculators. The common *light-emitting diode* (LED) display gives off a characteristic reddish glow when lit. LED displays do come in several colors other than red. The LED, LCD, and fluorescent displays are currently the most popular, but liquid-crystal displays are used in almost all solar-powered and battery-operated devices.

Because it is quite common and easy to use, the LED type of seven-segment display will be covered in greater detail. Figure 7-11a shows a 5-V power supply connected to a single LED. When the switch (SW1) is closed, current will flow in the circuit and light the LED. About 20 mA (milliamperes) of current would flow in this circuit, which is the typical current draw for an LED. The 150-Ω (ohm) resistor is placed in the circuit to limit the current to 20 mA. Without the resistor, the LED would burn out. LEDs typically can accept only about 1.7 V across their terminals. Being a diode, the LED is sensitive to polarity. The cathode (K) must be toward the negative (GND) of the power supply. The anode (A) must be toward the positive of the power supply.

A seven-segment LED display is shown in Fig. 7-11b. Each segment (*a* through *g*) contains an LED, as shown by the seven symbols. The display shown has all the anodes tied together and coming out the right side as a single connection (common anode). The inputs on the left go to the various segments of the display.

To understand how segments of the display are activated and lighted, consider the circuit in Fig. 7-11c. If switch *b* is closed, current will flow from GND through the limiting resistor to the *b* segment LED and out the common anode connection to the power supply. Only segment *b* will light.

Suppose you wanted the decimal number 7 to light on the display in Fig. 7-11c. Switches *a*, *b*, and *c* would be closed, which would light the LED segments *a*, *b*, and *c*. The decimal 7 would light on the display. Likewise, if the decimal 5 were to be lit, switches *a*, *c*, *d*, *f*, and *g* would be closed. Those five switches would ground the correct segments, and a decimal 5 would appear on the display. Note that it takes a GND voltage (LOW) to activate the LED segments on this display.

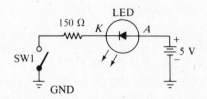

(a) Operation of a light-emitting diode (LED)

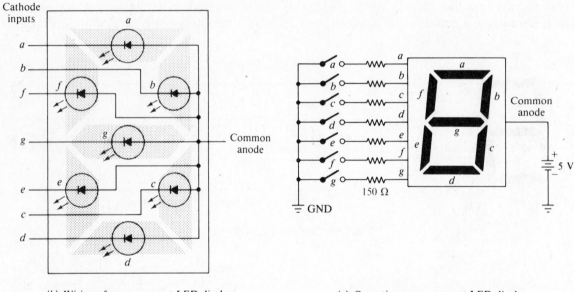

(b) Wiring of seven-segment LED display   (c) Operating seven-segment LED display

Fig. 7-11

Consider the common commercial decoder shown in Fig. 7-12a. This TTL device is called a 7447A BCD–to–seven-segment decoder/driver by the manufacturer. The input is a 4-bit BCD number shown on the left (inputs A, B, C, and D). The BCD number is decoded to form a seven-segment code that will light the appropriate segments on the LED display shown in Fig. 7-11b. Three extra inputs also are shown on the logic symbol. The lamp test input will turn all segments ON to see if all of them operate. Essentially, the blanking inputs turn all the segments OFF when activated. The lamp test and blanking inputs are activated by LOWs, as shown by the small bubbles at the inputs. The BCD inputs are activated by logical 1s. The 7447A decoder has active LOW outputs, as shown by the small bubbles at the outputs (a through g) of the logic symbol in Fig. 7-12a.

The operation of the 7447A decoder is detailed in the truth table furnished by Texas Instruments and shown in Fig. 7-12b. Consider line 1 on the truth table. To light the decimal 0 on the display, the BCD inputs (D, C, B, and A) must be at LLLL. This will activate or turn ON segments a, b, c, d, e, and f to form the decimal 0 on the seven-segment display. Note the invalid BCD inputs (decimals 10, 11, 12, 13, 14, and 15). They are not BCD numbers, however, they do generate a unique output, as shown in the truth table in Fig. 7-12b. Consider the decimal 10 line. With inputs of HLHL, the output column says that outputs d, e, and g are activated. This forms a small c. The unique outputs of this decoder for decimals 10 through 15 are shown as they would appear on the seven-segment display on the right side of Fig. 7-10b. Note that a decimal 15 will result in a *blank* display (all segments OFF).

A practical decoder system is shown in Fig. 7-13. A BCD number is entered at the left into the 7447A decoder. The decoder activates the proper outputs and allows the correct decimal number to appear on the display. Note that the display is a common-anode seven-segment LED display.

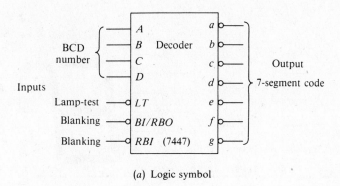

(a) Logic symbol

Decimal or function	Inputs						BI/RBO	Outputs							Note
	LT	RBI	D	C	B	A		a	b	c	d	e	f	g	
0	H	H	L	L	L	L	H	ON	ON	ON	ON	ON	ON	OFF	
1	H	X	L	L	L	H	H	OFF	ON	ON	OFF	OFF	OFF	OFF	
2	H	X	L	L	H	L	H	ON	ON	OFF	ON	ON	OFF	ON	
3	H	X	L	L	H	H	H	ON	ON	ON	ON	OFF	OFF	ON	
4	H	X	L	H	L	L	H	OFF	ON	ON	OFF	OFF	ON	ON	
5	H	X	L	H	L	H	H	ON	OFF	ON	ON	OFF	ON	ON	
6	H	X	L	H	H	L	H	OFF	OFF	ON	ON	ON	ON	ON	
7	H	X	L	H	H	H	H	ON	ON	ON	OFF	OFF	OFF	OFF	
8	H	X	H	L	L	L	H	ON	ON	ON	ON	ON	ON	ON	1
9	H	X	H	L	L	H	H	ON	ON	ON	OFF	OFF	ON	ON	
10	H	X	H	L	H	L	H	OFF	OFF	OFF	ON	ON	OFF	ON	
11	H	X	H	L	H	H	H	OFF	OFF	ON	ON	OFF	OFF	ON	
12	H	X	H	H	L	L	H	OFF	ON	OFF	OFF	OFF	ON	ON	
13	H	X	H	H	L	H	H	ON	OFF	OFF	ON	OFF	ON	ON	
14	H	X	H	H	H	L	H	OFF	OFF	OFF	ON	ON	ON	ON	
15	H	X	H	H	H	H	H	OFF	OFF	OFF	OFF	OFF	OFF	OFF	
BI	X	X	X	X	X	X	L	OFF	OFF	OFF	OFF	OFF	OFF	OFF	2
RBI	H	L	L	L	L	L	L	OFF	OFF	OFF	OFF	OFF	OFF	OFF	3
LT	L	X	X	X	X	X	H	ON	ON	ON	ON	ON	ON	ON	4

H = HIGH level, L = LOW level, X = irrelevant

Notes: 1. The blanking input (BI) must be open or held at a HIGH logic level when output functions 0 through 15 are desired. The ripple-blanking input (RBI) must be open or HIGH if blanking of a decimal zero is not desired.

2. When a LOW logic level is applied directly to the blanking input (BI), all segment outputs are OFF regardless of the level of any other input.

3. When ripple-blanking input (RBI) and inputs A, B, C, and D are at a LOW level with the lamp-test input HIGH, all segment outputs go OFF and the ripple-blanking output (RBO) goes to a LOW level (response condition).

4. When the blanking input/ripple-blanking output (BI/RBO) is open or held HIGH and a LOW is applied to the lamp-test input, all segment outputs are ON.

(b) Truth table (Reprinted by permission of Texas Instruments, Inc.)

Fig. 7-12 Commercial 7447 BCD–to–seven-segment decoder/driver

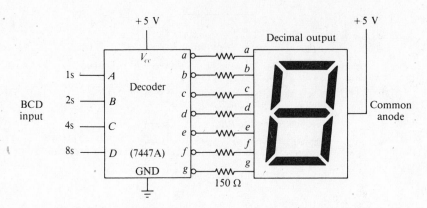

Fig. 7-13 Wiring of decoder and seven-segment LED display

Suppose the inputs to the decoder shown in Fig. 7-13 are LLLH (0001). This is the code for a decimal 1. According to the truth table for the 7447A decoder, this combination of inputs turns ON segments *b* and *c*. A decimal 1 is formed. Note that, when the truth table says ON, it means that the output of the 7447A decoder has gone to the LOW active state. It might also be said that the segment is being grounded through the decoder. The seven 150-Ω resistors limit the current from GND through the LED segment to a safe level. Recall that the 7447A was described as a decoder/*driver*. The driver description suggests that the current from the display LED flows directly through the 7447A IC. The decoder is directly *driving* the display. The 7447A IC is said to be *sinking* the current from the display.

It is assumed in Fig. 7-13 that the two blanking inputs (*RBI* and *BI/RBO*) plus the lamp test input are allowed to float HIGH. They are therefore not active and not shown on the logic symbol in Fig. 7-13.

Many CMOS display decoders are available. One example is the CMOS 74C48 BCD–to–seven-segment decoder, which is similar to the TTL 7447A IC. The 74C48 IC does need extra drive circuitry for most LED displays. Other examples of CMOS decoder ICs are the 4511 and 74HC4511 BCD–to–seven-segment latch/decoder/driver. The CMOS 4543 and 74HC4543 BCD–to–seven-segment latch/decoder/driver for liquid-crystal displays are also sold in convenient DIP IC form.

SOLVED PROBLEMS

7.13   Refer to Fig. 7-11*a*. Turn just the 5-V battery around and the LED will _____ (light, not light) as before.

Solution:

The LED will not light as before because it is sensitive to polarity.

7.14   Refer to Fig. 7-11*c*. A _____ (GND, +5-V) voltage is applied to the cathodes of the LED segments through the switches and limiting resistors.

Solution:

A GND voltage is applied to the cathodes of the LED segments when a switch shown in Fig. 7-11*c* is closed.

7.15   Refer to Fig. 7-11*c*. When switches *b*, *c*, *f*, and *g* are closed, a(n) _____ (decimal number) will be displayed on the seven-segment LED display.

Solution:

Lighting segments b, c, f, and g will form a 4 on the display.

7.16 Refer to Fig. 7-11c. When switches b and c are closed, a(n) _____ (decimal number) is displayed, and about _____ (1, 40) mA of current will be drawn by the LEDs.

Solution:

Lighting segments b and c will form a 1 on the display, which will cause about 40 mA of current to be drawn by the LEDs.

7.17 Refer to Fig. 7-12b. To display a decimal 2, the BCD inputs must be _____ _____ _____ _____ (H, L), which will turn ON segments _____ (list all ON segments).

Solution:

Displaying a decimal 2 requires a BCD input of LLHL, which turns on segments a, b, d, e, and g.

7.18 Invalid BCD inputs on the 7447A decoder produce _____ (OFF, unique) readings on the seven-segment display.

Solution:

Invalid BCD inputs (10, 11, 12, 13, 14, and 15) on the 7447A decoder produce unique readings on the display. See Fig. 7-10b.

7.19 List the *decimal* indication of the seven-segment display for each input pulse in Fig. 7-14.

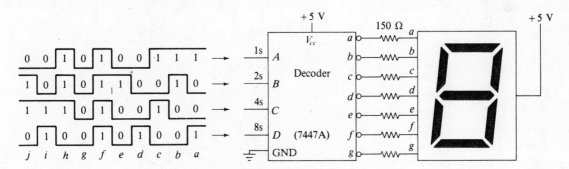

Fig. 7-14 Decoder-display pulse-train problem

Solution:

The decimal outputs for the various input pulses in Fig. 7-14 are as follows:

pulse $a = 9$ pulse $e = 2$ pulse $i = $ u (invalid BCD input)

pulse $b = 3$ pulse $f = $ display blank (invalid BCD input) pulse $j = 6$

pulse $c = 5$ pulse $g = 0$

pulse $d = 8$ pulse $h = 7$

7.20   List the segment of the seven-segment display that will be lit for each of the pulses in Fig. 7-14.

>   **Solution:**
>
>   The lit segments on the seven-segment display in Fig. 7-14 are as follows:
>
>   pulse $a = a, b, c, f, g$ pulse f = display blank (invalid BCD input)
>   pulse $b = a, b, c, d, g$ pulse $g = a, b, c, d, e, f$
>   pulse $c = a, c, d, f, g$ pulse $h = a, b, c$
>   pulse $d = a, b, c, d, e, f, g$ pulse $i = b, f, g$ (invalid BCD input)
>   pulse $e = a, b, d, e, g$ pulse $j = c, d, e, f, g$

7-5   LIQUID-CRYSTAL DISPLAYS

Most battery- or solar-powered electronic equipment use *liquid-crystal displays* (LCDs). The LCD in your calculator, wristwatch, portable telephone, or portable computer are but a few examples of the use of liquid-crystal displays. The main advantages of the liquid-crystal display is its extremely low power consumption and long life. The main disadvantage of the LCD is its slow switching time (on-off and off-on) which might be about 40 to 100 ms. Slow switching time becomes even more troublesome at low temperatures. A second disadvantage is the need for ambient light because the LCD reflects (controls) light but does not emit light like LED, VF, or incandescent displays.

A cutaway view of a typical *field-effect LCD* is detailed in Fig. 7-15. When a voltage is applied across the metalized segments on the top glass and the back plane, the segment changes to black on a silvery background. This is because the liquid crystal, or *nematic fluid*, sandwiched between the front and back pieces of glass transmits light differently when activated. The field-effect LCD uses polarized filters on the top and bottom of the display shown in Fig. 7-15. Each segment and the back plane are internally wired to contacts on the edge of the LCD package. The simplified diagram in Fig. 7-15 shows only three of many edge connectors.

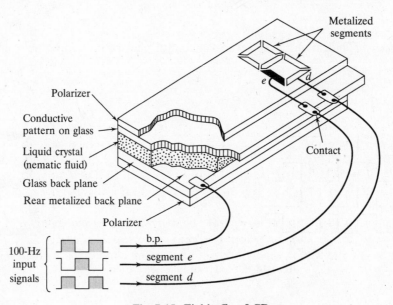

Fig. 7-15   Field-effect LCD

LCDs are driven by low-frequency (30 to 200 Hz) square-wave signals with a 50% duty cycle (50% of the time it's HIGH). Consider the signals entering the LCD in Fig. 7-15. Note that the signal entering the back plane (b.p.) is HLH (HIGH-LOW-HIGH). The square-wave signal applied to segment e is LHL (LOW-HIGH-LOW) which is 180° *out of phase* (inverted) with the back plane signal. An out-of-phase signal on a segment will activate the display as is the case with segment e in

Fig. 7-15. Next consider the signal applied to segment d of the LCD in Fig. 7-15. The signal goes HLH, which is a duplicate of the back plane signal, and they are said to be *in phase*. In-phase signals between the back plane and segment d produce *no voltage difference*, and segment d is not activated and remains invisible. In summary, in-phase signals do not activate the display while 180° out-of-phase signals activate an LCD segment.

A typical LCD is illustrated in Fig. 7-16. This unit comes in a 40-pin package ready for mounting in a printed circuit board. Notice that segments that can be activated can be manufactured in any shape, including numbers, symbols, and letters. Each segment, decimal point, word, and symbol is assigned a pin number. Only the back plane or common pin is noted on the drawing. Manufacturer's data sheets must be consulted for actual pin numbers. This is a commercial display that you might expect on a digital meter. In Fig. 7-16, note the construction of this field-effect LCD with nematic fluid sandwiched between glass plates and polarizers on the top and bottom. Plastic headers secure the glass plates of the LCD to the pins.

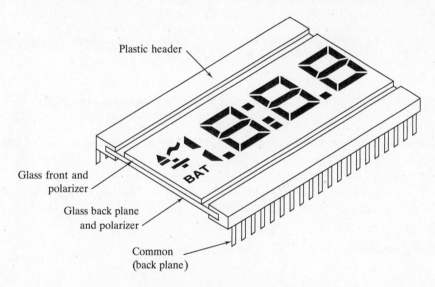

Fig. 7-16 Commercial $3\frac{1}{2}$ digit LCD

Care must be taken when using LCDs because they are made of glass and are somewhat fragile. Also, the driving signals should be generated by CMOS ICs for two reasons. CMOS ICs consume very little power like the LCD. The second reason is that signals from CMOS ICs do not have a dc voltage offset like that present when using TTL ICs. A dc voltage applied across the nematic fluid will *destroy the LCD* after a time.

An older type liquid-crystal display that produces frosty-white characters on a dark background is the *dynamic-scattering LCD*. The dynamic-scattering LCD uses a different nematic fluid and no polarizers. These must be viewed in very good light and consume more power than the more popular field-effect LCD.

SOLVED PROBLEMS

7.21 Digits appear _____ (black, silver) on a _____ (black, silver) background on a field-effect liquid-crystal display.

Solution:

When using a field-effect LCD, digits will appear black on a silver background.

7.22   List two advantages of LCDs over LED displays.

Solution:

Advantages of using an LCD are low-power consumption and long life.

7.23   List two disadvantages of LCD.

Solution:

An LCD has the disadvantage of slow switching times, especially at low temperatures. A second disadvantage is that the LCD cannot be viewed in darkness.

7.24   Refer to Fig. 7-15. When a voltage is applied across the nematic fluid in this LCD, the segment will be _____ (activated, deactivated).

Solution:

Voltage applied across the nematic fluid in an LCD activates the segment. An activated segment on the LCD in Fig. 7-15 will appear black on a silver background.

7.25   LCDs should be driven by _____ (low, high)-frequency square-wave signals.

Solution:

LCDs should be driven by low-frequency (30 to 200 Hz) square-wave signals with a 50% duty cycle.

7.26   When the signals applied to the back plane and segment of an LCD are 180° out of phase, the segment will be _____ (activated, deactivated).

Solution:

Out-of-phase signals activate LCD segment.

7.27   Refer to Fig. 7-16. What is sandwiched between the two glass plates in this LCD?

Solution:

Nematic fluid (liquid crystal) is sandwiched between the glass plates in the LCD pictured in Fig. 7-16.

7.28   LCDs can be damaged if driven by _____ (ac, dc) voltages.

Solution:

LCDs can be damaged if driven by dc voltages.

7-6 DRIVING LCDs

A block diagram of a simple LCD decoder/driver circuit is drawn in Fig. 7-17a. The input is in 8421 BCD code. The decoder converts the incoming BCD to seven-segment code. This decoder would operate much like the 7447 TTL decoder from Sec. 7-4 except it is a CMOS unit. Next, the LCD driver unit would take the 100-Hz square-wave signal from the free-running clock and send inverted (180° out-of-phase) signals to only the LCD segments that are to be activated. The LCD driver would send in-phase signals to the LCD segments that are inactive. The *free-running clock* is an *astable multivibrator* that continuously generates a string of square-wave pulses with a 50% duty cycle.

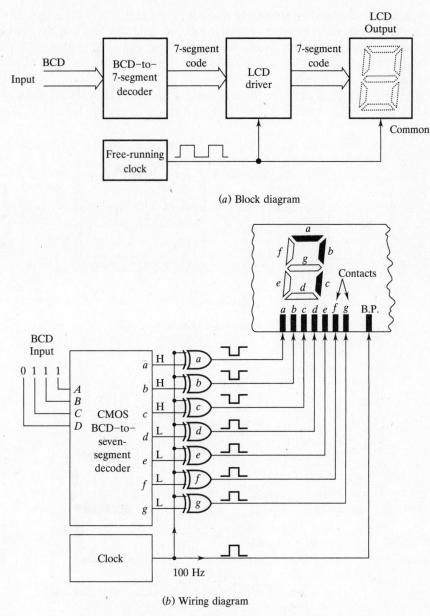

(a) Block diagram

(b) Wiring diagram

Fig. 7-17 Decoding/driving a seven-segment LCD

A more detailed diagram of the LCD decoder/driver is shown in Fig. 7-17b. In this example, the BCD input to the CMOS BCD–to–seven-segment decoder is 0111. The decoder translates the BCD input and activates outputs a, b, and c with HIGH, which is the proper seven-segment code to display a decimal 7. All other decoder outputs (d, e, f, and g) remain LOW or deactivated. Notice that the LCD driver section contains seven CMOS 2-input XOR gates. The 100-Hz square-wave signal drives the top input of each XOR gate. If the bottom input on an XOR gate is LOW, the signal passes through the gate with no change (in phase with clock signal). But if the bottom input of an XOR gate is driven HIGH, the signal is inverted as it passes through the gate (180° out of phase with clock signal). Refer back to Fig. 4-10 to verify the operation of an XOR gate. The out-of-phase signals in Fig. 7-17b are driving segments a, b, and c, which are activated and appear dark on a silver background on the LCD.

The clock signal is generated by an astable multivibrator in Fig. 7-17b. The 100-Hz signal is routed to both the common (back plane) of the LCD and each of the XOR gates in the driver section.

Two commercial CMOS ICs are available that perform the task of the LCD decoder/driver. These are the 4543 and 74HC4543 ICs, described by the manufacturer as a *BCD–to–seven-segment latch/decoder/driver for LCDs*.

A block diagram of an LCD decoder/driver circuit using the 74HC4543 IC is drawn in Fig. 7-18a. Note that the 74HC4543 chip contains the BCD–to–seven-segment decoder and LCD driver sections.

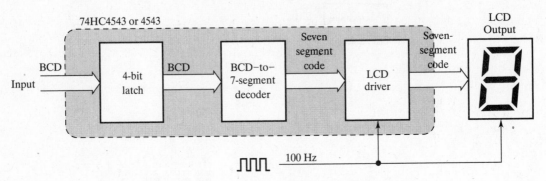

(a) Block diagram of 74HC4543 driving LCD

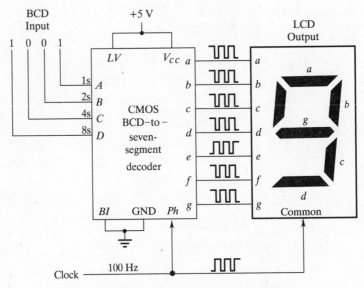

(b) Wiring diagram of 74HC4543 driving LCD

(c) Format of decimal numbers

Fig. 7-18 The 74HC4543 latch/decoder/driver CMOS IC

It also has an added 4-bit latch section to "lock in" the BCD input at a given time. Think of the latch as a memory unit that holds the four input bits on the input of the decoder section for a time.

A wiring diagram for the LCD decoder/driver circuit using the 74HC4543 IC is detailed in Fig. 7-18b. The BCD input is 1001 (decimal 9) in this example. The 1001 input is decoded into seven-segment code. The 100-Hz clock signal is routed to both the common (back plane) of the LCD and the *Ph* (phase) input of the 74HC4543 IC. Notice that the LCD driver section within the 74HC4543 IC inverts the signals to the segments that are to be activated. In this example segments *a*, *b*, *c*, *d*, *f*, and *g* are activated, displaying the decimal 9 on the LCD. The only in-phase signals passing to the LCD are those of the inactive segments. Only segment *e* is inactive in this example.

The format of the numbers generated by the 74HC4543 decoder is detailed in Fig. 7-18c. Note especially the numbers 6 and 9. These numbers are shaped differently from those generated by the 7447 decoder studied earlier in Sec. 7-4. Compare Fig. 7-18c with Fig. 7-10 to verify the different shapes for the numbers 6 and 9.

SOLVED PROBLEMS

7.29   Refer to Fig. 7-17a. What is the job of the decoder block?

Solution:

The decoder in Fig. 7-17a translates from a BCD number to seven-segment code.

7.30   Refer to Fig. 7-17a. What is the job of the LCD driver block?

Solution:

The LCD driver sends inverted signals to each segment that is to be activated and in-phase signals to each inactive segment of the liquid-crystal display.

7.31   Refer to Fig. 7-17a. The LCD driver block consists of _____ (NAND, XOR) gates.

Solution:

The LCD driver consists of XOR gates (see Fig. 7-17b).

7.32   Refer to Fig. 7-17b. If the input to the decoder were 0001_{BCD}, which XOR gates produce inverted outputs and which number does the LCD display.

Solution:

With an input of 0001, only XOR gates *a* and *b* will produce inverted signals at their outputs, activating segments *a* and *b* on the LCD (decimal 1 will show on display).

7.33   A free-running clock is also called a(n) _____ (astable, monostable) multivibrator.

Solution:

The free-running clock may also be called an astable multivibrator.

7.34   Refer to Fig. 7-19. What is the decimal reading on the LCD for each input pulse (*a* through *e*).

Solution:

Decimal outputs in Fig. 7-19 are as follows:
pulse $a = 2$
pulse $b = 4$
pulse $c = 8$
pulse $d = 5$
pulse $e = 6$

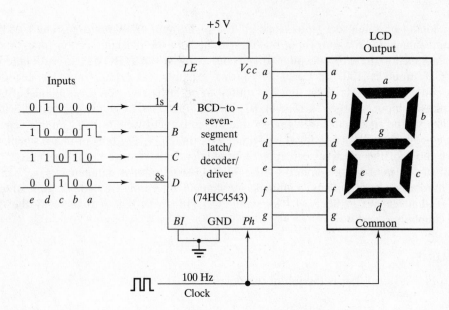

Fig. 7-19 Decoder pulse-train problem

7.35   Refer to Fig. 7-19. For input pulse *e* only, which drive line or lines to the LCD have *in-phase* signals appearing on them?

Solution:

Decimal 6 appears on the LCD during pulse *e* in Fig. 7-19. Only segment *b* is inactive, and therefore only drive line *b* has an in-phase signal. Also see Fig. 7-18*c* for the formation of decimal 6.

7.36   Refer to Fig. 7-19. For input pulse *b* only, which drive lines to the LCD have *out-of-phase* signals appearing on them?

Solution:

Decimal 4 appears on the LCD during pulse *b* in Fig. 7-19. Segments *b*, *c*, *f*, and *g* are activated, and therefore drive lines *b*, *c*, *f*, and *g* must have out-of-phase signals.

7-7   VACUUM FLUORESCENT DISPLAYS

The *vacuum fluorescent (VF) display* is a relative of the earlier triode vacuum tube. A schematic symbol for a triode vacuum tube is illustrated in Fig. 7-20. The parts of the triode tube are shown as the plate (*P*), control grid (*G*), and the cathode (*K*). The plate is sometimes referred to as the anode, while the cathode may be called the filament or heater. The cathode/filament is a fine wire that when coated with a material such as barium oxide will emit electrons when heated. The control grid is a screen placed between the cathode and plate.

Fig. 7-20 Schematic symbol of a triode vacuum tube

When the cathode/filament is heated, it "boils off" electrons into the vacuum surrounding the cathode. This is sometimes called thermionic emission. If the grid and plate are driven positive, the negatively charged electrons will be attracted and flow through the screenlike grid onto the plate. The triode is conducting current from cathode to anode.

To stop the triode from conducting, two methods can be employed. First, a negative charge can be placed on the control grid. This will repel the electrons and stop them from passing through the grid to the plate. Second, the voltage on the plate can be dropped from its normal positive to 0 volts. With no voltage on the plate, it will not attract electrons and the triode tube will not conduct. The VF display has parts closely resembling those in the triode vacuum tube.

Consider the schematic diagram of the vacuum fluorescent display shown in Fig. 7-21a. This schematic represents a single seven-segment digit having seven plates each coated with a zinc oxide fluorescent material. The VF display in Fig. 7-21a also has a single grid that controls the entire display. A single cathode/filament (K) is also shown, while the entire unit is enclosed in glass which contains a vacuum.

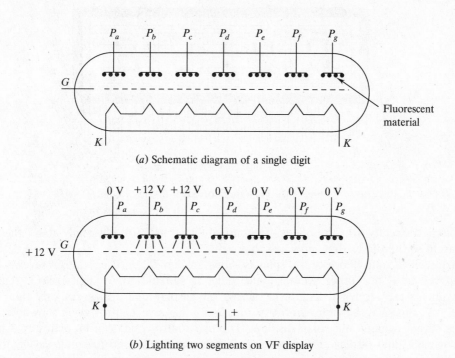

(a) Schematic diagram of a single digit

(b) Lighting two segments on VF display

Fig. 7-21 Seven-segment vacuum fluorescent display

Typical operation of a single digit in a VF display is illustrated in Fig. 7-21b. The cathode/filament is heated using a dc voltage. The control grid has +12 V applied, which "turns on" the entire display. In this example, only segments b and c are to be activated, so only plates P_b and P_c are energized with +12 V. Electrons flow from the cathode/filament *to only plates P_b and P_c* of the VF display. As the electrons strike the plates of the b and c segments, they will glow the characteristic blue-green color. In the example shown in Fig. 7-21b with only segment b and c energized, the number 1 appears on the seven-segment VF display. Also note in the example in Fig. 7-21b that the plates of the deactivated segments (P_a, P_d, P_e, P_f, and P_g) have 0 V applied to them. In summary, a plate voltage of +12 V lights a segment, whereas 0 V at a plate means the segment will not glow.

The physical layout of cathode/filament, grids, and plates in a VF display is shown in Fig. 7-22. Notice in Fig. 7-22a that the plates are shaped to form the individual segments of the seven-segment display. The cathode or heaters are thin wires stretched across the top. The grid is a screenlike panel

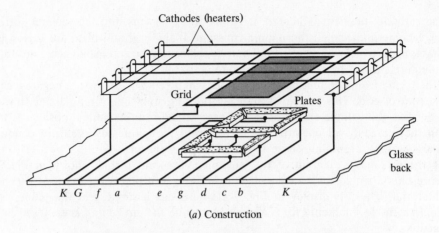

(a) Construction

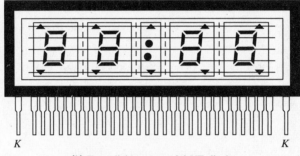

(b) Four-digit commercial VF display

Fig. 7-22 Vacuum fluorescent display

that is positioned directly over the plates. The cathodes and grid are physically above the plates but are transparent so the plates show up when they light.

A commercial vacuum fluorescent display is shown in Fig. 7-22b. It contains 4 seven-segment VF displays as well as a few other symbols that make it suitable for a digital clock readout. The cathode/filaments are stretched lengthwise across the display and appear as very thin wires in a commercial VF unit. The VF unit shown in Fig. 7-22b has five control grids shown as rectangles surrounding seven-segment and colon displays. The five control grids can be activated separately to "turn on" an individual display. The control grids are commonly used for *multiplexing* the displays (turning on seven-segment displays one at a time in rapid succession). The fluorescent-coated plates are shaped like the number segments, triangles, or colons.

Vacuum fluorescent displays are commonly used in readouts found in a wide variety of electronic equipment, especially those found in automobiles. A VF display has extremely long life, fast response times, operates at relatively low voltages (commonly 12 V), consumes little power, has good reliability, and is inexpensive. Although a VF display emits a blue-green color, filters can be used to display other colors. VF displays are compatible with the CMOS family of ICs.

SOLVED PROBLEMS

7.37 A vacuum fluorescent display glows with a _____ color when activated.

   Solution:

   Without a filter, the VF display glows with a blue-green color.

7.38 Refer to Fig. 7-21b. If +12 V was applied to the grid and all the plates of this VF display, which segments would glow and which number would appear?

Solution:

The +12 V at each plate will activate (light) all the segments (number 8 appears), while the +12 V on the grid turns on the entire seven-segment VF display.

7.39   Refer to Fig. 7-23. What are the parts labeled X, Y, and Z on the VF display?

Solution:

Part X = cathode, filament, or heater
Part Y = control grid
Part Z = plate or anode

7.40   Refer to Fig. 7-23. Which segments of this VF display will glow and which number will appear?

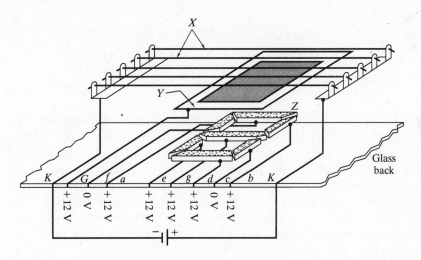

Fig. 7-23   VF display problem

Solution:

Segments a, b, d, e, and g are activated (+12 V at these plates) and will glow. The number 2 will show on the seven-segment VF display.

7.41   List a few of the advantages of the VF display.

Solution:

Advantages of VF displays include long life, fast response time, low power consumption, good reliability, compatibility with CMOS ICs, and ability to operate at relatively low voltages.

7-8   DRIVING VF DISPLAYS WITH CMOS

Consider the decoder/driver and VF display circuit drawn in Fig. 7-24. In this example the 0111_{BCD} is being decoded by the 4511 latch/decoder/driver IC, and the VF display reads decimal 7. Notice that only outputs a, b, and c are activated (HIGH) on the 4511 IC. These three HIGHs drive the plates of segments a, b, and c of the VF display to +12 V. The grid of the VF display in Fig. 7-24 is connected directly to positive of the 12-V power supply which activates the entire seven-segment display. The cathode (K) is connected in series with a limiting resistor (R_1) to heat the filament. The resistor limits the current through the filament (cathode) to a safe level. In this example, the inactive segments of the VF display (d, e, f, and g) have their plates held LOW (0 V), and they do not light.

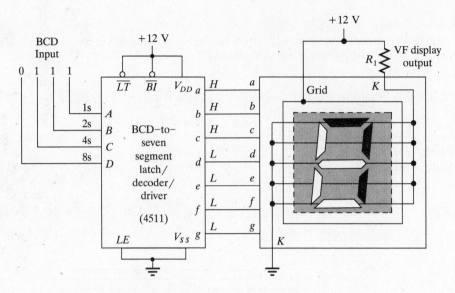

Fig. 7-24 Using the 4511 IC to drive a VF display

The block diagram for the 4511 BCD–to–seven-segment latch/decoder/driver IC is the same as for the 74HC4543 IC in Fig. 7-18*a*. The 4511 latch/decoder/driver has a 4-bit latch (memory unit). The latch section (*LE* input) of the 4511 IC is disabled in Fig. 7-24 by holding it LOW. With the latch disabled, data from the BCD input passes through to the decoder section of the 4511 IC. Note that a +12-V dc power supply is used for both the vacuum fluorescent display and the 4511 CMOS chip. The 4000 CMOS series is ideally suited for driving VF displays because this family of ICs can operate on a wide range of higher dc voltages up to +18 V. The decoder section of the 4511 translates from 8421 BCD code to seven-segment code. Figure 7-25*b* shows how the numbers are formed using the 4511 decoder. Note especially the formation of the 6 and 9 in Fig. 7-25*b*. The driver section of the

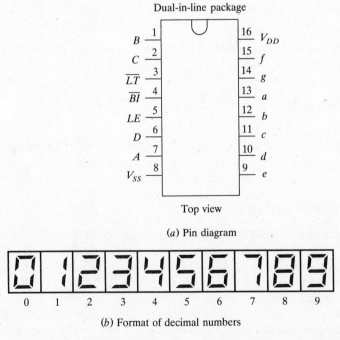

Fig. 7-25 The 4511 BCD–to–seven-segment latch/decoder/driver IC

4511 IC has its outputs connected directly to the plates (anodes) of the VF display. A HIGH at the output of the driver activates (lights) the segment on the seven-segment VF display (assuming the display's control grid is activated). A LOW at the output of the driver deactivates the segment of the VF display, and it does not light.

A pin diagram for the 4511 BCD–to–seven-segment latch/decoder/driver CMOS IC is reproduced in Fig. 7-25a. Recall that the power pins are labeled V_{DD} for positive (pin 16) and V_{SS} for negative (pin 7). The *LE* pin on the 4511 IC is a latch enable input. Latch enable is an active HIGH input and is shown disabled in the circuit in Fig. 7-24. To disable the latch means data will pass through the latch from BCD inputs to the decoder. The latch is said to be transparent when disabled. With the *LE* enabled (HIGH), four memory cells (or latches) hold current data on the input to the 4511 decoder. With the latches enabled, changes at the BCD inputs (labeled *A*, *B*, *C*, and *D*) to the 4511 IC will be disregarded. The 4511 IC has two active LOW inputs. When the $\overline{LT}$ (light test) input is activated with a LOW, all IC outputs go HIGH to test the attached display. When the $\overline{BI}$ (blanking input) is activated with a LOW, all outputs go LOW and all segments of the attached display go blank.

SOLVED PROBLEMS

7.42   Refer to Fig. 7-24. A _____ (5, 12)-V dc power supply is being used because the CMOS 4511 decoder/driver IC and the _____ (LCD, VF) display both operate at this voltage.

Solution:

In Fig. 7-24, a 12-V power supply is used because the CMOS 4511 IC and VF display both operate at this voltage.

7.43   Refer to Fig. 7-24. In this example, the control grid on the VF display is _____ (activated, deactivated) by being connected directly to +12 V.

Solution:

In this example, the control grid on the VF display is activated by being connected directly to +12 V.

7.44   Refer to Fig. 7-24. In this example, which plates of the VF display have +12 V (HIGH) applied to them.

Solution:

In this example, decimal 7 appears on the VF display, meaning that segments (plates) *a*, *b*, and *c* are activated (+12 V applied to them).

7.45   Refer to Fig. 7-24. What is the purpose of resistor R_1 in this circuit?

Solution:

Series resistor R_1 in Fig. 7-24 limits current through the filaments (cathode) to a safe level.

7.46   Refer to Fig. 7-24. If the BCD input was 0101, the decimal appearing on the VF display would be _____ .

Solution:

If the input was 0101_{BCD} in Fig. 7-24, the decimal appearing on the VF display would be 5.

7.47   Refer to Fig. 7-24. If the BCD input was 1000, the decimal appearing on the VF display would be _____ and segments (plates) _____ would be activated (HIGH).

Solution:

If the input in Fig. 7-24 was 1000_{BCD}, the VF display would show decimal 8 and all the segments (plates) would be activated.

7.48 Refer to Fig. 7-24. If the $\overline{LT}$ (lamp test) input is activated or goes _____ (HIGH, LOW), all segments of the seven-segment VF display would light.

Solution:

If the $\overline{LT}$ input in Fig. 7-24 was activated with a LOW, all segments of the VF display would light.

7.49 Refer to Fig. 7-25a. When connecting power to this 4000 series CMOS IC, the positive of the power supply is connected to the _____ (V_{DD}, V_{SS}) pin, while the negative is connected to the _____ (V_{DD}, V_{SS}) pin.

Solution:

On 4000 series CMOS ICs (see the 4511 in Fig. 7-25a), the V_{DD} pin is connected to the positive, while the V_{SS} pin is connected to the negative of the power supply.

Supplementary Problems

7.50 In a calculator, a(n) __(a)__ (decoder, encoder) would translate from decimal to binary while a(n) __(b)__ (decoder, encoder) would translate from binary to the decimal output.
Ans. (a) encoder (b) decoder

7.51 It is characteristic that a(n) _____ (decoder, encoder) has only one active input at any given time.
Ans. encoder

7.52 Refer to Fig. 7-26. This encoder has active _____ (HIGH, LOW) inputs and active _____ (HIGH, LOW) outputs.
Ans. The 74148 encoder has active LOW inputs and active LOW outputs, as shown by the small bubbles at the inputs and outputs of the logic symbol in Fig. 7-26.

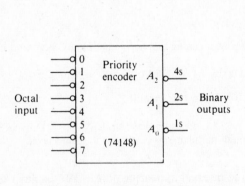

Inputs								Outputs		
0	1	2	3	4	5	6	7	A_2	A_1	A_0
H	H	H	H	H	H	H	H	H	H	H
X	X	X	X	X	X	X	L	L	L	L
X	X	X	X	X	X	L	H	L	L	H
X	X	X	X	X	L	H	H	L	H	L
X	X	X	X	L	H	H	H	L	H	H
X	X	X	L	H	H	H	H	H	L	L
X	X	L	H	H	H	H	H	H	L	H
X	L	H	H	H	H	H	H	H	H	L
L	H	H	H	H	H	H	H	H	H	H

H = HIGH, L = LOW, X = irrelevant

(a) Logic symbol   (b) Simplified truth table

Fig. 7-26 The 74148 octal-to-binary priority encoder

7.53 Refer to Fig. 7-26. If input 7 were *activated* with a _____ (HIGH, LOW), the output would be $A_2 = $ _____, $A_1 = $ _____, and $A_0 = $ _____ (HIGH, LOW).
Ans. If input 7 were activated with a LOW on the 74148 encoder, the outputs would be $A_2 = $ LOW, $A_1 = $ LOW, and $A_0 = $ LOW.

7.54   Refer to Fig. 7-26. If all inputs are HIGH, the outputs will be $A_2 =$ _____, $A_1 =$ _____, and $A_0 =$ _____ (HIGH, LOW).
    Ans.   If all inputs to the 74148 encoder are HIGH, the outputs will be all HIGH.

7.55   Refer to Fig. 7-26. If input 3 is activated, the outputs will be $A_2 =$ __(a)__ , $A_1 =$ __(b)__ , and $A_0 =$ __(c)__ (HIGH, LOW).
    Ans.   (*a*) HIGH (*b*) LOW (*c*) LOW

7.56   List the binary output indicator reading (3 bits) for each of the input pulses shown in Fig. 7-27.
    Ans.   pulse $a = 000$ pulse $c = 100$ pulse $e = 101$ pulse $g = 011$ pulse $i = 110$
             pulse $b = 010$ pulse $d = 111$ pulse $f = 010$ pulse $h = 001$ pulse $j = 111$

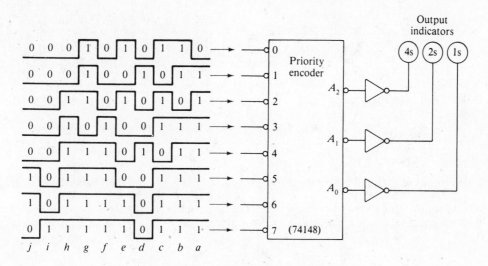

Fig. 7-27 Encoder pulse-train problem

7.57   Refer to Fig. 7-28. The 7443 decoder has active _____ (HIGH, LOW) inputs and active _____ (HIGH, LOW) outputs.
    Ans.   The 7443 decoder has active HIGH inputs and active LOW outputs, based on the logic symbol and truth table in Fig. 7-28.

7.58   Refer to Fig. 7-28. With an input of 0110, the __(a)__ (decimal number) output of the 7443 decoder is activated with a __(b)__ (HIGH, LOW).   *Ans.*   (*a*) 3 (*b*) LOW

7.59   Refer to Fig. 7-28. The invalid input 1111 generates all _____ (0s, 1s) at the outputs of the 7443 decoder.
    Ans.   1s

7.60   Refer to Fig. 7-28. The 7443 IC is a decoder that translates from __(a)__ (BCD, XS3) code to __(b)__ (decimals, hexadecimals).   *Ans.*   (*a*) XS3 (*b*) decimals

7.61   Refer to Fig. 7-28. When the outputs of the 7443 decoder are *deactivated*, they are at logical _____ (0, 1).   *Ans.*   1 (HIGH)

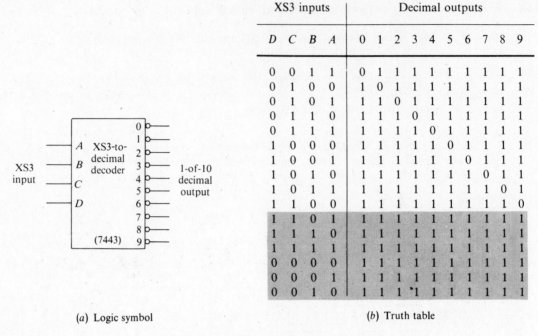

(a) Logic symbol

XS3 inputs				Decimal outputs									
D	C	B	A	0	1	2	3	4	5	6	7	8	9
0	0	1	1	0	1	1	1	1	1	1	1	1	1
0	1	0	0	1	0	1	1	1	1	1	1	1	1
0	1	0	1	1	1	0	1	1	1	1	1	1	1
0	1	1	0	1	1	1	0	1	1	1	1	1	1
0	1	1	1	1	1	1	1	0	1	1	1	1	1
1	0	0	0	1	1	1	1	1	0	1	1	1	1
1	0	0	1	1	1	1	1	1	1	0	1	1	1
1	0	1	0	1	1	1	1	1	1	1	0	1	1
1	0	1	1	1	1	1	1	1	1	1	1	0	1
1	1	0	0	1	1	1	1	1	1	1	1	1	0
1	1	0	1	1	1	1	1	1	1	1	1	1	1
1	1	1	0	1	1	1	1	1	1	1	1	1	1
1	1	1	1	1	1	1	1	1	1	1	1	1	1
0	0	0	0	1	1	1	1	1	1	1	1	1	1
0	0	0	1	1	1	1	1	1	1	1	1	1	1
0	0	1	0	1	1	1	1	1	1	1	1	1	1

(b) Truth table

Fig. 7-28 The 7443 XS3-to-decimal decoder

7.62 List the *decimal* output indicator activated for each pulse going into the 7443 decoder shown in Fig. 7-29.

  Ans. pulse $a = 6$ pulse $d = 9$ pulse $h = 3$

      pulse $b = 4$ pulse $e = 0$ pulse $i = 5$

      pulse c = all outputs deactivated   pulse $f = 8$ pulse j = all outputs deactivated

          (invalid XS3 input)   pulse $g = 1$ (invalid XS3 input)

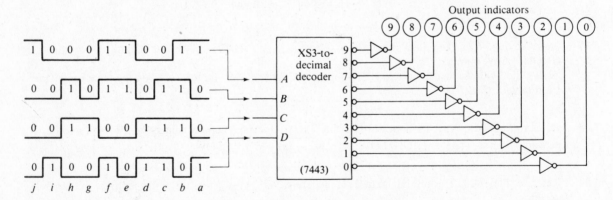

Fig. 7-29 Decoder pulse-train problem

7.63 Battery and solar-powered equipment most commonly use _____ (LCD, LED) displays.

  Ans. LCD

7.64 If an LED emits light, a liquid-crystal display is said to _____ (control, generate) light.

  Ans. control

7.65 What is the main disadvantage of liquid-crystal displays?
Ans.   Slow switching speeds or the need for ambient light.

7.66 Refer to Fig. 7-15. On an LCD, only segments that are driven by _____ (in-phase, out-of-phase) square-wave signals are activated and visible on the display.   *Ans.*   out-of-phase

7.67 LCDs that show _____ (black, frosty white) segments on a silvery background are referred to as field-effect liquid-crystal displays.   *Ans.*   black

7.68 Square-wave signals are required when driving _____ (liquid-crystal, VF) displays.
Ans.   liquid-crystal

7.69 The liquid-crystal sandwiched between glass plates on an LCD is called _____ fluid.   *Ans.*   nematic

7.70 Refer to Fig. 7-17a. The decoder and LCD driver are _____ (CMOS, TTL) devices.   *Ans.*   CMOS

7.71 Refer to Fig. 7-17a. The LCD driver section consists of seven _____ (AND, XOR) gates.
Ans.   XOR

7.72 Refer to Fig. 7-30. What is the decimal reading on the LCD for each input pulse (*a* through *d*).
Ans.   pulse *a* = 0
pulse *b* = 9
pulse *c* = 3
pulse *d* = 6

7.73 Refer to Fig. 7-30. For input pulse *c* only, which drive lines to the LCD have out-of-phase signals appearing on them?
Ans.   Out-of-phase signals are activated on segments *a*, *b*, *c*, *d*, and *g*.

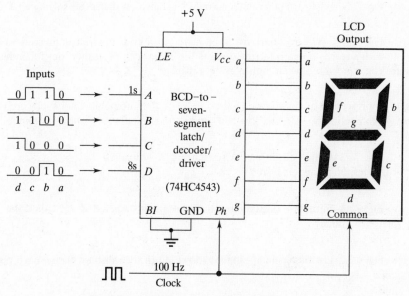

Fig. 7-30 Decoder/driver pulse-train problem

7.74 Refer to Fig. 7-30. For input pulse *b* only, which drive lines to the LCD have in-phase signals appearing on them?
Ans.   Segment *e* is deactivated with an in-phase signal (decimal 9 appears).

7.75   A VF display has parts comparable to a _____ (diode, triode) vacuum tube.   *Ans.*   triode

7.76   The plates of a VF display are coated with a _____ (barium oxide, zinc chloride) fluorescent material that glows when bombarded by electrons.   *Ans.*   barium oxide

7.77   Refer to Fig. 7-31. List the plates (anodes) that are activated on this seven-segment VF display.
   Ans.   Activated plates $= P_b$, P_c, P_f, and P_g

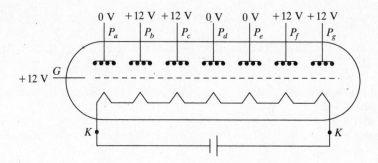

Fig. 7-31   VF display problem

7.78   Refer to Fig. 7-31. What decimal number would be shown on the seven-segment VF display?   *Ans.*   4

7.79   Refer to Fig. 7-22*b*. Why are there five separate control grids in this commercial VF display?
   Ans.   Each figure has a control grid so digits (or colon) can be turned on/off individually. The control grids are commonly used when multiplexing a display.

7.80   Vacuum fluorescent displays are widely used in _____ (automobiles, solar-powered equipment) because of voltage compatibility, long life, low cost, and good reliability.   *Ans.*   automobiles

7.81   Refer to Fig. 7-23. In the VF display, wires labeled X are called the _____.
   Ans.   cathodes, filaments, or heaters

7.82   Refer to Fig. 7-23. In the VF display, the shaped segments labeled Z are called the _____.
   Ans.   plates or anodes

7.83   Refer to Fig. 7-32. List the decimal number shown on the VF display during each pulse *a* through *d*.
   Ans.   pulse $a = 8$
        pulse $b = 9$
        pulse $c = 5$
        pulse $d = 2$

7.84   Refer to Fig. 7-32. During pulse *a* only, what is the logic level at each output (*a* through *g*) of the 4511 IC?
   Ans.   All outputs (*a* through *g*) are HIGH or activated (decimal 8 is displayed).

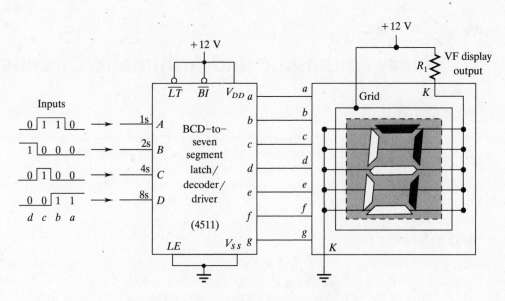

Fig. 7-32 Decoder/driver pulse-train problem

7.85 Refer to Fig. 7-32. During pulse *c* only, what is the logic level at each output (*a* through *g*) of the 4511 IC?

    Ans.   Decimal 5 is displayed.
            pulse *a* = HIGH output
            pulse *b* = LOW output
            pulse *c* = HIGH output
            pulse *d* = HIGH output
            pulse *e* = LOW output
            pulse *f* = HIGH output
            pulse *g* = HIGH output

Chapter 8

Binary Arithmetic and Arithmetic Circuits

8-1 INTRODUCTION

The general public thinks of digital devices as fast and accurate calculating machines. The calculator and digital computer are probably the reason for that. Arithmetic circuits are common in many digital systems. It will be found that rather simple combinational logic circuits (several gates connected) will add, subtract, multiply, and divide. This chapter will cover binary arithmetic and the way it is performed by logic circuits.

8-2 BINARY ADDITION

Adding binary numbers is a very simple task. The rules (addition tables) for binary addition using two bits are shown in Fig. 8-1. The first three rules are obvious. Rule 4 says that, in binary, $1 + 1 = 10$ (decimal 2). The 1 in the sum must be carried to the next column as in regular decimal addition.

	Sum	Carry out
Rule 1	$0 + 0 = 0$	
Rule 2	$0 + 1 = 1$	
Rule 3	$1 + 0 = 1$	
Rule 4	$1 + 1 = 0$ and carry $1 = 10$	

+ symbol means add

Fig. 8-1 Rules for binary addition

Two sample binary addition problems are shown below:

```
                                           carries
                                         1↙  1↙  1↙
          1  0  0     4                     1   0   1      5
        +0  1  0   +2                +     0   1   1    +3
(sum)   ――――――――   ――   (decimal)  (sum)  ―――――――――    ――
         1  1  0     6                    1  0  0  0      8   (decimal)
```

It is now possible to design a gating circuit that will perform addition. Looking at the left two columns of Fig. 8-1 reminds one of a two-variable truth table. The binary rules are reproduced in truth table form in Fig. 8-2. The inputs to be added are given the letters A and B. The sum output is often

Inputs		Outputs	
A	B	Sum	Carry out
0	0	0	0
0	1	1	0
1	0	1	0
1	1	0	1
$A + B$		Σ	Co

Fig. 8-2 Half adder truth table

170

given the *summation symbol* (Σ). The carry-out output column is often just represented with the *Co* symbol.

The truth table in Fig. 8-2 is that of a *half adder* circuit. A block diagram for a half adder might be drawn as in Fig. 8-3*a*. Note the two inputs *A* and *B* on the symbol in Fig. 8-3*a*. The outputs are labeled Σ (sum) and *Co* (carry out). It is common to label the half adder with HA as shown on the block symbol.

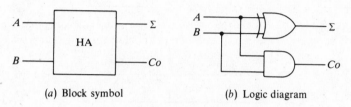

(*a*) Block symbol   (*b*) Logic diagram

Fig. 8-3 Half adder

Looking at the sum (Σ) output column of the truth table in Fig. 8-2, note that it takes an XOR function to produce the Σ output. The carry-out column will use an AND function. A complete logic circuit for the half adder with two inputs (*A* and *B*) and two outputs (Σ and *Co*) is shown in Fig. 8-3*b*. Composed only of gates (XOR and AND), the half adder is classified as a combinational logic circuit.

Consider the binary addition problem in Fig. 8-4*a*. The 1s column is $1 + 1$, and it follows rule 4 in Fig. 8-1. The sum is 0 with a carry of 1 to the 2s column. The 2s column must now be added. In the 2s column we have $1 + 1 + 1$. This is a new situation. It equals binary 11 (decimal 3). The 1 is placed below the 2s column in the sum position. A 1 is carried to the 4s column. The single 1 at the top of the 4s column is added to the 0s with a result of 1, which is written in the sum position. The result is a sum of 110.

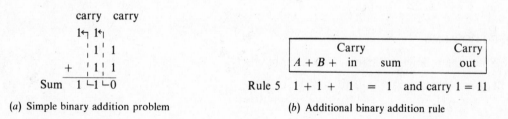

(*a*) Simple binary addition problem   (*b*) Additional binary addition rule

Fig. 8-4

Rule 5 for binary addition is formally written in Fig. 8-4*b*. Note the three inputs (*A*, *B*, and carry in). The outputs are the usual sum and carry out. Rule 5 suggests that a half adder will *not* work if a *carry-in* situation arises. Half adders will add only two inputs (*A* and *B*), as in the 1s column of an addition problem. When the 2s column or the 4s column is added, a new circuit is needed. The new circuit is called a *full adder*. A block diagram of a full adder is shown in Fig. 8-5*a*.

The full adder circuit has three inputs which are added. The inputs shown in the block diagram in Fig. 8-5 are *A*, *B*, and *Cin* (carry in). The outputs from the full adder are the customary Σ (sum) and *Co* (carry out). Note the use of the letters FA to symbolize full adder in the block diagram. To repeat, the half adder is used in only the 1s place when larger binary numbers are added. Full adders are used for adding all other columns (2s, 4s, 8s, and so forth).

A full adder circuit can be constructed from half adders and an OR gate. A full adder circuit is diagrammed in Fig. 8-5*b*. The half adder becomes a basic building block in constructing other adders. A truth table for the full adder is detailed in Fig. 8-5*c*.

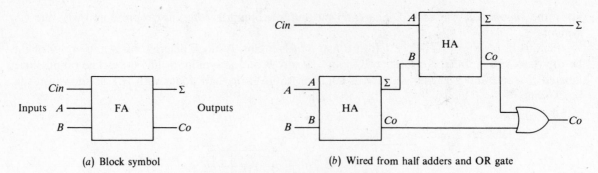

(a) Block symbol   (b) Wired from half adders and OR gate

Inputs			Outputs	
A	B	Cin	Σ	Co
0	0	0	0	0
0	0	1	1	0
0	1	0	1	0
0	1	1	0	1
1	0	0	1	0
1	0	1	0	1
1	1	0	0	1
1	1	1	1	1
$A + B + C_{in}$			Sum	Carry out

(c) Truth table

Fig. 8-5 Full adder

SOLVED PROBLEMS

8.1 Solve the following binary addition problems:

(a)   100   (b)   1010   (c)   1001
      +11   +110   +101

Solution:

Refer to Figs. 8-1 and 8-2. The sums in the problems are as follows:
(a) 111,   (b) 10000,   (c) 1110.

8.2 Find the binary sums in the following problems:

(a)   1110   (b)   1011   (c)   1111
      +11   +111   +111

Solution:

Refer to Figs. 8-1 and 8-2. The binary sums in the problems are as follows:
(a) 10001,   (b) 10010,   (c) 10110.

8.3   A half adder circuit has _____ input(s) and _____ output(s).

   Solution:

   A half adder circuit has 2 inputs and 2 outputs.

8.4   A full adder circuit has _____ input(s) and _____ output(s).

   Solution:

   A full adder circuit has 3 inputs and 2 outputs.

8.5   Draw a block diagram of a half adder and label the inputs and outputs.

   Solution:

   See Fig. 8-3a.

8.6   Draw a block diagram of a full adder and label the inputs and outputs.

   Solution:

   See Fig. 8-5a.

8.7   A half adder circuit is constructed from what two logic gates?

   Solution:

   A half adder circuit is constructed from a 2-input XOR gate and a 2-input AND gate.

8.8   A full adder circuit can be constructed by using two _____ (FAs, HAs) and a 2-input _____ (AND, OR) gate.

   Solution:

   A full adder circuit can be constructed by using two HAs and a 2-input OR gate.

8.9   An HA circuit is used to add the bits in the _____ (1s, 2s) column of a binary addition problem.

   Solution:

   An HA adds the 1s column in a binary addition problem.

8.10   Draw the logic diagram of a full adder using AND, XOR, and OR gates.

   Solution:

   See Fig. 8-6.

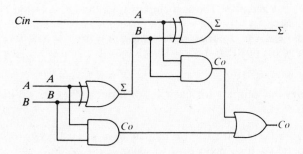

Fig. 8-6 Logic diagram of a full adder

8.11 List the HA *sum* outputs for each set of input pulses shown in Fig. 8-7.

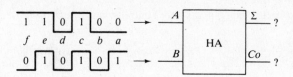

Fig. 8-7 Half adder pulse-train problem

Solution:

Based on the truth table in Fig. 8-2, the sum outputs from the half adder in Fig. 8-7 are as follows:

pulse $a = 1$ pulse $c = 0$ pulse $e = 0$
pulse $b = 0$ pulse $d = 0$ pulse $f = 1$

8.12 List the half adder *carry-out* outputs for each set of input pulses shown in Fig. 8-7.

Solution:

Based on the truth table in Fig. 8-2 the carry-out outputs from the half adder in Fig. 8-7 are as follows:

pulse $a = 0$ pulse $c = 1$ pulse $e = 1$
pulse $b = 0$ pulse $d = 0$ pulse $f = 0$

8.13 List the full adder *sum* outputs for each set of input pulses shown in Fig. 8-8.

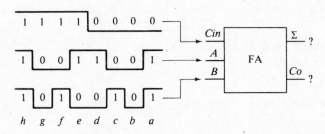

Fig. 8-8 Full adder pulse-train problem

Solution:

Refer to Figs. 8-1 and 8-4b. The sum outputs from the FA shown in Fig. 8-8 are as follows:

pulse $a = 0$ pulse $c = 1$ pulse $e = 0$ pulse $g = 1$
pulse $b = 0$ pulse $d = 1$ pulse $f = 0$ pulse $h = 1$

8.14 List the FA *carry-out* outputs for each set of input pulses shown in Fig. 8-8.

Solution:

Refer to Figs. 8-1 and 8-4b. The Co outputs from the full adder shown in Fig. 8-8 are as follows:

pulse $a = 1$ pulse $c = 0$ pulse $e = 1$ pulse $g = 0$
pulse $b = 0$ pulse $d = 0$ pulse $f = 1$ pulse $h = 1$

8-3 BINARY SUBTRACTION

Half subtractors and full subtractors will be explained in this section. The rules for the binary subtraction of two bits are given in Fig. 8-9. The top number in a subtraction problem is called the *minuend*. The bottom number is called the *subtrahend*, and the answer is called the *difference*. Rule 1 in Fig. 8-9 is obvious. Rule 2 (Fig. 8-9) concerns 1 being subtracted from the smaller number 0. In Fig. 8-10, note that, in the 1s column of the binary number, 1 is subtracted from 0. A 1 must be borrowed from the binary 2s column, leaving a 0 in that column. Now the subtrahend 1 is subtracted from the minuend 10 (decimal 2). This leaves a difference of 1 in the 1s column. The binary 2s column uses rule 1 $(0 - 0)$ and is equal to 0. Therefore, rule 2 is $0 - 1 = 1$ with a borrow of 1. Rules 3 and 4 are also rather obvious.

	Minuend		Subtrahend		Difference	Borrow out
Rule 1	0	−	0	=	0	
Rule 2	0	−	1	=	1	and borrow 1
Rule 3	1	−	0	=	1	
Rule 4	1	−	1	=	0	

Fig. 8-9 Rules for binary subtraction

	Binary		Decimal
	borrow		
Minuend	10	10	2
Subtrahend	− 0	1	− 1
Difference	0	1	1

Fig. 8-10 Binary subtraction problem showing a borrow

The subtraction rules given in Fig. 8-9 look somewhat like a truth table. These rules have been reproduced in truth table form in Fig. 8-11. Consider the difference (Di) output in the truth table. Note that this output represents the XOR function. The logic function for the difference output in a subtractor is the same as that for the sum output in a half adder circuit. Now consider the borrow (Bo) column in the truth table. The logic function for this column can be represented by the Boolean expression $\overline{A} \cdot B = Y$. It can be implemented by using an inverter and a 2-input AND gate.

Inputs		Outputs	
Minuend A	Subtrahend B	Difference	Borrow
0	0	0	0
0	1	1	1
1	0	1	0
1	1	0	0
$A - B$		Di	Bo

Fig. 8-11 Half subtractor truth table

The truth table in Fig. 8-11 represents a logic circuit called a *half subtractor*. The Boolean expression for the difference output is $A \oplus B = Di$. The Boolean expression for the borrow (Bo) output is $\overline{A} \cdot B = Bo$. A half subtractor would be wired from logic gates as shown in Fig. 8-12a. Input

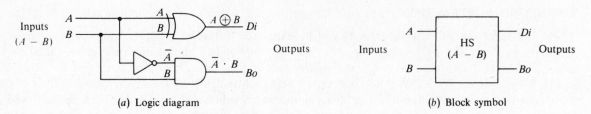

(a) Logic diagram   (b) Block symbol

Fig. 8-12 Half subtractor

A is the minuend and B is the subtrahend. The Di output is the difference; Bo is the borrow. A simplified block diagram for a half subtractor is in Fig. 8-12b.

Compare the half subtractor logic diagram in Fig. 8-12a with the half adder in Fig. 8-3b. The only difference in the logic circuits is that the half subtractor has one added inverter at the A input of the AND gate.

Consider the subtraction problem in Fig. 8-13. Several borrows are evident in this problem. If six subtractor circuits are used for the six binary places, the borrows must be considered. A half subtractor may be used for the 1s place. Full subtractors must be used in the 2s, 4s, 8s, 16s, and 32s columns of this problem.

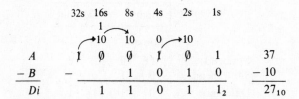

Fig. 8-13 Sample binary subtraction problem showing borrows

A block diagram of a full subtractor (FS) is in Fig. 8-14a. The inputs are A (minuend), B (subtrahend), and Bin (borrow input). The outputs are Di (difference) and Bo (borrow output). The Bo and Bin lines are connected from subtractor to subtractor to keep track of the borrows.

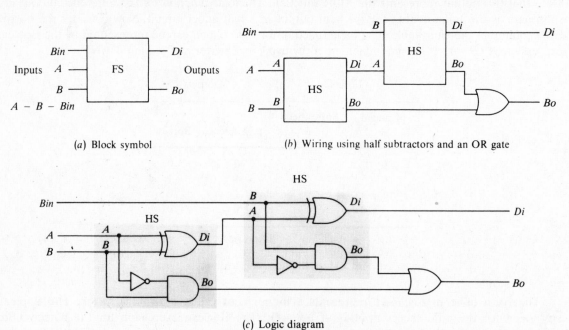

(a) Block symbol   (b) Wiring using half subtractors and an OR gate

(c) Logic diagram

Fig. 8-14 Full subtractor

The diagram in Fig. 8-14*b* shows how to wire two half subtractors (HS) and an OR gate together to form a full subtractor (FS) circuit. Note that the wiring pattern is similar to that used for adders. Finally, Fig. 8-14*c* shows how gates could be wired to form a full subtractor circuit. Remember that full subtractors must be used to subtract all columns except the 1s column in binary subtraction.

The truth table for the full subtractor is in Fig. 8-15. The inputs are labeled as minuend (A), subtrahend (B), and borrow in (Bin). The outputs are the customary difference (Di) and borrow out (Bo).

	Inputs			Outputs	
Line	Minuend (A)	Subtrahend (B)	Borrow in (Bin)	Difference	Borrow out (Bo)
1	0	0	0	0	0
2	0	0	1	1	1
3	0	1	0	1	1
4	0	1	1	0	1
5	1	0	0	1	0
6	1	0	1	0	0
7	1	1	0	0	0
8	1	1	1	1	1
		$A - B - Bin$		Di	Bo

Fig. 8-15 Truth table for the full subtractor

The binary subtraction problem in Fig. 8-16 will aid understanding of the full subtractor truth table. *Follow as this problem is solved, using only the truth tables in Figs. 8-11 and 8-15.* Look at the 1s column of the problem in Fig. 8-16. The 1s place uses a half subtractor. Find this situation in the truth table in Fig. 8-11. You find that line 3 of the half subtractor truth table gives an output of 1 for Di (difference) and 0 for borrow out (Bo). This is recorded below the 1s column in Fig. 8-16.

	64s	32s	16s	8s	4s	2s	1s	
A	1	1	1	0	1	0	1	117
$-B$	−0	0	1	1	1	0	0	−28
Di	1	0	1	1	0	0	1	89

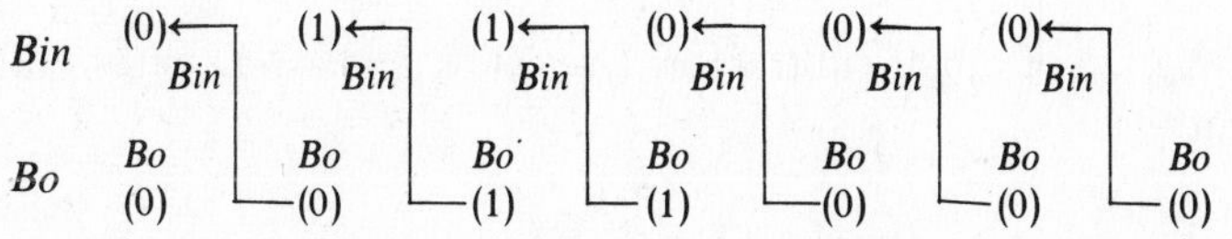

Fig. 8-16 Solving a binary subtraction problem using truth tables

Consider the 2s column in Fig. 8-16. The 2s column uses a full subtractor. On the full subtractor truth table, look for the situation where $A = 0$, $B = 0$, and $Bin = 0$. This is line 1 in Fig. 8-15. According to the truth table, both outputs (Di and Bo) are 0. This is recorded below the 2s column in Fig. 8-16.

Next consider the 4s column in Fig. 8-16. The inputs to this full subtractor will be $A = 1$, $B = 1$, and $Bin = 0$. Looking at the input side of the truth table in Fig. 8-15, it appears that line 7 shows this situation. The outputs (Di and Bo) are both 0 according to the truth table and are written as such on Fig. 8-16 under the 4s column.

Look at the 8s column in Fig. 8-16. The inputs to the full subtractor will be $A = 0$, $B = 1$, and $Bin = 0$. Line 3 of the truth table (Fig. 8-15) shows this situation. The outputs (Di and Bo) in line 3 are both 1s and are recorded in the 8s column in Fig. 8-16.

The 16s column in Fig. 8-16 has inputs of $A = 1$, $B = 1$, and $Bin = 1$. This corresponds with line 8 in the truth table. Line 8 generates an output of $Di = 1$ and $Bo = 1$. These 1s are recorded under the 16s column in the problem.

The 32s column has inputs of $A = 1$, $B = 0$, and $Bin = 1$. This corresponds to line 6 in the truth table in Fig. 8-15. Line 6 generates outputs of $Di = 0$ and $Bo = 0$. These 0s are recorded in the 32s column of the problem.

Finally consider the 64s column in Fig. 8-16. The inputs to the full subtractor are $A = 1$, $B = 0$, and $Bin = 0$. This input combination is shown in line 5 in the truth table. Line 5 generates an output of $Di = 1$ and $Bo = 0$. Figure 8-16 illustrates how binary 11100 is subtracted from binary 1110101 using truth tables. The borrows are shown below the problem. This procedure is quite cumbersome for humans, but electronic circuits can accurately perform this subtraction in microseconds.

SOLVED PROBLEMS

8.15 Solve the following binary subtraction problems:

(a)	110	(b)	1111	(c)	10110	(d)	10001	(e)	110001
	-100		-1010		-1100		-110		-111

Solution:

By a procedure similar to that illustrated in Fig. 8-13, the differences for the problems are found to be as follows:
(a) 010,   (b) 101,   (c) 1010,   (d) 1011,   (e) 101010.

8.16 Draw a block diagram of a half subtractor and label inputs and outputs.

Solution:

See Fig. 8-12b.

8.17 Draw a block diagram of a full subtractor and label inputs and outputs.

Solution:

See Fig. 8-14a.

8.18 Draw the logic diagram of a half subtractor. Use XOR and AND gates plus an inverter. Label inputs and outputs.

Solution:

See Fig. 8-12a.

8.19 When half adders and subtractors are compared, it is found that the half subtractor logic circuit contains one extra logic element which is an _____ (AND, inverter, OR).

Solution:

A HS contains one inverter more than an HA logic circuit.

8.20 List the difference (Di) outputs from the half subtractor shown in Fig. 8-17.

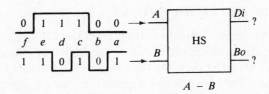

Fig. 8-17 Half subtractor pulse-train problem

Solution:

Refer to the truth table in Fig. 8-11. The Di outputs from the HS (Fig. 8-17) are as follows:

pulse $a = 1$ pulse $c = 0$ pulse $e = 0$
pulse $b = 0$ pulse $d = 1$ pulse $f = 1$

8.21 List the borrow-out (Bo) outputs from the half subtractor shown in Fig. 8-17.

Solution:

Refer to the truth table in Fig. 8-11. The Bo outputs from the HS (Fig. 8-17) are as follows:

pulse $a = 1$ pulse $c = 0$ pulse $e = 0$
pulse $b = 0$ pulse $d = 0$ pulse $f = 1$

8.22 List the difference (Di) outputs from the full subtractor shown in Fig. 8-18.

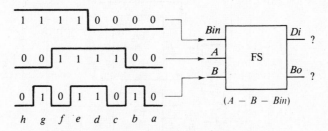

Fig. 8-18 Full subtractor pulse-train problem

Solution:

Refer to the truth table in Fig. 8-15. The Di outputs of the FS (Fig. 8-18) are as follows:

pulse $a = 0$ pulse $c = 1$ pulse $e = 1$ pulse $g = 0$
pulse $b = 1$ pulse $d = 0$ pulse $f = 0$ pulse $h = 1$

8.23 List the borrow-out (Bo) outputs from the full subtractor shown in Fig. 8-18.

Solution:

Refer to the truth table in Fig. 8-15. The Bo outputs from the FS (Fig. 8-18) are as follows:

pulse $a = 0$ pulse $c = 0$ pulse $e = 1$ pulse $g = 1$
pulse $b = 1$ pulse $d = 0$ pulse $f = 0$ pulse $h = 1$

8.24 When 2-bit numbers are subtracted, an _____ (FS, HS) is used for the 1s column and an _____ (FS, HS) is used for the 2s column.

Solution:

In binary subtraction, an HS is used for the 1s column and an FS is used for the 2s column.

8-4   PARALLEL ADDERS AND SUBTRACTORS

Binary addition can be accomplished in two different ways. Either *parallel* or *serial adders* can be used. A serial adder operates in much the same way as addition by hand. It first adds the 1s column, then the 2s column plus the carry, then the 4s column plus the carry, and so forth. Serial addition takes a fair amount of time when long binary numbers are added. Parallel addition, however, is very fast. In parallel addition, all the binary words (a *word* is a group of bits of a given length, such as 4, 8, or 16) to be added are applied to the inputs and the sum is almost immediate. Serial adders are simpler but slower. Parallel adders are faster, but they have more complicated logic circuits.

A *4-bit parallel adder* is shown in Fig. 8-19. A single half adder (HA) and three full adder (FA) circuits are used. Note that the top HA adds the 1s column (A_1 and B_1). The 2s column uses a full adder. The 2s FA adds the A_2 and B_2 plus the carry from the 1s adder. Note that the carry line runs from the Co of the 1s adder to the Cin of the 2s adder. The 4s and 8s adders also are full adders. The sum (Σ) output of each adder is connected to a sum indicator at the lower right in Fig. 8-19. The Co of the 8s FA is an overflow and forms the 16s place in the sum.

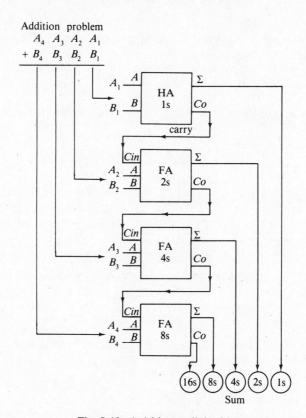

Fig. 8-19 A 4-bit parallel adder

Suppose the task were to add binary 1111 to 1111 with the parallel adder shown in Fig. 8-19. As soon as these numbers were applied to the eight inputs on the left, the output of 11110 (decimal 30) would appear on the output sum indicators. This parallel adder is limited to four input bits. More full adders could be attached to the circuit for the 16s place, 32s place, and so forth.

As with addition, subtraction can be done serially or by parallel subtractors. Figure 8-20 is a diagram of a familiar-looking *4-bit parallel subtractor*. Its wiring is quite similar to that of the 4-bit parallel adder that was just studied. The two 4-bit numbers are shown in the problem section at the upper left. Note that $B_4B_3B_2B_1$ (subtrahend) is subtracted from $A_4A_3A_2A_1$ (minuend). The

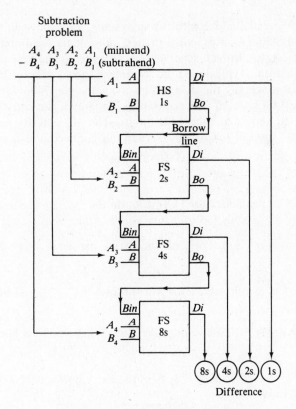

Fig. 8-20 A 4-bit parallel subtractor

difference between these numbers will appear on the difference output indicators at the lower right in Fig. 8-20.

The 1s column in Fig. 8-20 uses a half subtractor (HS). The 2s, 4s, and 8s columns use full subtractors (FS). Each of the Di outputs of the subtractors is connected to an output indicator to show the difference. The borrow lines connect the Bo output of one subtractor to the Bin input of the next most significant bit. The borrow lines keep track of the many borrows in binary subtraction. If greater than 4-bit numbers were to be subtracted, more full subtractors would be added to the circuit. FSs would be added in the pattern shown in Fig. 8-20. This parallel subtractor acts on the inputs and gives the difference almost immediately.

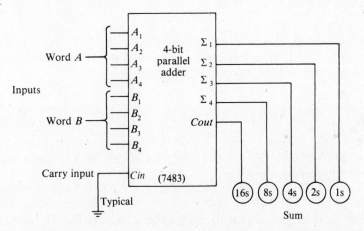

Fig. 8-21 Logic symbol for a commercial 7483 4-bit parallel-adder IC

By comparing the 4-bit parallel adder with the subtractor, it can be seen that the circuits are very similar (see Figs. 8-19 and 8-20). As a practical matter, full adders are purchased in IC form rather than being wired from logic gates. In fact, several more complicated adders and arithmetic logic units (ALUs) are available in IC form. Typically, an adder unit is shown as a block symbol like the one in Fig. 8-21. This logic symbol is actually the diagram for a commercial 7483 *4-bit full adder* IC. It could also be the symbol for the 4-bit parallel adder shown in Fig. 8-19 if the carry input (*Cin*) were left off the symbol. The A_1 and B_1 inputs are the LSB inputs. The A_4 and B_4 connections are the MSB inputs. It is typical to GND the *Cin* (carry input) when not connected to a preceding parallel adder.

SOLVED PROBLEMS

8.25   Refer to Fig. 8-19. The top adder (1s HA) will add the _____ (LSBs, MSBs).

Solution:

   The top adder shown in Fig. 8-19 will add the LSBs (least significant bits).

8.26   Refer to Fig. 8-20. The 8s full subtractor will subtract the _____ (LSBs, MSBs).

Solution:

   The 8s FS shown in Fig. 8-20 will subtract the MSBs (most significant bits).

8.27   Refer to Fig. 8-19. If $A_1 = 1$ and $B_1 = 1$, then the carry line between the 1s and 2s adders will be _____ (HIGH, LOW).

Solution:

   According to line 4 in the truth table in Fig. 8-2, with A_1 and B_1 equal to 1, the carry line between the 1s and 2s adders shown in Fig. 8-19 will be HIGH, indicating a carry.

8.28   Draw a diagram of a 6-bit parallel adder using one HA and five FAs.

Solution:

   See Fig. 8-22.

8.29   When the 6-bit unit in Prob. 8.28 adds 111111 and 111111, the sum is a binary _____, which equals _____ in decimal.

Solution:

   When the 6-bit unit in Prob. 8.28 adds binary 111111 and 111111, the sum is binary 1111110 by a procedure similar to that illustrated in Fig. 8-4a. The sum (1111110) is then converted to its decimal equivalent of 126 by means of the procedure shown in Fig. 1-2.

8.30   Refer to Fig. 8-19. When 1100 and 0011 are added, which carry lines will be HIGH?

Solution:

   When 1100 and 0011 are added with the adder shown in Fig. 8-19, *no* carry lines are HIGH because no carries occur in this addition problem.

8.31   Refer to Fig. 8-19. If all eight inputs to the parallel adder are HIGH, the binary output will be _____, which equals _____ in decimal.

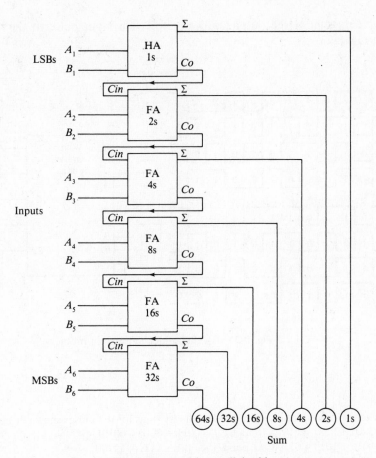

Fig. 8-22 A 6-bit parallel adder

Solution:

If all inputs to the parallel adder shown in Fig. 8-19 are HIGH, the binary output (sum) will be

$$1111_2 + 1111_2 = 11110_2 \text{ (sum)}$$

The sum 11110_2 is equal to a decimal 30 according to the procedure shown in Fig. 1-2.

8.32 Refer to Fig. 8-20. The _____ (bottom, top) subtractor is subtracting the least significant bits.

Solution:

The top subtractor is subtracting the LSBs in the problem in Fig. 8-20.

8.33 When 0011 is subtracted from 1101 in Fig. 8-20, the borrow line between the _____ (1s, 2s, 4s) subtractor and the _____ (2s, 4s, 8s) subtractor is HIGH.

Solution:

When 0011 is subtracted from 1101 in Fig. 8-20, the borrow line between the 2s and 4s subtractors is HIGH. This is shown by noting the borrow between the 4s and 2s places in the binary subtraction problem

$$
\begin{array}{cccc}
 & & \nearrow 1\cancel{0} & \\
1 & 1\cancel{0} & \cancel{0} & 1 \\
-0 & 0 & 1 & 1 \\
\hline
1 & 0 & 1 & 0
\end{array}
$$

8.34 List the binary sum at the output indicator for each input pulse to the 4-bit parallel adder shown in Fig. 8-23.

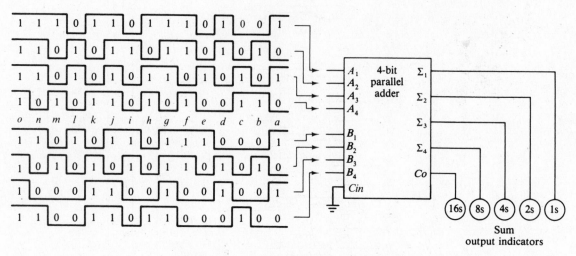

Fig. 8-23 Parallel-adder pulse-train problem

Solution:

Refer to the procedure in Fig. 8-4a. The binary sums for the pulses shown in Fig. 8-23 are as follows:

pulse $a = 0101 + 0101 = 01010$	pulse $i = 0011 + 0010 = 00101$
pulse $b = 0010 + 1010 = 01100$	pulse $j = 1101 + 1111 = 11100$
pulse $c = 1000 + 1100 = 10100$	pulse $k = 1110 + 1001 = 10111$
pulse $d = 0110 + 0011 = 01001$	pulse $l = 0001 + 0110 = 00111$
pulse $e = 0001 + 0100 = 00101$	pulse $m = 0010 + 1001 = 01011$
pulse $f = 0011 + 1011 = 01110$	pulse $n = 1001 + 0111 = 10000$
pulse $g = 1111 + 0111 = 10110$	pulse $o = 1111 + 1111 = 11110$
pulse $h = 1000 + 1101 = 10101$	

8-5 USING FULL ADDERS

A 4-bit parallel adder using three full adders and one half adder was studied in Fig. 8-19. To standardize circuitry and to do more complicated arithmetic, a somewhat different 4-bit adder is used. This new 4-bit adder is diagrammed in Fig. 8-24. Note that four full adders are used in this revised circuit. To make the 1s FA operate like a half adder, the Cin input to the FA is grounded (LOW). The revised circuit shown in Fig. 8-24 will operate exactly like the older version shown in Fig. 8-19.

The full adder truth table in Fig. 8-25 has been rearranged somewhat to help show that the full adder can be converted to a half adder by holding the Cin input LOW. Consider the upper half of the full adder truth table in Fig. 8-25. Note that $Cin = 0$ for each line of the unshaded section of the truth table. The B, A, Σ, and Co columns now *correspond exactly* with the half adder truth table in Fig. 8-2.

The 4-bit full adder shown in Fig. 8-24 is a block diagram of the 7483 full adder IC introduced in Fig. 8-21. Four-bit full adder ICs can be connected to form 8-, 12-, 16-, or even 32-bit parallel adders. An 8-bit parallel adder is featured in Fig. 8-26. Note the Cin input to the top 7483 IC is grounded (LOW). As in Fig. 8-24, this LOW at Cin converts the 1s full adder to a half adder. The Co output of

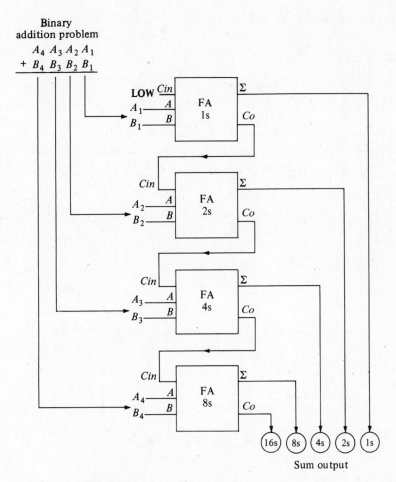

Fig. 8-24 Parallel adder using four full adders

Inputs			Outputs	
Cin	B	A	Σ	Co
0	0	0	0	0
0	0	1	1	0
0	1	0	1	0
0	1	1	0	1
1	0	0	1	0
1	0	1	0	1
1	1	0	0	1
1	1	1	1	1

Fig. 8-25 Full adder truth table

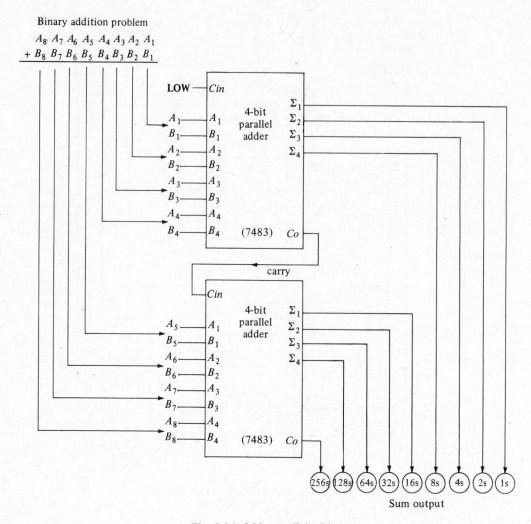

Fig. 8-26 8-bit parallel adder

the top 7483 IC is connected to the *Cin* input of the bottom unit. This handles carries from the 8s to the 16s places. Other carries are handled internally in the 7483 parallel adder ICs.

SOLVED PROBLEMS

8.35 Draw a diagram of an 8-bit parallel adder by using eight FAs.

Solution:

See Fig. 8-27.

8.36 Refer to Fig. 8-24. The 1s FA is converted to function as a _____ by grounding the *Cin* input.

Solution:

The 1s FA in Fig. 8-24 is converted to a half adder by grounding the *Cin* input.

8.37 Refer to Fig. 8-24. The conductors leading from the *Co* of one FA to the *Cin* of the next FA are known as _____ lines.

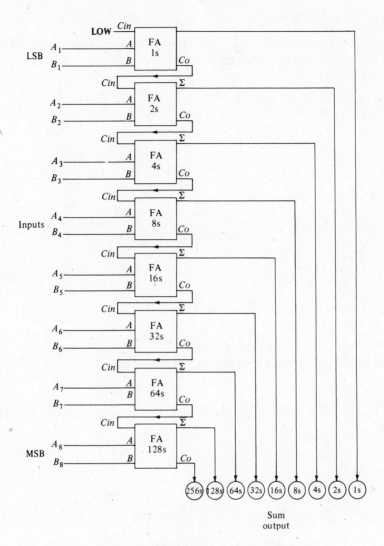

Fig. 8-27 8-bit parallel adder circuit

Solution:

Conductors leading from *Co* to *Cin* (Fig. 8-24) are called carry lines. These lines pass carries from one FA to the next.

8.38 Refer to Fig. 8-24. When 0001 and 0001 are added, the carry line between the 1s and 2s FAs is _____ (HIGH, LOW).

Solution:

Binary addition problem:

$$0001 + 0001 = 0010$$

When 0001 and 0001 are added (Fig. 8-24), a carry occurs between the 1s and 2s place and that carry line will be HIGH.

8.39 Refer to Fig. 8-24. Which carry lines are HIGH when binary 0111 and 0101 are added?

Solution:

See Fig. 8-28. When 0111 and 0101 are added in the device diagrammed in Fig. 8-24, three carry lines will be HIGH. They are the carry lines between:

1. 1s FA and 2s FA

2. 2s FA and 4s FA

3. 4s FA and 8s FA

$$\begin{array}{c} 1 \leftharpoonup\ 1 \leftharpoonup\ 1 \leftharpoonup \\ 0\ \vdots\ 1\ \vdots\ 1\ \vdots\ 1 \\ +0\ \vdots\ 1\ \vdots\ 0\ \vdots\ 1 \\ \hline 1\ \llcorner 1\ \llcorner 0\ \llcorner 0 \end{array}$$

Fig. 8-28 Binary addition problem

8.40   The block diagram in Fig. _____ (8-19, 8-24) most accurately describes the 7483 4-bit parallel adder IC.

Solution:

The block diagram in Fig. 8-24 describes the 7483 adder IC.

8.41   Refer to Fig. 8-26. What is the sum when the binary numbers 10011000 and 10101111 are added?

Solution:

See Fig. 8-29.

$$\begin{array}{r} ^{1\,1\,1} \\ 1001\ 1000 \\ +\quad 1010\ 1111 \\ \hline 10100\ 0111 \quad \text{Sum} \end{array}$$

Fig. 8-29 Binary addition problem

8-6   USING ADDERS FOR SUBTRACTION

With minor changes, parallel adders can be used to perform binary subtraction. The 4-bit parallel adder shown in Fig. 8-24 can be modified slightly to form a subtractor circuit. A *4-bit parallel subtractor* circuit is diagrammed in Fig. 8-30. Note that four full adders (FAs) are used. Note also that data entering each full adder's *B* input is inverted. Finally, note that the *Cin* input to the 1s FA (top full adder shown in Fig. 8-30) is held HIGH. The 4-bit parallel subtractor circuit shown in Fig. 8-30 will subtract the subtrahend ($B_4B_3B_2B_1$) from the minuend ($A_4A_3A_2A_1$).

The theory of operation of the circuit shown in Fig. 8-30 is based on a special mathematical technique outlined in Fig. 8-31. The problem given in Fig. 8-31 is to subtract binary 0111 from 1110. The problem is solved across the top by using traditional decimal and binary subtraction. The three steps below detail how the subtraction problem would be solved by using adders and a 2s complement subtrahend.

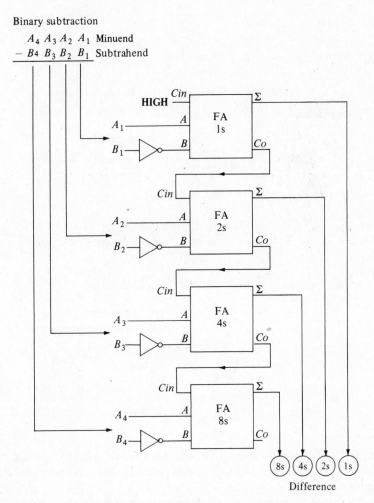

Fig. 8-30 4-bit subtractor using full adders

Follow the steps in Fig. 8-31 when solving the sample problem.

Step 1. *Change the subtrahend to its 2s complement form*. Only the subtrahend must be converted to its 2s complement equivalent. First the binary number 0111 is changed to its 1s complement form (1000), and then 1 is added to form the 2s complement (1000 + 1 = 1001).

Step 2. *Add the minuend to the 2s complement subtrahend*. The original minuend is added to the 2s complement subtrahend to get a temporary result (1110 + 1001 = 10111 in this example).

Step 3. *Discard the overflow*. The MSB is discarded, and the remaining 4 bits are equal to the binary difference. In this example the difference is binary 0111.

The reason why the circuit shown in Fig. 8-30 will work as a subtractor can now be explained. The four inverters change the binary subtrahend to its 1s complement form (each 1 is changed to 0 and each 0 to 1). The HIGH at the *Cin* input to the 1s FA is the same as adding +1 to the subtrahend. The minuend and 2s complement subtrahend are added. The *Co* terminal of the 8s FA is the *overflow* output. The *Co* output which discards the overflow is not displayed.

	Decimal	Binary	
	14	1110	Minuend
Problem:	−7	− 0111	Subtrahend
	7	0111	Difference

(a) Traditional decimal and binary subtraction

Step ① *Change subtrahend to 2s complement form.*

Binary   1s complement   2s complement

$$0111 \xrightarrow[\text{1s complement}]{\text{Form}} 1000 \xrightarrow{\text{Add 1}} 1001$$

Step ② *Add minuend to 2s complement subtrahend.*

$$
\begin{array}{r}
^{1}\\
1110 \quad \text{Minuend}\\
+\ 1001 \quad \text{2s complement subtrahend}\\
\hline
10111
\end{array}
$$

Step ③ *Discard overflow. The difference is 0111 in this example.*

$$
\begin{array}{r}
1110 \quad \text{Minuend}\\
+\quad 1001 \quad \text{2s complement subtrahend}\\
\hline
①\ 0111 \quad \text{Difference}
\end{array}
$$

Discard ↙
overflow

(b) Special technique subtraction using 2s complement subtrahend and addition

Fig. 8-31

If the 4-bit parallel adder and subtractor circuits from Figs. 8-24 and 8-30 are compared, they are seen to be almost identical. These circuits can be combined to form an adder/subtractor circuit. Such a circuit is diagrammed in Fig. 8-32.

The 4-bit parallel adder/subtractor circuit shown in Fig. 8-32 has an additional input called the mode control. If the mode control input is LOW (logical 0), the four XOR gates have no effect on the data at the B inputs (data passes through the XORs and is not inverted). The *Cin* input to the 1s FA is held LOW, which makes the FA function as a half adder. A 4-bit sum will appear at the output indicators at the lower right.

When the mode control of the adder/subtractor circuit shown in Fig. 8-32 is HIGH (logical 1), the four XOR gates act as inverters. The subtrahend $(B_4 B_3 B_2 B_1)$ is inverted. The *Cin* input to the 1s FA is HIGH which is like adding $+1$ to the 1s complement subtrahend. The difference will appear at the lower right in Fig. 8-32 in binary form.

SOLVED PROBLEMS

8.42   List three modifications that must be made to convert the 4-bit adder shown in Fig. 8-24 to a 4-bit parallel subtractor.

Solution:

See Fig. 8-30 for the solution. The modifications to the adder shown in Fig. 8-24 are (1) adding four inverters, (2) connecting *Cin* input of the 1s FA to HIGH, and (3) leaving the *Co* output from the 8s FA disconnected.

8.43   Refer to Fig. 8-32. The XOR gates act as inverters when the mode control is _____.

Solution:

The XOR gates in Fig. 8-32 act as inverters when the mode control is HIGH (subtract mode).

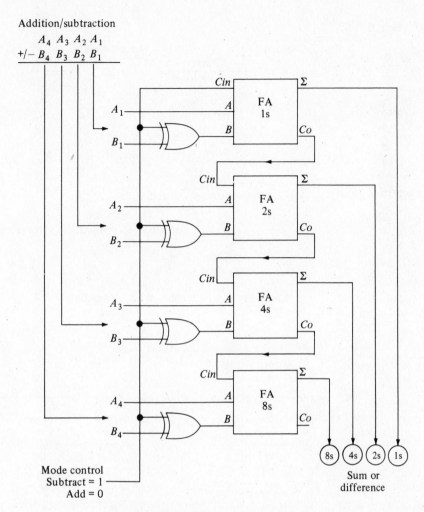

Fig. 8-32 4-bit adder/subtractor circuit

8.44   Refer to Fig. 8-32. This circuit acts as a 4-bit parallel _____ when the mode control input is LOW.

Solution:

   The circuit shown in Fig. 8-32 acts as a 4-bit parallel adder when the mode control input is LOW.

8.45   Use the special technique shown in Fig. 8-31 to subtract binary 0110 from 1111.

Solution:

   See Fig. 8-33.

8.46   Draw a diagram of a 4-bit parallel subtractor circuit by using a 7483 4-bit parallel adder IC and four inverters.

Solution:

   See Fig. 8-34.

8.47   Refer to Fig. 8-30. When binary 0101 is subtracted from 1100, the difference is _____ .

Solution:

   When binary 0101 is subtracted from 1100, the difference is 0111 (decimal $12 - 5 = 7$).

Step ①

Binary		1s complement		2s complement
	Form		Add 1	
0110	$\xrightarrow{\text{1s complement}}$	1001	$\xrightarrow{}$	1010

Step ②

$$\begin{array}{r} {\scriptstyle 1\,1} \\ 1111 \\ +\ 1010 \\ \hline 11001 \end{array}$$

Step ③

$$\begin{array}{rl} 1111 & \text{Minuend} \\ +\ 1010 & \text{2s complement subtrahend} \\ \hline \textcircled{1}\ 1001 & \text{Difference} \end{array}$$

Discard
overflow

Fig. 8-33 Special technique for subtraction

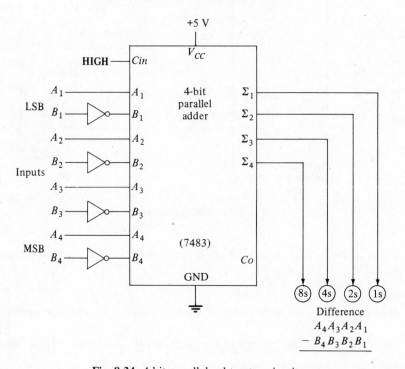

Fig. 8-34 4-bit parallel subtractor circuit

8.48 Refer to Fig. 8-30. List the B inputs to the four FAs when 0101 is subtracted from 1100.

Solution:

The 1s complement of the subtrahend will appear at the B inputs to the FAs shown in Fig. 8-30. If the subtrahend equals 0101, the 1s complement of the subtrahend will be 1010.

8.49 Refer to Fig. 8-30. When 0101 is subtracted from 1100, which carry lines will be HIGH?

Solution:

The FAs add 1100 (minuend) to 1011 (2s complement subtrahend), which means a carry occurs only at Co of the 8s FA. All the carry lines in Fig. 8-30 will be LOW in this operation.

8-7 2S COMPLEMENT ADDITION AND SUBTRACTION

The 2s complement method of representing numbers is widely used in microprocessors. Until now, the numbers added or subtracted were positive numbers. However, microprocessors must add and subtract both positive and negative numbers. Using 2s complement numbers makes adding and subtracting signed numbers possible. A review of 2s complement numbers and their use in representing positive and negative values is given in Sec. 1-4.

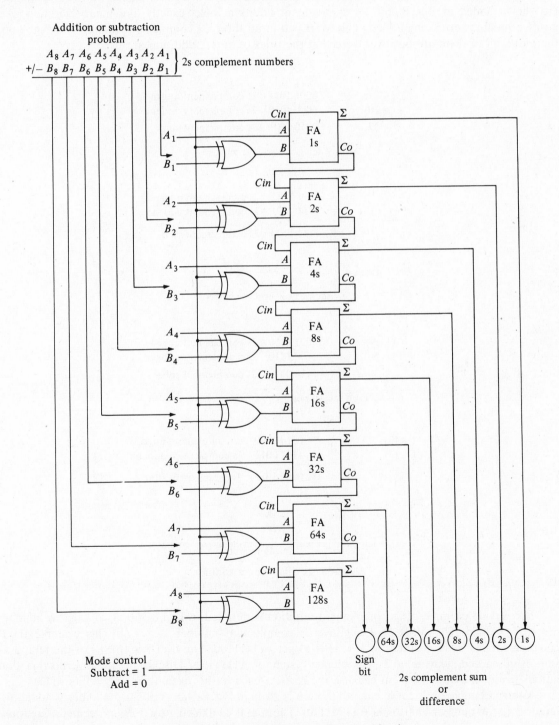

Fig. 8-35 8-bit parallel adder/subtractor circuit

A circuit for adding and subtracting signed numbers in 2s complement notation is diagrammed in Fig. 8-35. This is an 8-bit parallel adder/subtractor that will add or subtract signed numbers. *All inputs and outputs are in 2s complement form.*

Note that the 8-bit 2s complement adder/subtractor shown in Fig. 8-35 is an extension of the 4-bit adder/subtractor shown in Fig. 8-32. If the mode control shown in Fig. 8-35 is LOW, the circuit adds. However, if the mode control is HIGH, the circuit performs as an 8-bit parallel subtractor.

Four examples of adding 2s complement numbers are shown in Fig. 8-36. Two positive numbers are shown added in Fig. 8-36a. 2s complement addition looks exactly like binary addition when positive numbers are being added. The MSB is 0 in all three 2s complement numbers in Fig. 8-36a; therefore, all of them are positive. Note that the rules of binary addition are used.

```
                          11  1
    (+27)          0001 1011    2s complement augend
  +(+10)        +  0000 1010    2s complement addend
  ──────          ─────────
   +37₁₀           0010 0101    2s complement sum
```

(a) Adding two positive numbers

```
                       11111 111
    (−1)           1111 1111    2s complement augend
  +(−3)         +  1111 1101    2s complement addend
  ──────          ─────────
   −4₁₀        (1) 1111 1100    2s complement sum
                Discard
```

(b) Adding two negative numbers

```
                         11  1
    (+20)          0001 0100    2s complement augend
  +(−50)        +  1100 1110    2s complement addend
  ──────          ─────────
   −30₁₀           1110 0010    2s complement sum
```

(c) Adding a smaller positive to a larger negative number

```
                        111
    (+40)          0010 1000    2s complement augend
  +(−13)        +  1111 0011    2s complement addend
  ──────          ─────────
   +27₁₀        (1) 0001 1011    2s complement sum
                Discard
```

(d) Adding a larger positive to a smaller negative number

Fig. 8-36 2s complement addition

The second example of 2s complement addition is detailed in Fig. 8-36b. Two negative numbers are being added. The MSB of a negative 2s complement number is a 1. In this example the 2s complement 11111111 is added to 11111101 to get 1 11111100. The overflow (MSB) of the temporary sum is discarded, leaving a 2s complement sum of 11111100. Discarding the overflow is done automatically in a digital system because the register used in this example is only 8 bits wide.

A third example of 2s complement addition is given in Fig. 8-36c. A positive number is added to a larger negative number (00010100 + 11001110). The sum is 11100010, or a −30 in decimal. The fourth example adds a positive number to a smaller negative number. When 00101000 is added to 11110011, the result is 1 00011011. The overflow (MSB) is discarded, leaving the sum of 00011011.

Four examples of 2s complement subtraction are shown in Fig. 8-37. Two positive numbers are subtracted in Fig. 8-37a. The $+41$ is converted to its 2s complement form (00101001), and then it is 2s complemented again to get a subtrahend of 11010111. The minuend and subtrahend are then *added* to get 1 00100010. The overflow (MSB) is discarded, leaving a *2s complement difference* of 00100010, or $+34$ in decimal.

$(+75)$			0100 1011	Minuend
$-(+41)$	0010 1001	Form 2s complement and add →	+ 1101 0111	Subtrahend
$+34_{10}$			① 0010 0010	2s complement difference
			↙ Discard	

(a) Subtracting two positive numbers

			1 1	
(-80)			1011 0000	Minuend
$-(-30)$	1110 0010	Form 2s complement and add →	+ 0001 1110	Subtrahend
-50_{10}			1100 1110	2s complement difference

(b) Subtracting two negative numbers

			1	
$(+24)$			0001 1000	Minuend
$-(-20)$	1110 1100	Form 2s complement and add →	+ 0001 0100	Subtrahend
$+44_{10}$			0010 1100	2s complement difference

(c) Subtracting a negative from a positive number

			1 1	
(-60)			1100 0100	Minuend
$-(+15)$	0000 1111	Form 2s complement and add →	+ 1111 0001	Subtrahend
-75_{10}			① 1011 0101	2s complement difference
			↙ Discard	

(d) Subtracting a positive from a negative number

Fig. 8-37 2s complement subtraction

The second example of 2s complement subtraction is detailed in Fig. 8-37b. Two negative numbers are subtracted. The minuend (-80) is converted to its 2s complement form (10110000). The subtrahend (-30) is 2s complemented twice to get first 11100010 and finally 00011110. The 2s complement difference is 11001110 (-50) when the minuend and subtrahend are *added*.

The third example of 2s complement subtraction is explained in Fig. 8-37c. A -20 is subtracted from a $+24$. The -20 is 2s complemented twice to get the temporary 11101100 and the final subtrahend of 00010100. The subtrahend (00010100) is then added to the minuend (00011000) to get the 2s complement difference of 00101100, or $+44$ in decimal.

The final example of 2s complement subtraction is given in Fig. 8-37d. A $+15$ is subtracted from -60. The minuend (-60) is converted to its 2s complement form (11000100). The subtrahend ($+15$) is 2s complemented twice to get the temporary 00001111 and the final subtrahend of 11110001. The minuend (11000100) and subtrahend (11110001) are added to get 1 10110101. The overflow (MSB) is discarded, leaving a 2s complement difference of 10110101, or -75 in decimal.

All the sample problems can be tested by using the 8-bit parallel adder/subtractor shown in Fig. 8-35. Remember that both the inputs to and outputs from the adder/subtractor circuit shown in Fig. 8-35 must be in 2s complement notation.

SOLVED PROBLEMS

8.50   Why are 2s complement numbers used in digital systems?
   Solution:
   2s complement numbers are used to represent signed numbers.

8.51   Refer to Fig. 8-35. This circuit can add or subtract _____ numbers.
   Solution:
   The circuit shown in Fig. 8-35 can add or subtract signed numbers in 2s complement notation.

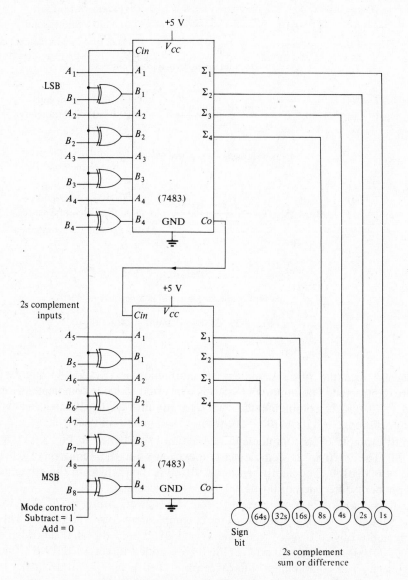

Fig. 8-38   8-bit adder/subtractor circuit

8.52 Draw a diagram of an 8-bit parallel adder/subtractor by using two 7483 ICs and eight XOR gates. Use Figs. 8-34 and 8-35 as a guide.

Solution:

See Fig. 8-38.

8.53 Refer to Fig. 8-35. The numbers input into the adder/subtractor must be in what code?

Solution:

The numbers input into the adder/subtractor shown in Fig. 8-35 must be in 2s complement notation.

8.54 Refer to Fig. 8-35. What do inputs A_8 and B_8 represent?

Solution:

Inputs A_8 and B_8 in Fig. 8-35 represent the signs of the numbers. If the sign bit is 0, the number is positive; if the sign bit is 1, the number is negative.

8.55 Add $+83$ and $+17$ by using 2s complement numbers. Use the procedure shown in Fig. 8-36.

Solution:

See Fig. 8-39.

```
              1   11
(+83)       0101 0011   2s complement augend
+(+17)    + 0001 0001   2s complement addend
─────       ─────────
+100₁₀      0110 0100   2s complement sum
```

Fig. 8-39 Solution to 2s complement addition problem

8.56 Add $+119$ and -13 by using 2s complement numbers. Use the procedure illustrated in Fig. 8-36.

Solution:

See Fig. 8-40.

```
              111  111
(+119)      0111 0111   2s complement augend
+ (-13)   + 1111 0011   2s complement addend
─────    ① 0110 1010   2s complement sum
+106₁₀
         Discard
```

Fig. 8-40 Solution to 2s complement addition problem

8.57 Subtract $+26$ from $+64$ by using 2s complement numbers. Use the procedure illustrated in Fig. 8-37.

Solution:

See Fig. 8-41.

$$
\begin{array}{r}
(+64) \\
-(+26) \\
\hline
+38_{10}
\end{array}
\quad \xrightarrow{\text{2s complement}} \quad
0001\ 1010
\quad \xrightarrow[\text{and add}]{\text{2s complement}} \quad
$$

$$
\begin{array}{r}
\overset{1}{0100}\ 0000 \quad \text{Minuend} \\
+\ \ 1110\ 0110 \quad \text{Subtrahend} \\
\hline
\textcircled{1}\ 0010\ 0110 \quad \text{2s complement difference}
\end{array}
$$

Discard

Fig. 8-41 Solution to 2s complement subtraction problem

8.58 Subtract -23 from -53 by using 2s complement numbers. Use the procedure illustrated in Fig. 8-37.

Solution:

See Fig. 8-42.

$$
\begin{array}{r}
(-53) \\
-(-23) \\
\hline
-30_{10}
\end{array}
\quad \xrightarrow{\text{2s complement}} \quad
1110\ 1001
\quad \xrightarrow[\text{and add}]{\text{2s complement}} \quad
$$

$$
\begin{array}{r}
\overset{11\ 111}{1100}\ 1011 \quad \text{Minuend} \\
+\ \ 0001\ 0111 \quad \text{Subtrahend} \\
\hline
1110\ 0010 \quad \text{2s complement difference}
\end{array}
$$

Fig. 8-42 Solution to 2s complement subtraction problem

Supplementary Problems

8.59 Solve the following binary addition problems:

(a) 1111   (b) 11110   (c) 10111   (d) 0111   (e) 11111   (f) 10001
    +1011   +10101   + 1111   +1100   +   1   +11011

Ans. (a) 11010   (b) 110011   (c) 100110   (d) 10011   (e) 100000   (f) 101100

8.60 Give the letter symbol for the following inputs to and outputs from a half adder (HA):
(a) top input,   (b) bottom input,   (c) sum output,   (d) carry output.
Ans. (a) top input $= A$ (b) bottom input $= B$ (c) sum output $= \Sigma$ (d) carry output $= Co$

8.61 Give the letter symbol for the following inputs to and outputs from a full adder (FA):
(a) carry input,   (b) top data input,   (c) bottom data input,   (d) sum output,   (e) carry output.
Ans. (a) carry input $= Cin$ (c) bottom data input $= B$ (e) carry output $= Co$
      (b) top data input $= A$ (d) sum output $= \Sigma$

8.62 Draw a logic diagram of an HA circuit using gates. Label the inputs and outputs.
Ans. See Fig. 8-3b.

8.63 Draw a logic diagram of an FA circuit by using XOR and NAND gates only. Label the inputs and outputs. Use Fig. 8-6 as a guide.   *Ans.* See Fig. 8-43.

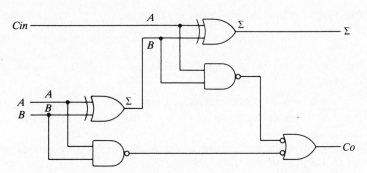

Fig. 8-43 Full adder logic diagram using XOR and NAND gates

8.64   An HA will add two variables, and an FA will add _____ input variables.   *Ans.*   three

8.65   List the full adder Σ outputs for each set of input pulses shown in Fig. 8-44.
   Ans.   pulse $a = 1$ pulse $c = 1$ pulse $e = 0$ pulse $g = 1$ pulse $i = 0$
            pulse $b = 0$ pulse $d = 1$ pulse $f = 1$ pulse $h = 0$ pulse $j = 0$

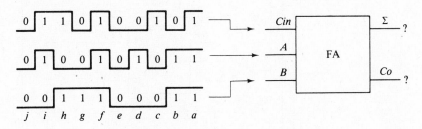

Fig. 8-44 Full adder pulse-train problem

8.66   List the full adder *Co* outputs for each set of input pulses shown in Fig. 8-44.
   Ans.   pulse $a = 1$ pulse $c = 0$ pulse $e = 0$ pulse $g = 0$ pulse $i = 1$
            pulse $b = 1$ pulse $d = 0$ pulse $f = 1$ pulse $h = 1$ pulse $j = 0$

8.67   Solve the following binary subtraction problems:

(a) $\begin{array}{r} 11011 \\ -01110 \\ \hline \end{array}$ (b) $\begin{array}{r} 11100 \\ -01110 \\ \hline \end{array}$ (c) $\begin{array}{r} 11001 \\ -01010 \\ \hline \end{array}$ (d) $\begin{array}{r} 10000 \\ -01001 \\ \hline \end{array}$ (e) $\begin{array}{r} 10111 \\ -10001 \\ \hline \end{array}$

Ans. (a) 1101 (b) 1110 (c) 1111 (d) 0111 (e) 0110

8.68   Give the names of the following inputs to and outputs from a half subtractor:
   (a) *A,* (b) *B,* (c) *Di,* (d) *Bo.*
   Ans. (a) *A* = minuend input (c) *Di* = difference output
            (b) *B* = subtrahend input (d) *Bo* = borrow output

8.69   Give the letter symbols for the following inputs to and outputs from the FS:
   (a) borrow input, (b) minuend input, (c) subtrahend input, (d) difference output,
   (e) borrow output.
   Ans. (a) borrow input = *Bin* (c) subtrahend input = *B* (e) borrow output = *Bo*
            (b) minuend input = *A* (d) difference output = *Di*

8.70   Refer to Fig. 8-19. The A_2 and B_2 inputs are from the _____ (1s, 2s, 4s, 8s) column of the addition
problem.   *Ans.*   2s

8.71   Refer to Fig. 8-20. The A_3 and B_3 inputs are from the _____ (1s, 2s, 4s, 8s) column of the subtraction
problem.   *Ans.*   4s

8.72   Refer to Fig. 8-20. If all inputs to the 4s FS are 1, the output from this FS will be $Di =$ _____ and
$Bo =$ _____ .
Ans.   When all inputs to the 4s FS shown in Fig. 8-20 are HIGH, the outputs will be $Di = 1$ and $Bo = 1$.
This is based on line 8 of the FS truth table in Fig. 8-15.

8.73   Refer to Fig. 8-45. The outputs from the 1s HS are $Di =$ __(a)__ and $Bo =$ __(b)__ according to line
__(c)__ in the truth table in Fig. 8-11.

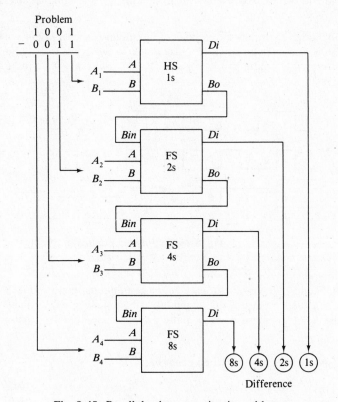

Fig. 8-45   Parallel-subtractor circuit problem

8.74   Refer to Fig. 8-45. The inputs to the 2s FS are $A =$ _____, $B =$ _____, and $Bin =$ _____ with
outputs of $Di =$ _____ and $Bo =$ _____ according to line _____ in the truth table in Fig. 8-15.
Ans.   The inputs to the 2s FS (Fig. 8-45) are $A = 0$, $B = 1$, and $Bin = 0$ with outputs of $Di = 1$ and
$Bo = 1$ according to line 3 in Fig. 8-15.

8.75   Refer to Fig. 8-45. The inputs to the 4s FS are $A =$ _____, $B =$ _____, and $Bin =$ _____ with
outputs of $Di =$ _____ and $Bo =$ _____ according to line _____ in the truth table in Fig. 8-15.
Ans.   The inputs to the 4s FS (Fig. 8-45) are $A = 0$, $B = 0$, and $Bin = 1$ with outputs of $Di = 1$ and $Bo = 1$
according to line 2 in Fig. 8-15.

8.76 Refer to Fig. 8-45. The inputs to the 8s FS are $A =$ _____, $B =$ _____, and $Bin =$ _____ with outputs of $Di =$ _____ and $Bo =$ _____ according to line _____ in the truth table in Fig. 8-15.
Ans. The inputs to the 8s FS in Fig. 8-45 are $A = 1$, $B = 0$, and $Bin = 1$ with outputs of $Di = 0$ and $Bo = 0$ according to line 6 in Fig. 8-15.

8.77 Refer to Fig. 8-45. The difference showing on the indicators is a binary _____.   *Ans.*   0110.

8.78 Refer to Fig. 8-45. This unit is a _____ -bit _____ (parallel, serial) _____ (adder, subtractor).
Ans.   4-bit parallel subtractor

8.79 List the binary differences at the output indicators of the 4-bit parallel subtractor circuit shown in Fig. 8-46.
Ans.   The differences for the pulses shown in Fig. 8-46 are as follows:
pulse $a = 0010$ pulse $c = 0100$ pulse $e = 0011$ pulse $g = 0001$ pulse $i = 0011$
pulse $b = 1000$ pulse $d = 1001$ pulse $f = 0011$ pulse $h = 0111$ pulse $j = 1101$

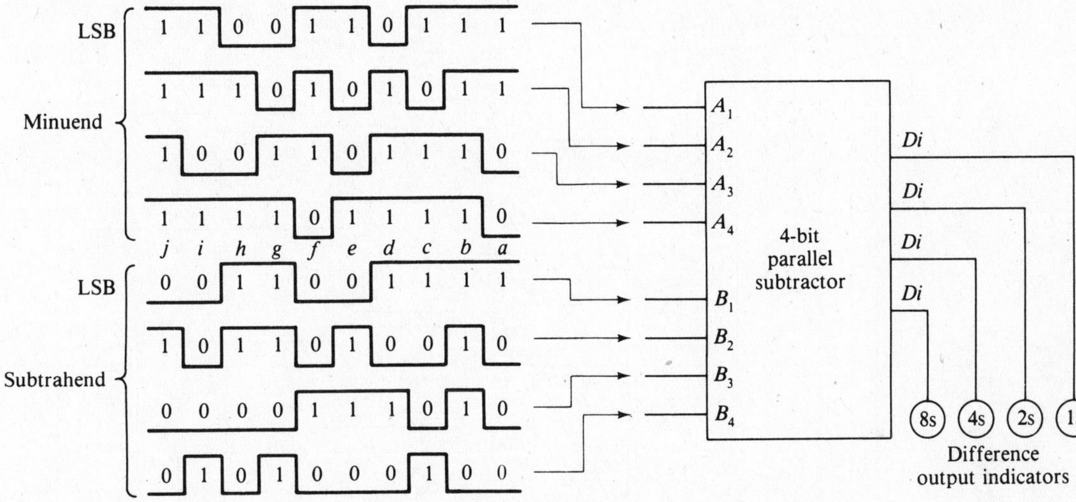

Fig. 8-46 Parallel-subtractor pulse-train problem

8.80 Refer to Fig. 8-46. The subtractor probably contains   *(a)*   HS(s) and   *(b)*   FSs.
Ans. *(a)* one   *(b)* three

8.81 Refer to Fig. 8-46. The subtractor circuit is classified as a _____ (combinational, sequential) logic circuit.   *Ans.* combinational

8.82 Refer to Fig. 8-24. What is the effect of grounding the 1s full adder Cin input?
Ans. Grounding the 1s FA Cin input shown in Fig. 8-24 has the effect of converting the 1s full adder to a half adder.

8.83 The 7483 TTL IC is described as a 4-bit _____ (parallel, serial) adder DIP integrated circuit.
Ans. parallel

8.84 Refer to Fig. 8-26. What is the sum when the binary numbers 11101010 and 01001110 are added?
Ans. sum = 100111000

8.85   Refer to Fig. 8-26. What is the highest sum that could be generated by the 8-bit parallel adder?
           Ans. $11111111 + 11111111 = 1\ 11111110_2\ (255 + 255 = 510_{10})$.

8.86   Refer to Fig. 8-32. The XOR gates act like _____ (AND gates, inverters) when the mode control is
           HIGH.   *Ans.*   inverters

8.87   Refer to Fig. 8-32. This circuit acts as a 4-bit parallel _____ when the mode control input is HIGH.
           Ans.   subtractor

8.88   Draw a diagram of an 8-bit parallel adder/subtractor using eight FAs and eight XOR gates.
           Ans.   See Fig. 8-47.

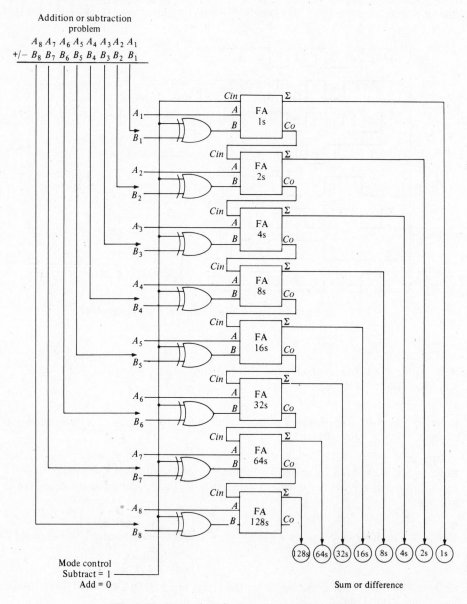

Fig. 8-47 8-bit parallel adder/subtractor circuit

8.89 Refer to Fig. 8-35. The output from the adder/subtractor is in what code?
Ans.   2s complement notation

8.90 Refer to Fig. 8-35. Why does this adder/subtractor circuit specify the use of 2s complement numbers?
Ans.   2s complement notation is one method of representing *signed numbers* in digital circuits.

8.91 Refer to Fig. 8-35. The MSB in the result (sum or difference) is the _____ bit.
Ans.   sign (0 = positive or 1 = negative)

8.92 Add $+18$ to -55 by using 2s complement numbers. Use the procedure shown in Fig. 8-36.
Ans.   See Fig. 8-48.

$$
\begin{array}{rll}
(+18) & 0001\ 0010 & \text{2s complement augend} \\
+(-55) & +\ 1100\ 1001 & \text{2s complement addend} \\
\hline
-37_{10} & 1101\ 1011 & \text{2s complement sum}
\end{array}
$$

Fig. 8-48 Solution to 2s complement addition problem

8.93 Subtract -14 from $+47$ by using 2s complement numbers. Use the procedure shown in Fig. 8-37.
Ans.   See Fig. 8-49.

$$
\begin{array}{rll}
(+47) & \overset{1\ \ 11}{0010\ 1111} & \text{Minuend} \\
-(-14) \longrightarrow 1111\ 0010 \longrightarrow & +\ 0000\ 1110 & \text{Subtrahend} \\
\hline
+61_{10} & 0011\ 1101 & \text{2s complement difference}
\end{array}
$$

Fig. 8-49 Solution to 2s complement subtraction problem

Chapter 9

Flip-Flops and Other Multivibrators

9-1 INTRODUCTION

Logic circuits are classified in two broad categories. The groups of gates described thus far have been wired as *combinational logic circuits*. In this chapter a valuable type of circuit will be introduced: the *sequential logic circuit*. The basic building block of combinational logic is the logic gate. The basic building block of the sequential logic circuit is the *flip-flop* circuit. Sequential logic circuits are extremely valuable because of their *memory characteristic*.

Several types of flip-flops will be detailed in this chapter. Flip-flops are also called "latches," "bistable multivibrators," or "binaries." The term "flip-flop" will be used in this book. Useful flip-flops can be wired from logic gates, such as NAND gates, or bought in IC form. Flip-flops are interconnected to form sequential logic circuits for data storage, timing, counting, and sequencing.

Besides the bistable multivibrator (flip-flop), two other types of multivibrators (MVs) are introduced in this chapter. The *astable multivibrator* is also called a *free-running MV*. The astable MV produces a continuous series of square-wave pulses and is commonly used as a clock in a digital system. The *monostable multivibrator* is also called a *one-shot MV* in that it produces a single pulse when triggered by an external source.

9-2 *RS* FLIP-FLOP

The most basic flip-flop is called the *RS flip-flop*. A block logic symbol for the *RS* flip-flop is shown in Fig. 9-1. The logic symbol shows two inputs, labeled *set* (S) and *reset* (R), on the left. The *RS* flip-flop in this symbol has active LOW inputs, shown as small bubbles at the S and R inputs. Unlike logic gates, flip-flops have two complementary outputs. The outputs are typically labeled Q and $\overline{Q}$ (say "not Q or Q not"). The Q output is considered the "normal" output and is the one most used. The other output ($\overline{Q}$) is simply the complement of output Q, and it is referred to as the complementary output. Under normal conditions these outputs are always complementary. Hence, if $Q = 1$, then $\overline{Q} = 0$; or if $Q = 0$, then $\overline{Q} = 1$.

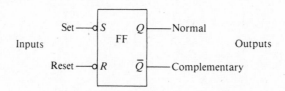

Fig. 9-1 Logic symbol for *RS* flip-flop

The *RS* flip-flop can be constructed from logic gates. An *RS* flip-flop is shown wired from two NAND gates in Fig. 9-2*a*. Note the characteristic feedback from the output of one NAND gate into the input of the other gate. As with logic gates, a truth table defines the operation of the flip-flop. Line 1 of the truth table in Fig. 9-2*b* is called the *prohibited state* in that it drives both outputs to 1, or HIGH. This condition is not used on the *RS* flip-flop. Line 2 of the truth table shows the *set* condition of the flip-flop. Here a LOW, or logical 0, activates the set (S) input. That sets the normal Q output to HIGH, or 1, as shown in the truth table. This set condition works out if the NAND circuit shown in Fig. 9-2*a* is analyzed. A 0 at gate 1 generates a 1 at output Q. This 1 is fed back to gate 2. Gate 2 now has two 1s applied to its inputs, which forces the output to 0. Output $\overline{Q}$ is therefore 0, or LOW. Line 3 in Fig. 9-2*b* is the *reset* condition. The LOW, or 0, activates the reset input. This clears (or resets) the normal Q output to 0. The fourth line of the table shows the disabled, or *hold*, condition of the *RS*

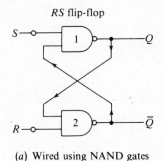

RS flip-flop

(a) Wired using NAND gates

Mode of operation	Inputs		Outputs	
	S	R	Q	Q̄
Prohibited	0	0	1	1
Set	0	1	1	0
Reset	1	0	0	1
Hold	1	1	no change	

(b) Truth table

Fig. 9-2 *RS* flip-flop

flip-flop. The outputs remain *as they were* before the hold condition existed. There is no change in the outputs from their previous states.

Note that, when the truth table in Fig. 9-2b refers to the *set* condition, it means setting output Q to 1. Likewise, the *reset* condition means resetting (clearing) output Q to 0. The operating conditions therefore refer to the normal output. Note that the complementary output ($\overline{Q}$) is exactly the opposite. Because of its typical function of holding data temporarily, the *RS* flip-flop is often called the *RS latch*. *RS* latches can be wired from gates or purchased in IC form. Think of the *RS* flip-flop as a memory device that will hold a single bit of data.

SOLVED PROBLEMS

9.1   Refer to Fig. 9-1. This *RS* flip-flop has active _____ (HIGH, LOW) inputs.

Solution:
As indicated by the small bubbles at the inputs of the logic symbol in Fig. 9-1, the *RS* flip-flop has active LOW inputs.

9.2   If the normal output of the *RS* flip-flop is HIGH, then output $Q =$ _____ (0, 1) and $\overline{Q} =$ _____ (0, 1).

Solution:
If the normal output of the *RS* flip-flop is HIGH, then output $Q = 1$ and $\overline{Q} = 0$.

9.3   Activating the reset input with a _____ (HIGH, LOW) effectively _____ (clears, sets) output Q to a logical _____ (0, 1).

Solution:
Activating the reset input with a LOW clears output Q to 0.

9.4   List the *binary* outputs at the normal output (Q) of the *RS* flip-flop shown in Fig. 9-3.

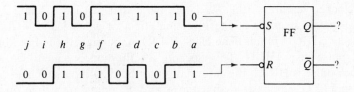

Fig. 9-3 *RS* flip-flop pulse-train problem

Solution:

The binary outputs at output Q shown in Fig. 9-3 are as follows:

pulse $a = 1$ pulse $d = 0$ pulse $g = 1$ pulse $j = 0$
pulse $b = 1$ pulse $e = 0$ pulse $h = 1$
pulse $c = 0$ pulse $f = 0$ pulse $i = 1$ (prohibited state)

9.5   List the *binary* outputs at output $\overline{Q}$ of the *RS* flip-flop shown in Fig. 9-3.

Solution:

The binary outputs at output $\overline{Q}$ (Fig. 9-3) are as follows:

pulse $a = 0$ pulse $d = 1$ pulse $g = 0$ pulse $j = 1$
pulse $b = 0$ pulse $e = 1$ pulse $h = 0$
pulse $c = 1$ pulse $f = 1$ pulse $i = 1$ (prohibited state)

9.6   List the mode of operation of the *RS* flip-flop for each input pulse shown in Fig. 9-3.

Solution:

The modes of operation of the *RS* flip-flop (Fig. 9-3) are as follows:

pulse a = set   pulse d = hold   pulse g = set   pulse j = reset
pulse b = hold   pulse e = reset   pulse h = hold
pulse c = reset   pulse f = hold   pulse i = prohibited

9-3 CLOCKED *RS* FLIP-FLOP

The basic *RS* latch is an *asynchronous* device. It does *not* operate in step with a clock or timing device. When an input (such as the set input) is activated, the normal output is immediately activated just as in combinational logic circuits. Gating circuits and *RS* latches operate asynchronously.

The clocked *RS flip-flop* adds a valuable *synchronous* feature to the *RS* latch. The clocked *RS* flip-flop *operates in step with the clock* or timing device. In other words, it operates synchronously. A logic symbol for the clocked *RS* flip-flop is shown in Fig. 9-4. It has the set (S) and reset (R) inputs and the added clock (CLK) input. The clocked *RS* flip-flop has the customary normal output (Q) and complementary output ($\overline{Q}$).

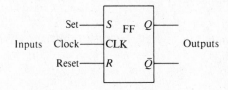

Fig. 9-4 Logic symbol for clocked *RS* flip-flop

The clocked *RS* flip-flop can be implemented with NAND gates. Figure 9-5a illustrates two NAND gates being added to the *RS* latch (flip-flop) to form the clocked *RS* flip-flop. NAND gates 3 and 4 add the clocked feature to the *RS* latch. Note that just gates 1 and 2 form the *RS* latch, or flip-flop. Note also that, because of the inverting effect of gates 3 and 4, the set (S) and reset (R) inputs are now active HIGH inputs. The clock (CLK) input triggers the flip-flop (enables the flip-flop) when the clock pulse goes HIGH. The clocked *RS* flip-flop is said to be a *level-triggered* device. Anytime the clock pulse is HIGH, the information at the data inputs (R and S) will be transferred to the outputs. It should be emphasized that the S and R inputs are active during the entire time the clock pulse level is HIGH. The HIGH of the clock pulse may be thought of as an enabling pulse.

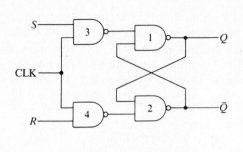

Mode of operation	Inputs			Outputs	
	CLK	S	R	Q	$\bar{Q}$
Hold	_⊓_	0	0	no change	
Reset	_⊓_	0	1	0	1
Set	_⊓_	1	0	1	0
Prohibited	_⊓_	1	1	1	1

_⊓_ = positive clock pulse

(a) Wired using NAND gates   (b) Truth table

Fig. 9-5 Clocked *RS* flip-flop

The truth table in Fig. 9-5*b* details the operation of the clocked *RS* flip-flop. The hold mode of operation is described in line 1 of the truth table. When a clock pulse arrives at the CLK input (with 0s at the *S* and *R* inputs), the outputs *do not change*. The outputs stay the same as they were before the clock pulse. This mode might also be described as the *disabled* condition of the flip-flop. Line 2 is the reset mode. The normal output (*Q*) will be cleared or reset to 0 when a HIGH activates the *R* input and a clock pulse arrives at the CLK input. It will be noted that just placing *R* = 1 and *S* = 0 does not immediately reset the flip-flop. The flip-flop waits until the clock pulse goes from LOW to HIGH, and then the flip-flop resets. This unit operates synchronously, or in step with the clock. Line 3 of the truth table describes the set condition of the flip-flop. A HIGH activates the *S* input (with *R* = 0 and a HIGH clock pulse), setting the *Q* output to 1. Line 4 of the truth table is a prohibited combination (all inputs 1) and is not used because it drives both outputs HIGH.

Waveforms, or *timing diagrams*, are widely used and are quite useful for working with flip-flops and sequential logic circuits. Figure 9-6 is a timing diagram for the clocked *RS* flip-flop. The top three lines represent the binary signals at the clock, set, and reset inputs. Only a single output (*Q*) is shown across the bottom. Beginning at the left, clock pulse 1 arrives but has no effect on *Q* because inputs *S* and *R* are in the hold mode. Output *Q* therefore stays at 0. At point *a* on the timing diagram, the set input is activated to a HIGH. After a time at point *b*, output *Q* is set to 1. Note that the flip-flop waited until clock pulse 2 started to go from LOW to HIGH before output *Q* was set. Pulse 3 senses the inputs (*R* and *S*) in the hold mode, and therefore the output does not change. At point *c* the reset input is activated with a HIGH. A short time later at point *d*, output *Q* is cleared, or reset to 0. Again this happens on the LOW-to-HIGH transition of the clock pulse. Point *e* senses the set input activated, which sets output *Q* to 1 at point *f* on the timing diagram. Input *S* is deactivated and *R* is activated before pulse 6, which causes output *Q* to go to LOW, or to the reset condition. Pulse 7 shows that output *Q* follows inputs *S* and *R* the entire time the clock is HIGH. At point *g* on the timing diagram in Fig. 9-6, the set input (*S*) goes HIGH and the *Q* output follows by going HIGH. Input *S* then goes LOW. Next the reset input (*R*) is activated by a HIGH at point *h*. That causes

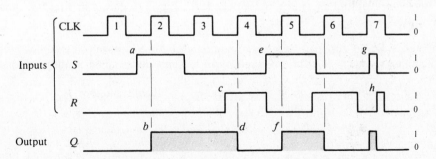

Fig. 9-6 Waveform diagram for clocked *RS* flip-flop

output Q to reset, or go LOW. Input R then returns to LOW, and finally clock pulse 7 ends with a HIGH-to-LOW transition. During clock pulse 7, the output was *set* to HIGH and then *reset* to LOW. Note that between pulses 5 and 6 it appears that both inputs S and R are at 1. The condition of inputs R and S both being HIGH would normally be considered the prohibited state for the flip-flop. In this case it is acceptable for both R and S to be HIGH because the clock pulse is LOW and the flip-flop is not activated.

SOLVED PROBLEMS

9.7   Refer to Fig. 9-4. The reset and set inputs on the clocked RS flip-flop are said to be active _____ (HIGH, LOW) inputs.

Solution:

The R and S inputs are active HIGH inputs on the clocked RS flip-flop shown in Fig. 9-4.

9.8   A flip-flop that operates in step with the clock is said to operate _____ (asynchronously, synchronously).

Solution:

A flip-flop that operates in step with the clock operates synchronously.

9.9   The RS latch operates _____ (asynchronously, synchronously).

Solution:

The RS latch operates asynchronously.

9.10   The clocked RS flip-flop operates _____ (asynchronously, synchronously).

Solution:

The clocked RS flip-flop operates synchronously.

9.11   Draw the logic symbol of a clocked RS flip-flop by using NAND gates.

Solution:

See Fig. 9-5a.

9.12   List the *binary* output at $\overline{Q}$ for the clocked RS flip-flop shown in Fig. 9-6 during the input clock pulses.

Solution:

The binary outputs at $\overline{Q}$ in this flip-flop are the opposite of those at the Q output. They are as follows:

pulse 1 = 1   pulse 3 = 0   pulse 5 = 0   pulse 7 = 1, then 0, and then 1
pulse 2 = 0   pulse 4 = 1   pulse 6 = 1

9.13   List the *binary* output at Q for the flip-flop of Fig. 9-7 during the eight clock pulses.

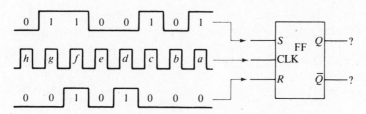

Fig. 9-7 Clocked *RS* flip-flop pulse-train problem

Solution:

The binary outputs at *Q* for the clocked *RS* flip-flop of Fig. 9-7 are as follows:

pulse *a* = 1   pulse *c* = 1   pulse *e* = 0   pulse *g* = 1
pulse *b* = 1   pulse *d* = 0   pulse *f* = 1 (prohibited condition)   pulse *h* = 1

9.14 List the mode of operation of the flip-flop of Fig. 9-7 during the eight clock pulses (use terms: hold, reset, set, prohibited).

Solution:

The operational modes for the clocked *RS* flip-flop of Fig. 9-7 are as follows:

pulse *a* = set   pulse *c* = set   pulse *e* = hold   pulse *g* = set
pulse *b* = hold   pulse *d* = reset pulse *f* = prohibited   pulse *h* = hold

9.15 Refer to Fig. 9-6. The clocked *RS* flip-flop is level-triggered, which means the unit is enabled during the entire _____ (HIGH, LOW) portion of the clock pulse.

Solution:

The flip-flop of Fig. 9-6 is level-triggered, which means it is enabled during the entire HIGH portion of the clock pulse.

9-4 *D* FLIP-FLOP

The logic symbol for a common type of flip-flop is shown in Fig. 9-8. The *D flip-flop* has only a single *data* input (*D*) and a clock input (CLK). The customary *Q* and $\overline{Q}$ outputs are shown on the right side of the symbol. The *D* flip-flop is often called a *delay flip-flop*. This name accurately describes the unit's operation. Whatever the input at the data (*D*) point, it is *delayed* from getting to the normal output (*Q*) *by one clock pulse*. Data is transferred to the output on the LOW-to-HIGH transition of the clock pulse.

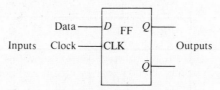

Fig. 9-8 Logic symbol for *D* flip-flop

The clocked *RS* flip-flop can be converted to a *D* flip-flop by adding an inverter. That conversion is shown in the diagram in Fig. 9-9*a*. Note that the *R* input to the clocked *RS* flip-flop has been inverted.

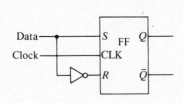

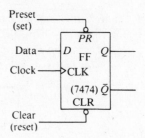

(a) D flip-flop wired from clocked RS flip-flop   (b) Logic symbol for 7474 D flip-flop with asynchronous inputs

Fig. 9-9

A commercial D flip-flop is shown in Fig. 9-9b. The D flip-flop in Fig. 9-9b is a TTL device described by the manufacturers as a 7474 IC. The logic symbol for the 7474 D flip-flop shows the regular D and CLK inputs. Those inputs are called the *synchronous* inputs, for they operate in step with the clock. The extra two inputs are the *asynchronous* inputs, and they operate just as in the RS flip-flop discussed previously. The asynchronous inputs are labeled preset (PR) and clear (CLR). The preset (PR) input can be activated by a LOW, as shown by the small bubble on the logic symbol. When the preset (PR) is activated, it *sets* the flip-flop. In other words, it places a 1 at the normal output (Q). It presets Q to 1. The clear (CLR) input can be activated by a LOW, as shown by the small bubble on the logic symbol. When the clear (CLR) input to the D flip-flop is activated, the Q output is reset, or cleared to 0. The *asynchronous inputs override the synchronous inputs* on this D flip-flop.

A truth table for the 7474 D flip-flop is in Fig. 9-10. The modes of operation are given on the left and the truth table on the right. The first three lines are for asynchronous operation (preset and clear inputs). Line 1 shows the preset (PR) input activated with a LOW. That sets the Q output to 1. Note the X's under the synchronous inputs (CLK and D). The X's mean that these inputs are irrelevant because the asynchronous inputs override them. Line 2 shows the clear (CLR) input activated with a LOW. This results in output Q being reset, or cleared to 0. Line 3 shows the prohibited asynchronous input (both PR and CLR at 0). The synchronous inputs (D and CLK) will operate when both asynchronous inputs are disabled ($PR = 1, \text{CLR} = 1$). Line 4 shows a 1 at the data (D) input and a rising clock pulse (shown with the upward arrow). The 1 at input D is transferred to output Q on the clock pulse. Line 5 shows a 0 at the data (D) input being transferred to output Q on the LOW-to-HIGH clock transition.

Mode of operation	Inputs				Outputs	
	Asynchronous		Synchronous			
	PR	CLR	CLK	D	Q	$\bar{Q}$
Asynchronous set	0	1	X	X	1	0
Asynchronous reset	1	0	X	X	0	1
Prohibited	0	0	X	X	1	1
Set	1	1	↑	1	1	0
Reset	1	1	↑	0	0	1

0 = LOW, 1 = HIGH, X = irrelevant, ↑ = LOW-to-HIGH transition of the clock pulse.

Fig. 9-10 Truth table for 7474 D flip-flop

Only the bottom two lines of the truth table in Fig. 9-10 are needed if the *D* flip-flop does not have the asynchronous inputs. *D* flip-flops are widely used in data storage. Because of this use, it is sometimes also called a *data flip-flop*.

Look at the *D* flip-flop symbols shown in Figs. 9-8 and 9-9*b*. Note that the clock (CLK) input in Fig. 9-9*b* has a small > inside the symbol, meaning that this is an *edge-triggered* device. This edge-triggered flip-flop transfers data from input *D* to output *Q* on the LOW-to-HIGH transition of the clock pulse. In edge triggering, it is the *change of the clock* from LOW to HIGH (or H to L) that transfers data. Once the clock pulse is HIGH on the edge-triggered flip-flop, a change in input *D* will have no effect on the outputs.

Figures 9-8 and 9-9*a* show a *D* flip-flop that is *level-triggered* (as opposed to edge-triggered). The absence of the small > inside the symbol at the clock input indicates a level-triggered device. On a level-triggered flip-flop, a certain voltage level will cause the data at input *D* to be transferred to output *Q*. The problem with the level-triggered device is that the output will follow the input if the input changes while the clock pulse is HIGH. Level triggering, or clocking, can be a problem if input data changes while the clock is HIGH.

SOLVED PROBLEMS

9.16   What two other names are given to the *D* flip-flop?

Solution:

The *D* flip-flop is also called the *delay* and the *data* flip-flop.

9.17   Draw a logic diagram of a clocked *RS* flip-flop and inverter wired as a *D* flip-flop.

Solution:

See Fig. 9-9*a*.

9.18   Draw the logic symbol for a *D* flip-flop. Label inputs as *D*, CLK, *PR*, and CLR. Label outputs as *Q* and $\overline{Q}$.

Solution:

See Fig. 9-9*b*.

9.19   The data bit at the *D* input of the 7474 *D* flip-flop is transferred to output _____ (Q, $\overline{Q}$) on the _____ (H-to-L, L-to-H) transition of the clock pulse.

Solution:

The data at the *D* input of a *D* flip-flop is transferred to output *Q* on the L-to-H transition of the clock pulse.

9.20   Refer to Fig. 9-10. An X in the truth table stands for an _____ (extra, irrelevant) input.

Solution:

An X in the truth table stands for an irrelevant input. An X input can be either 0 or 1 and has no effect on the output.

9.21   List the binary outputs at the complementary output ($\overline{Q}$) of the *D* flip-flop of Fig. 9-11 after each of the clock pulses.

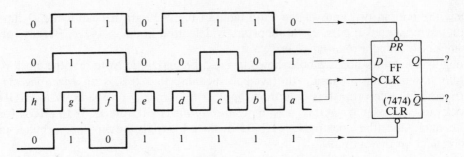

Fig. 9-11 *D* flip-flop pulse-train problem

Solution:

Refer to the truth table in Fig. 9-10. The binary outputs at $\overline{Q}$ of the *D* flip-flop (Fig. 9-11) are as follows:

pulse $a = 0$ pulse $c = 0$ pulse $e = 0$ pulse $g = 0$
pulse $b = 1$ pulse $d = 1$ pulse $f = 1$ pulse $h = 1$ (prohibited state)

9.22   Refer to Fig. 9-11. Which input has control of the flip-flop during pulse *a*?

Solution:

The preset (*PR*) input is activated during pulse *a* and overrides all other inputs. It sets the *Q* output to 1.

9.23   Refer to Fig. 9-11. Just before pulse *b*, output *Q* is _____ (HIGH, LOW); during pulse *b*, output *Q* is _____ (HIGH, LOW); on the H-to-L clock-pulse transition, output *Q* is _____ (HIGH, LOW).

Solution:

Just before pulse *b*, output *Q* is HIGH; during pulse *b*, output *Q* is LOW; on the H-to-L clock-pulse transition, output *Q* is LOW.

9-5   *JK* FLIP-FLOP

The logic symbol for a *JK* flip-flop is shown in Fig. 9-12. This device might be considered the universal flip-flop; other types can be made from it. The logic symbol shown in Fig. 9-12 has three synchronous inputs (*J*, *K*, and CLK). The *J* and *K* inputs are data inputs, and the clock input transfers data from the inputs to the outputs. The logic symbol shown in Fig. 9-12 also has the customary normal output (*Q*) and complementary output ($\overline{Q}$).

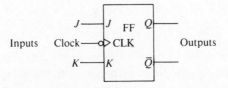

Fig. 9-12 Logic symbol for *JK* flip-flop

A truth table for the *JK* flip-flop is in Fig. 9-13. The modes of operation are given on the left and the truth table is on the right. Line 1 of the truth table shows the hold, or disabled, condition. Note that both data inputs (*J* and *K*) are LOW. The reset, or clear, condition of the flip-flop is shown in

Mode of operation	Inputs			Outputs	
	CLK	J	K	Q	$\bar{Q}$
Hold	⎍	0	0	no change	
Reset	⎍	0	1	0	1
Set	⎍	1	0	1	0
Toggle	⎍	1	1	opposite state	

Fig. 9-13 Truth table for pulse-triggered JK flip-flop

line 2 of the truth table. When $J = 0$ and $K = 1$ and a clock pulse arrives at the CLK input, the flip-flop is reset ($Q = 0$). Line 3 shows the set condition of the JK flip-flop. When $J = 1$, $K = 0$, and a clock pulse is present, output Q is set to 1. Line 4 illustrates a very useful condition of the JK flip-flop that is called the *toggle* position. When both inputs J and K are HIGH, the output will go to the opposite state when a pulse arrives at the CLK input. With repeated clock pulses, the Q output might go LOW, HIGH, LOW, HIGH, LOW, and so forth. This LOW-HIGH-LOW-HIGH idea is called *toggling*. The term "toggling" comes from the ON-OFF nature of a toggle switch.

Note in the truth table in Fig. 9-13 that an entire clock pulse is shown under the clock (CLK) input heading. Many JK flip-flops are *pulse-triggered*. It *takes the entire pulse to transfer data* from the input to the outputs of the flip-flop. With the clock input in the truth table, it is evident that the JK flip-flop is a synchronous flip-flop.

The JK is considered the universal flip-flop. Figure 9-14a shows how a JK flip-flop and an inverter would be wired to form a D flip-flop. Note the single D input at the far left and the clock input. This wired D flip-flop would trigger on the HIGH-to-LOW transition of the clock pulse, as shown by the bubble at the CLK input.

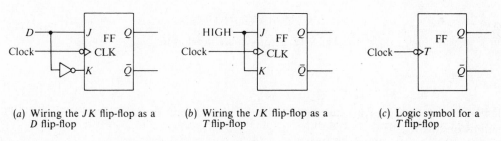

(*a*) Wiring the JK flip-flop as a D flip-flop

(*b*) Wiring the JK flip-flop as a T flip-flop

(*c*) Logic symbol for a T flip-flop

Fig. 9-14

A useful *toggle flip-flop* (T-type flip-flop) is shown wired in Fig. 9-14b. A JK flip-flop is shown being used in its toggle mode. Note that the J and K inputs are simply tied to a HIGH, and the clock is fed into the CLK input. As the repeated clock pulses feed into the CLK input, the outputs will simply toggle.

The toggle operation is widely used in sequential logic circuits. Because of its wide use, a special symbol is sometimes used for the toggle (T-type) flip-flop. Figure 9-14c shows the logic symbol for the toggle flip-flop. The single input (labeled T) is the clock input. The customary Q and $\bar{Q}$ outputs are shown on the right of the symbol. The T flip-flop has only the toggle mode of operation.

One commercial JK flip-flop is detailed in Fig. 9-15. This is described by the manufacturer as a *7476 TTL dual JK flip-flop*. A pin diagram of the 7476 IC is reproduced in Fig. 9-15a. Note that the IC contains two separate JK flip-flops. Each flip-flop has asynchronous preset (*PR*) and clear (*CLR*) inputs. The synchronous inputs are shown as J, K, and CLK (clock). The customary normal (Q) and

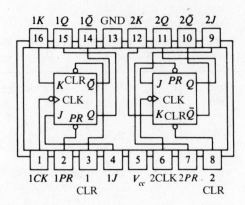

(a) Pin diagram (*Reprinted by permission of Texas Instruments, Inc.*)

Mode of operation	Inputs					Outputs	
	PR	CLR	CLK	J	K	Q	$\bar{Q}$
Asynchronous set	0	1	X	X	X	1	0
Asynchronous clear	1	0	X	X	X	0	1
Prohibited	0	0	X	X	X	1	1
Hold	1	1	⊓	0	0	no change	
Reset	1	1	⊓	0	1	0	1
Set	1	1	⊓	1	0	1	0
Toggle	1	1	⊓	1	1	opposite state	

X = irrelevant   ⊓ = positive clock pulse

(b) Mode-select truth table

Fig. 9-15 The 7476 *JK* flip-flop IC

complementary ($\bar{Q}$) outputs are available. Pins 5 and 13 are the +5-V (V_{cc}) and GND power connections on this IC.

A truth table for the 7476 *JK* flip-flop is shown in Fig. 9-15b. The top three lines detail the operation of the asynchronous inputs preset (*PR*) and clear (CLR). Line 3 of the truth table shows the prohibited state of the asynchronous inputs. Lines 4 through 7 detail the conditions of the synchronous inputs for the hold, reset, set, and toggle modes of the 7476 *JK* flip-flop. The manufacturer describes the 7476 as a *master-slave JK flip-flop* using positive-pulse triggering. The data at the outputs changes on the H-to-L transition of the clock pulse, as symbolized by the small bubble and > symbol at the CLK input on the flip-flop logic diagram in Fig. 9-15a.

Most commercial *JK* flip-flops have asynchronous input features (such as *PR* and CLR). Most *JK* flip-flops are pulse-triggered devices like the 7476 IC, but they can also be purchased as edge-triggered units.

Flip-flops are the fundamental building blocks of sequential logic circuits. Therefore, IC manufacturers produce a variety of flip-flops using both the TTL and CMOS technologies. Typical TTL flip-flops are the 7476 *JK* flip-flop with preset and clear, 7474 dual positive-edge-triggered *D* flip-flop with preset and clear, and the 7475 4-bit bistable latch. Typical CMOS flip-flops include the 4724 8-bit addressable latch, 40175 quad *D* flip-flop, and the 74C76 *JK* flip-flop with preset and clear.

SOLVED PROBLEMS

9.24 Draw the logic symbol for a *JK* flip-flop with pulse triggering. Label inputs as *J*, *K*, and CLK. Label outputs as *Q* and $\overline{Q}$.

Solution:

See Fig. 9-12.

9.25 List the four synchronous modes of operation of the *JK* flip-flop.

Solution:

The synchronous modes of operation for the *JK* flip-flop are hold, reset, set, and toggle.

9.26 When the output of a flip-flop goes LOW, HIGH, LOW, HIGH on repeated clock pulses, it is in what mode of operation?

Solution:

If a flip-flop's output alternates states (LOW, HIGH, LOW) on repeated clock pulses, it is in the toggle mode.

9.27 List the *binary* output at output *Q* of the *JK* flip-flop of Fig. 9-16 after each of the eight clock pulses.

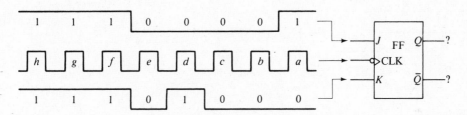

Fig. 9-16 *JK* flip-flop pulse-train problem

Solution:

Refer to the truth table in Fig. 9-13. Based on the truth table, the binary output (at *Q*) (Fig. 9-16) after each clock pulse is as follows:

pulse *a* = 1   pulse *c* = 1   pulse *e* = 0   pulse *g* = 0
pulse *b* = 1   pulse *d* = 0   pulse *f* = 1   pulse *h* = 1

9.28 List the mode of operation of the *JK* flip-flop during each of the eight clock pulses shown in Fig. 9-16.

Solution:

Refer to the mode-select truth table in Fig. 9-13. Based on the table, the mode of the *JK* flip-flop during each clock pulse shown in Fig. 9-16 is as follows:

pulse *a* = set   pulse *c* = hold   pulse *e* = hold   pulse *g* = toggle
pulse *b* = hold   pulse *d* = reset   pulse *f* = toggle   pulse *h* = toggle

9.29 List the asynchronous inputs to the 7476 *JK* flip-flop.

Solution:

The asynchronous inputs to the 7476 *JK* flip-flop are preset (*PR*) and clear (CLR).

9.30 The asynchronous inputs to the 7476 *JK* flip-flop have active _____ (HIGH, LOW) inputs.

Solution:

   The asynchronous inputs to the 7476 *JK* flip-flop have active LOW inputs.

9.31 Both asynchronous inputs to the 7476 IC must be _____ (HIGH, LOW); the *J* and *K* inputs must be _____ (HIGH, LOW); and a clock pulse must be present for the flip-flop to toggle.

Solution:

   Both asynchronous inputs to the 7476 IC must be HIGH; the *J* and *K* inputs must be HIGH, and a clock pulse must be present for the flip-flop to toggle.

9.32 List the mode of operation of the 7476 *JK* flip-flop during each of the seven clock pulses shown in Fig. 9-17.

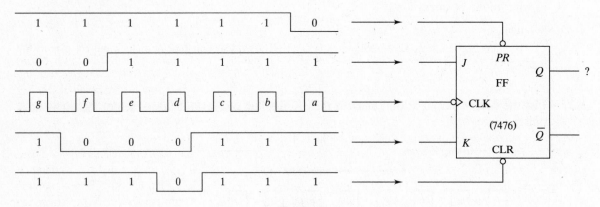

Fig. 9-17

Solution:

   Refer to the mode-select truth table in Fig. 9-15*b*. Based on the table, the mode of the *JK* flip-flop during each clock pulse shown in Fig. 9-17 is as follows:
pulse *a* = asynchronous set
pulse *b* = toggle
pulse *c* = toggle
pulse *d* = asynchronous clear (reset)
pulse *e* = set
pulse *f* = hold
pulse *g* = reset

9.33 Refer to Fig. 9-17. List the *binary* output at *Q* of the *JK* flip-flop after each of the seven clock pulses.

Solution:

   Refer to the truth table in Fig. 9-15*b*. Based on the table, the binary output (at *Q*) after each clock pulse is as follows:
pulse *a* = 1
pulse *b* = 0
pulse *c* = 1
pulse *d* = 0
pulse *e* = 1
pulse *f* = 1
pulse *g* = 0

9-6   TRIGGERING OF FLIP-FLOPS

Most complicated digital equipment operates as a synchronous sequential system. This suggests that a master clock signal is sent to all parts of the system to coordinate system operation. A typical clock pulse train is shown in Fig. 9-18. Remember that the horizontal distance on the waveform is *time* and the vertical distance is *voltage*. The clock pulses shown in the figure are for a TTL device because of the +5-V and GND voltages. Other digital circuits use clocks, but the voltages might be different.

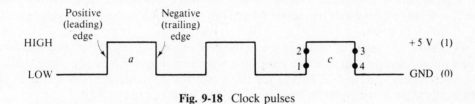

Fig. 9-18   Clock pulses

Start at the left of the waveform in Fig. 9-18. The voltage is at first LOW, or at GND. That is also called a logical 0. Pulse *a* shows the *leading edge* (also called the *positive edge*) of the waveform going from GND voltage to +5 V. This edge of the waveform may also be called the LOW-to-HIGH (L-to-H) edge of the waveform. On the right side of pulse *a*, the waveform drops from +5 V to GND voltage. This edge is called the HIGH-to-LOW (H-to-L) edge of the clock pulse. It is also called the *negative-going* edge or the *trailing* edge of the clock pulse.

Some flip-flops transfer data from input to output on the positive (leading) edge of the clock pulse. These flip-flops are referred to as *positive-edge-triggered flip-flops*. An example of such a flip-flop is shown in Fig. 9-19. The clock input is shown by the middle waveform. The top waveform shows the Q output when the positive-edge-triggered flip-flop is in its toggle mode. Note that each leading edge (positive-going edge) of the clock toggles the flip-flop.

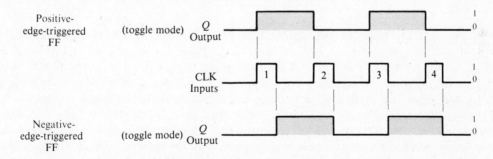

Fig. 9-19   Triggering of positive- and negative-edge flip-flops

Other flip-flops are classed as *negative-edge-triggered flip-flops*. The operation of a negative-edge-triggered flip-flop is shown by the bottom two waveforms in Fig. 9-19. The middle waveform is the clock input. The bottom waveform is the Q output when the flip-flop is in its toggle mode. Note that this flip-flop toggles to its opposite state only on the trailing edge (negative-going edge) of the clock pulse. It is important to note the difference in timing of the positive- and negative-edge-triggered flip-flops shown in Fig. 9-19. The timing difference is of great importance in some applications.

Many *JK* flip-flops are *pulse-triggered* units. These pulse-triggered devices are *master-slave JK flip-flops*. A master-slave *JK* flip-flop is actually several gates and flip-flops wired together to use the entire clock pulse to transfer data from input to output. In Fig. 9-18, pulse *c* will be used to help explain how pulse triggering works with a master-slave *JK* flip-flop. The following events happen, during the pulse-triggering sequence, at the numbered points in Fig. 9-18:

1. The input and output of the flip-flop are isolated.
2. Data is entered from the J and K inputs, but it is not transferred to the output.
3. The J and K inputs are disabled.
4. Previously entered data from J and K is transferred to the output.

Note that the data actually appears at the outputs at point 4 (trailing edge) of the waveform in Fig. 9-18. The logic symbol for a pulse-triggered flip-flop has a small bubble attached to the clock (CLK) input (see Fig. 9-15a) to show that the actual transfer of data to the output takes place on the H-to-L transition of the clock pulse.

The waveforms in Fig. 9-20 will aid understanding of the operation of the master-slave JK flip-flop and pulse triggering. Start at the left of the waveform diagram. The top three waveforms are the synchronous inputs J, K, and CLK. The top line describes the mode of operation during the clock pulse. The bottom line is the resultant output of the JK flip-flop at output Q.

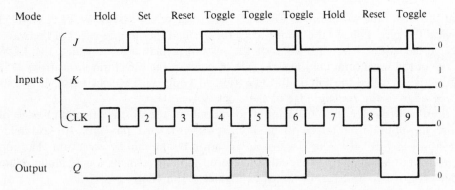

Fig. 9-20 Waveform diagram for a master-slave JK flip-flop

Look at clock (CLK) pulse 1 shown in Fig. 9-20. Both J and K inputs are LOW. This is the hold condition, and so the Q output stays at 0, as it was before pulse 1. Look at clock (CLK) pulse 2. Inputs J and K are in the set mode ($J = 1, K = 0$). On the trailing edge of pulse 2, output Q goes to a logical 1, or HIGH. Pulse 3 sees the inputs in the reset mode ($J = 0, K = 1$). On the trailing edge of clock pulse 3, output Q is reset, or cleared to 0. Pulse 4 sees the inputs in the toggle mode ($J = 1, K = 1$). On the trailing edge of clock pulse 4, output Q toggles to a logical 1, or HIGH. Pulse 5 sees the inputs in the toggle mode again. On the trailing edge of pulse 5, output Q toggles to a logical 0 or LOW.

Pulse 6 (Fig. 9-20) will show an unusual characteristic of the master-slave JK flip-flop. Note that, on the leading edge of clock pulse 6, input $K = 1$ and $J = 0$. When clock pulse 6 is HIGH, input K goes from 1 to 0 while J goes from 0 to 1 to 0. On the trailing edge of clock pulse 6, both inputs (J and K) are LOW. However, as strange as it may seem, the flip-flop still *toggles* to a HIGH. The master-slave JK flip-flop *remembers any or all HIGH inputs while the clock pulse is HIGH*. During pulse 6, both input J and K were HIGH for a short time when the clock input was HIGH. The flip-flop therefore regarded this as the toggle condition.

Next refer to clock pulse 7 (Fig. 9-20). Pulse 7 sees inputs J and K in the hold mode ($J = 0, K = 0$). Output Q remains in its present state (stays at 1). Pulse 8 sees input K at 1 for a short time and input J at 0. The flip-flop interprets this as the reset mode. Output Q therefore resets output Q to 0 on the trailing edge of clock pulse 8.

Refer to clock pulse 9 (Fig. 9-20). The master-slave JK flip-flop sees both J and K inputs at a LOW on the positive edge of clock pulse 9. When the pulse is HIGH, input K goes HIGH for a short time and then input J goes HIGH for a short time. Inputs J and K are *not* HIGH at the same time, however. On the trailing edge of clock pulse 9, both inputs (J and K) are LOW. The flip-flop interprets this as the toggle mode. Output Q changes states and goes from a 0 to a 1.

It should be noted that not all JK flip-flops are of the master-slave type. Some JK flip-flops are edge-triggered. Manufacturers' data manuals will specify if the flip-flop is edge-triggered or pulse-triggered.

SOLVED PROBLEMS

9.34 Flip-flops are classified as either edge-triggered or _____-triggered units.

   Solution:

   Flip-flops are classified as either edge-triggered or pulse-triggered units.

9.35 A positive-edge-triggered flip-flop transfers data from input to output on the _____ (leading, trailing) edge of the clock pulse.

   Solution:

   A positive-edge-triggered flip-flop transfers data from input to output on the leading edge of the clock pulse.

9.36 A negative-edge-triggered flip-flop transfers data from input to output on the _____ (H-to-L, L-to-H) transition of the clock pulse.

   Solution:

   A negative-edge-triggered flip-flop transfers data from input to output on the H-to-L transition of the clock pulse.

9.37 The master-slave JK flip-flop is an example of a _____ (positive-edge-, pulse-) triggered unit.

   Solution:

   The master-slave JK flip-flop is an example of a pulse-triggered unit.

9.38 Refer to Fig. 9-20. List the *binary* output (at $\overline{Q}$) *after* each of the nine clock pulses.

   Solution:

   The $\overline{Q}$ output is always the complement of the Q output on a flip-flop. Therefore the binary outputs (at $\overline{Q}$) in Fig. 9-20 after each clock pulse are as follows:
   pulse 1 = 1   pulse 3 = 1   pulse 5 = 1   pulse 7 = 0   pulse 9 = 0
   pulse 2 = 0   pulse 4 = 0   pulse 6 = 0   pulse 8 = 1

9.39 List the *binary* output (at Q) of the master-slave JK flip-flop of Fig. 9-21 *after* each of the eight clock pulses.

   Solution:

   Refer to the truth table in Fig. 9-13. According to this table, the binary output (at Q) of the master-slave JK flip-flop (Fig. 9-21) after each clock pulse is as follows:
   pulse a = 1   pulse c = 1   pulse e = 0   pulse g = 0
   pulse b = 0   pulse d = 0   pulse f = 1   pulse h = 1

9.40   List the mode of operation for the master-slave *JK* flip-flop of Fig. 9-21 for each clock pulse.

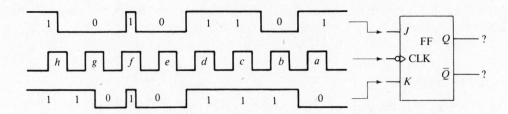

Fig. 9-21 *JK* flip-flop pulse-train problem

Solution:

Refer to the mode-select truth table in Fig. 9-13. According to the table, the modes of operation for the master-slave *JK* flip-flop (Fig. 9-21) for each clock pulse are as follows:

pulse *a* = set   pulse *c* = toggle   pulse *e* = hold   pulse *g* = reset
pulse *b* = reset   pulse *d* = toggle   pulse *f* = toggle   pulse *h* = toggle

9.41   Refer to Fig. 9-21. Assume the *JK* flip-flop is a negative-edge-triggered unit. List the *binary* output (at *Q*) of the edge-triggered flip-flop after each of the eight clock pulses.

Solution:

Refer to the truth table in Fig. 9-13, but remember that this is a negative-edge-triggered *JK* flip-flop (it triggers on the H-to-L transition of the clock pulse). The binary output (at *Q*) for the negative-edge-triggered *JK* flip-flop after each clock pulse is as follows:

pulse *a* = 1   pulse *c* = 1   pulse *e* = 0   pulse *g* = 0
pulse *b* = 0   pulse *d* = 0   pulse *f* = 0   pulse *h* = 1

9-7   ASTABLE MULTIVIBRATORS—CLOCKS

Introduction

A multivibrator (MV) is a pulse generator circuit which produces a rectangular-wave output. Multivibrators are classified as *astable*, *bistable*, or *monostable*.

An astable MV is also called a *free-running multivibrator*. The astable MV generates a continuous flow of pulses as depicted in Fig. 9-22*a*.

A bistable multivibrator is also called a *flip-flop*. The bistable MV is always in one of two stable states (set or reset). The basic idea of a bistable MV is diagrammed in Fig. 9-22*b*, where the input pulse triggers a change in output from LOW to HIGH.

A monostable MV is also called a *one-shot multivibrator*. When the one-shot is triggered, as shown in Fig. 9-22*c*, the MV generates a single short pulse.

Astable Multivibrator

The versatile 555 Timer IC can be used to implement an astable, bistable, or monostable multivibrator. The 555 timer is shown wired as a free-running (astable) multivibrator in Fig. 9-23*a*. If

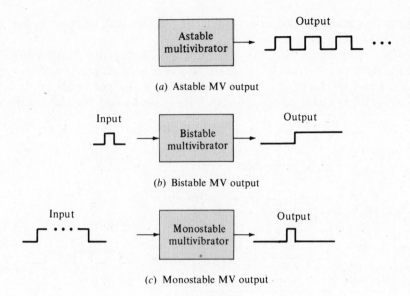

(a) Astable MV output

(b) Bistable MV output

(c) Monostable MV output

Fig. 9-22

both resistors (R_A and R_B) = 4.7 kΩ (kilohms) and $C = 100 \ \mu F$, the output will be a string of TTL level pulses at a frequency of 1 Hz.

The output frequency of the MV shown in Fig. 9-23a can be increased by *decreasing* the value of the resistors and/or capacitor. For example, if the resistors (R_A and R_B) = 330 Ω and $C = 0.1 \ \mu F$, then the output frequency will rise to about 10 kHz.

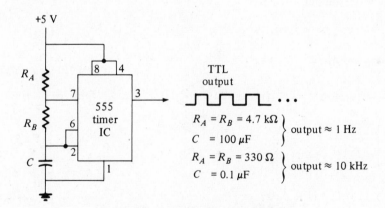

(a) 555 timer IC wired as an astable MV

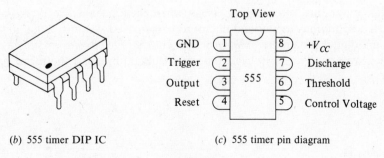

(b) 555 timer DIP IC

(c) 555 timer pin diagram

Fig. 9-23

The 555 timer is commonly sold in an 8-pin DIP IC like that pictured in Fig. 9-23*b*. The pin functions for the 555 timer IC are shown in Fig. 9-23*c*.

Another astable multivibrator circuit is shown in Fig. 9-24. This free-running MV uses two CMOS inverters from the 4069 hex inverter IC. Note the use of a 10-V dc power source, which is common (but not standard) in CMOS circuits. The frequency of the output is about 10 kHz. The output frequency can be varied by changing the value(s) of the resistors and capacitor in the circuit.

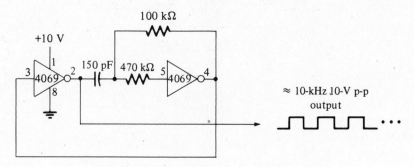

Fig. 9-24 An astable MV using CMOS inverters

Another astable multivibrator circuit using CMOS inverters is diagrammed in Fig. 9-25. This free-running MV contains a crystal-controlled oscillator (4049*a* and 4049*b*) with inverters (4049*c* and 4049*d*) used to square up the waveform. The output frequency is controlled by the natural frequency of the crystal, which is 100 kHz in this circuit. The frequency is very stable. The square-wave output is at CMOS voltage levels (about 10 V p-p).

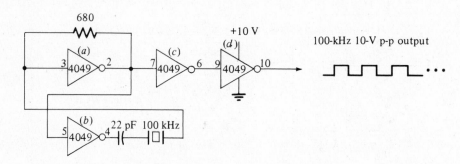

Fig. 9-25 Crystal-controlled astable MV

Astable multivibrators are often called *clocks* when they are used in digital systems. A system clock is used in all synchronous digital and microprocessor-based systems. Some important characteristics of a clock in a digital system are *frequency*, *clock cycle time*, *frequency stability*, *voltage stability*, and *shape* of the waveform. The clock cycle time is calculated by using the formula

$$T = \frac{1}{f}$$

where T = time, s
f = frequency, Hz

Clocks require square-wave pulses with fast rise times and fast fall times.

SOLVED PROBLEMS

9.42 List three classes of multivibrators.

   Solution:

   Multivibrators are classified as astable, bistable, or monostable.

9.43 Another name for an astable multivibrator is _____.

   Solution:

   An astable MV is also called a free-running multivibrator.

9.44 Another name for a bistable multivibrator is _____.

   Solution:

   A bistable MV is also called a flip-flop.

9.45 Another name for a monostable multivibrator is _____.

   Solution:

   A monostable MV is also called a one-shot multivibrator.

9.46 Increasing the value of both resistors and the capacitor in the MV circuit shown in Fig. 9-23a
   will _____ (decrease, increase) the output frequency.

   Solution:

   Increasing the value of the resistors and capacitor in the free-running MV of Fig. 9-23a will decrease
   the output frequency.

9.47 Pin number _____ is next to the dot on the 8-pin DIP IC shown in Fig. 9-23b.

   Solution:

   Pin 1 is located next to the dot on top of the 8-pin DIP IC of Fig. 9-23b.

9.48 The clock pulses from the MV shown in Fig. 9-23a _____ (are, are not) compatible with TTL.

   Solution:

   The clock pulses from the 555 timer shown in Fig. 9-23a are at TTL voltage levels. (LOW = 0 V and
   HIGH = about +4.5 V.)

9.49 The 4069 inverters used in the MV shown in Fig. 9-24 are _____ (CMOS, TTL) ICs.

   Solution:

   The 4069 is a CMOS hex inverter IC.

9.50 The astable MV shown in Fig. _____ (9-24, 9-25) would have the greatest frequency stability.

   Solution:

   The free-running MV shown in Fig. 9-25 would have great frequency stability. The oscillator's
   frequency is crystal-controlled.

9.51   The output from the crystal-controlled free-running MV shown in Fig. 9-25 is compatible with
a _____ (CMOS, TTL) circuit.

Solution:

The 10-V output from the MV shown in Fig. 9-25 means it is compatible with a CMOS circuit. TTL
voltage levels must range from 0 to +5.5 V only.

9.52   The clock cycle time for the MV shown in Fig. 9-25 is _____ s.

Solution:

The formula is $T = 1/f$, so $T = 1/100,000 = 0.00001$. The clock cycle time for the MV shown in Fig.
9-25 is 0.00001 s, or 10 μs.

9-8   MONOSTABLE MULTIVIBRATORS

The *one-shot* or *monostable multivibrator* generates an output pulse of fixed duration each time
its input is triggered. The basic idea of the monostable MV is shown graphically in Fig. 9-22c. The
input trigger may be an entire pulse, an L-to-H transition of the clock, or an H-to-L transition of the
trigger pulse depending on the one-shot. The output pulse may be either a positive or a negative
pulse. The designer can adjust the time duration of the output pulse by using different resistor-capaci-
tor combinations.

The adaptable 555 timer IC is shown wired as a one-shot MV in Fig. 9-26. A short negative input
pulse causes the longer positive output pulse. The time duration t of the output pulse is calculated by
using the formula

$$t = 1.1 R_A C$$

where R_A is equal to the value of the resistor in ohms, C is equal to the value of the capacitor in
farads, and t is equal to the time duration of the output pulse in seconds. By calculating the output
pulse time duration t for the one-shot shown in Fig. 9-26, we have

$$t = 1.1 \times 10,000 \times 0.0001 = 1.1 \text{ s}$$

The calculated time duration t of the output pulse for the one-shot MV shown in Fig. 9-26 is 1.1 s.

The one-shot MV shown in Fig. 9-26 is *nonretriggerable*. This means that when the one-shot's
output is HIGH, it will disregard any input pulse. Retriggerable monostable MVs also are available.

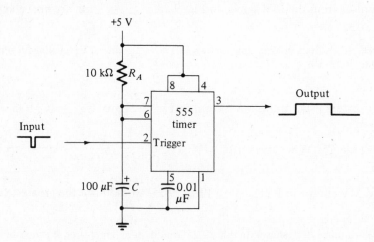

Fig. 9-26 555 timer IC wired as a monostable MV

In Fig. 9-27 the TTL 74121 one-shot IC is shown being used to generate *single* TTL level pulses when a mechanical switch is pressed. Many digital trainers used in technical education and design work use circuits of this type to generate single clock pulses. Both positive and negative clock pulses are available from the normal (Q) and complementary ($\overline{Q}$) outputs of the 74121 one-shot IC.

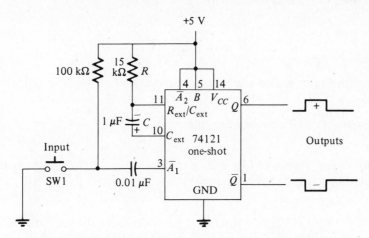

Fig. 9-27 74121 IC wired to generate single clock pulses

The duration of the output pulse can be adjusted by varying the values of the resistor R and capacitor C. To calculate the time duration of the output pulse, use the formula

$$t = 0.7RC$$

where R equals the value of the resistor in ohms, C equals the value of the capacitor in farads, and t is the duration of the output pulse in seconds. By calculating the duration of the output pulse in the one-shot circuit in Fig. 9-27, we have

$$t = 0.7 \times 15,000 \times 0.000001 = 0.0105 \text{ s}$$

The calculated time duration t for the output pulse from the one-shot shown in Fig. 9-27 is 0.0105 s, or about 10 ms.

A pin diagram and truth table for the 74121 one-shot IC are reproduced in Fig. 9-28. Note that the 74121 one-shot has three separate trigger inputs ($\overline{A}_1$, $\overline{A}_2$, and B). Typically, only a single input

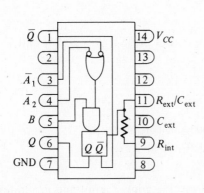

	Inputs		Outputs	
$\overline{A}_1$	$\overline{A}_2$	B	Q	$\overline{Q}$
L	X	H	L	H
X	L	H	L	H
X	X	L	L	H
H	H	X	L	H
H	↓	H	⊓	⊔
↓	H	H	⊓	⊔
↓	↓	H	⊓	⊔
L	X	↑	⊓	⊔
X	L	↑	⊓	⊔

H = HIGH voltage level
L = LOW voltage level
X = don't care
↑ = LOW-to-HIGH transition
↓ = HIGH-to-LOW transition

(*a*) Pin diagram   (*b*) Truth table (*Courtesy of Signetics Corporation*)

Fig. 9-28

would be used at a time. In the application shown in Fig. 9-27, input $\overline{A}_1$ (pin 3) serves as the trigger input. This matches the situation in line 6 of the truth table (Fig. 9-28b). Inputs $\overline{A}_2$ and B are HIGH, and trigger input $\overline{A}_1$ reacts to a HIGH-to-LOW transition of the trigger pulse.

Monostable multivibrators are useful for timing applications when precision is not critical. One-shots are also used to introduce delays in digital systems.

SOLVED PROBLEMS

9.53   A _____ multivibrator is also referred to as a one-shot.

Solution:

A monostable MV is also called a one-shot.

9.54   The one-shot circuit shown in Fig. 9-26 generates a _____ (negative, positive) output pulse when triggered by a negative input pulse.

Solution:

The one-shot shown in Fig. 9-26 generates a positive output pulse when triggered by a negative input pulse.

9.55   What is the calculated time duration t of the output pulse from the one-shot shown in Fig. 9-26 if $R_A = 9.1$ kΩ and $C = 10$ μF?

Solution:

The formula is $t = 1.1R_AC$; then

$$t = 1.1 \times 9100 \times 0.00001$$

The calculated time duration of the output pulse from the one-shot would be 0.1 s.

9.56   An _____ (H-to-L, L-to-H) transition of the trigger pulse shown in Fig. 9-27 causes the one-shot to generate an output pulse.

Solution:

In Fig. 9-27, it is the HIGH-to-LOW transition of the trigger pulse that causes the one-shot to generate an output pulse.

9.57   In Fig. 9-27, what will be the time duration of the output pulse if $R = 30$ kΩ and $C = 100$ μF?

Solution:

The formula is

$$t = 0.7RC$$

so

$$t = 0.7 \times 30,000 \times 0.0001$$

The time duration of the output pulse from the 74121 IC will be 2.1 s.

9.58   Pressing the switch SW1 shown in Fig. 9-27 activates the trigger input and generates a _____ (negative, positive) pulse from the complementary ($\overline{Q}$) output.

Solution:

Pressing SW1 in Fig. 9-27 triggers the 74121 one-shot and generates a negative pulse from the complementary ($\overline{Q}$) output. Also see truth table in Fig. 9-28b.

Supplementary Problems

9.59   If it is said that "the flip-flop is set," then output Q is _____ (HIGH, LOW).   *Ans.*   HIGH

9.60   A(n) _____ (clocked RS, RS) flip-flop is an example of a synchronously operated device.
Ans.   clocked RS

9.61   A(n) _____ (D, RS) flip-flop does *not* have a clock input.   *Ans.* RS

9.62   Combinational logic circuits and the RS latch operate _____ (asynchronously, synchronously).
Ans.   asynchronously

9.63   The normal output of a flip-flop is the _____ (Q, $\overline{Q}$) output.   *Ans.* Q

9.64   List the *binary* output (at Q) of the RS latch shown in Fig. 9-29 for each of the eight pulses.
Ans.   pulse $a = 0$ pulse $c = 1$ pulse $e = 1$ pulse $g = 1$
         pulse $b = 0$ pulse $d = 0$ pulse $f = 1$ (prohibited)   pulse $h = 1$

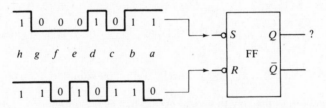

Fig. 9-29 RS flip-flop pulse-train problem

9.65   List the mode of operation of the RS flip-flop shown in Fig. 9-29 for each of the eight pulses.
Ans.   pulse a = reset   pulse c = set   pulse e = set   pulse g = set
         pulse b = hold   pulse d = reset   pulse f = prohibited condition   pulse h = hold

9.66   List the binary output at $\overline{Q}$ for the clocked RS flip-flop shown in Fig. 9-7 for each of the eight clock pulses.
Ans.   pulse $a = 0$ pulse $c = 0$ pulse $e = 1$ pulse $g = 0$
         pulse $b = 0$ pulse $d = 1$ pulse $f = 1$ pulse $h = 0$

9.67   Refer to Fig. 9-30. The clocked RS flip-flop is triggered by the _____ (leading, trailing) edge of the clock pulse.   *Ans.*   leading

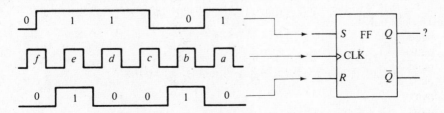

Fig. 9-30 Clocked RS flip-flop pulse-train problem

9.68 List the *binary* output (at Q) of the clocked *RS* flip-flop shown in Fig. 9-30 for each of the six clock pulses.
Ans.   pulse $a = 1$ pulse $c = 0$ pulse $e = 1$ (prohibited condition)
         pulse $b = 0$ pulse $d = 1$ pulse $f = 1$

9.69 List the mode of operation for the clocked *RS* flip-flop shown in Fig. 9-30 as each pulse triggers the unit.
Ans.   pulse $a =$ set   pulse $d =$ set
         pulse $b =$ reset   pulse $e =$ prohibited condition
         pulse $c =$ hold (S and R are both 0   pulse $f =$ set ($S = 1$, $R = 0$ on the
                on leading edge)   leading edge)

9.70 List the *binary* output (at Q) for the D flip-flop shown in Fig. 9-11 after each of the eight clock pulses.
Ans.   pulse $a = 1$ pulse $c = 1$ pulse $e = 1$ pulse $g = 1$
         pulse $b = 0$ pulse $d = 0$ pulse $f = 0$ pulse $h = 1$ (prohibited condition)

9.71 Refer to Fig. 9-11. Which input has control of the flip-flop during pulse e?
Ans.   The preset (PS) input is activated during pulse e and overrides all other inputs. It sets the Q output to 1.

9.72 Refer to Fig. 9-11. Which input has control of the flip-flop during pulse f?
Ans.   The clear (CLR) input is activated during pulse f and overrides all other inputs. It resets the Q output to 0.

9.73 A delay flip-flop is also called a _____ (D, T)-type flip-flop.   *Ans.* D

9.74 On a D flip-flop, the data bit at input D is delayed _____ (0, 1, 2, 3, 4) clock pulse(s) from getting to output _____ (Q, $\overline{Q}$).   *Ans.*   1; Q

9.75 A T-type flip-flop is also called a _____ (toggle, truth-table) flip-flop.   *Ans.*   toggle

9.76 Draw a logic diagram showing how to wire a JK flip-flop as a T flip-flop.   *Ans.*   See Fig. 9-14b.

9.77 Draw a logic diagram showing how to wire a JK flip-flop and an inverter as a D flip-flop.
Ans.   See Fig. 9-14a.

9.78 List the *binary* output (at $\overline{Q}$) fo the JK flip-flop shown in Fig. 9-16 after each of the eight clock pulses.
Ans.   pulse $a = 0$ pulse $c = 0$ pulse $e = 1$ pulse $g = 1$
         pulse $b = 0$ pulse $d = 1$ pulse $f = 0$ pulse $h = 0$

9.79 Refer to Fig. 9-16. The _____ (asynchronous, synchronous) inputs to the JK flip-flop are shown being used on this unit.
Ans.   The J, K, and CLK inputs are synchronous inputs.

9.80 Refer to Fig. 9-17. Which input has control of the JK flip-flop during pulse a?
Ans.   PR (Preset input activated with LOW sets output Q to 1.)

9.81 Refer to Fig. 9-17. Which input has control of the JK flip-flop during pulse d?
Ans.   CLR (Clear input activated with LOW resets output Q to 0.)

9.82 Refer to Fig. 9-17. List the *binary* output at $\overline{Q}$ (complementary output) of the JK flip-flop *after* each of the seven clock pulses.
Ans.   pulse $a = 0$
         pulse $b = 1$
         pulse $c = 0$
         pulse $d = 1$
         pulse $e = 0$
         pulse $f = 0$
         pulse $g = 1$

9.83   A negative-edge-triggered flip-flop transfers data from input to outputs on the _____ (leading, trailing) edge of the clock pulse.   *Ans.*   trailing

9.84   A positive-edge-triggered flip-flop transfers data from input to outputs on the _____ (H-to-L, L-to-H) transition of the clock pulse.   *Ans.*   L-to-H

9.85   Refer to Fig. 9-21. List the *binary* output (at $\overline{Q}$) of the master-slave *JK* flip-flop *after* each of the eight clock pulses.

Ans.   pulse $a = 0$ pulse $c = 0$ pulse $e = 1$ pulse $g = 1$
             pulse $b = 1$ pulse $d = 1$ pulse $f = 0$ pulse $h = 0$

9.86   Refer to Fig. 9-21. List the mode of operation of the *negative-edge-triggered JK* flip-flop for each of the clock pulses.

Ans.   pulse $a =$ set   pulse $c =$ toggle   pulse $e =$ hold   pulse $g =$ reset
             pulse $b =$ reset   pulse $d =$ toggle   pulse $f =$ hold (J and $K = 0$ pulse $h =$ toggle
                                                                                 during H-to-L
                                                                                 pulse)

9.87   A flip-flop is also referred to as a(n) _____ multivibrator.   *Ans.*   bistable

9.88   A free-running clock is also referred to as a(n) _____ multivibrator.   *Ans.*   astable

9.89   One-shots are also called _____.   *Ans.*   monostable multivibrators

9.90   The main advantage of a crystal-controlled clock is its _____ stability.   *Ans.*   frequency

9.91   If a free-running MV has a clock cycle time of 0.000001 s, the frequency of the clock is _____. *Ans.*   1 MHz

9.92   The output of the astable MV shown in Fig. 9-25 _____ (is, is not) TTL-compatible. *Ans.*   is not (voltage is too high.)

Chapter 10

Counters

10-1 INTRODUCTION

Counters are important digital electronic circuits. They are sequential logic circuits because *timing* is obviously important and because they need a *memory* characteristic. *Digital counters* have the following important characteristics:

1. Maximum number of counts (modulus of counter)
2. Up or down count
3. Asynchronous or synchronous operation
4. Free-running or self-stopping

As with other sequential circuits, flip-flops are used to construct counters.

Counters are extremely useful in digital systems. Counters can be used to count events such as a number of clock pulses in a given time (measuring frequency). They can be used to divide frequency and store data as in a digital clock, and they can also be used in sequential addressing and in some arithmetic circuits.

10-2 RIPPLE COUNTERS

Digital counters will count only in binary or in binary codes. Figure 10-1 shows the counting sequence in binary from 0000 to 1111 (0 to 15 in decimal). A digital counter that would count from binary 0000 to 1111 as shown in the table might be called a *modulo-16 counter*. The *modulus* of a counter is the number of counts the counter goes through. The term "modulo" is sometimes shortened to "mod." This counter might thus be called a *mod-16 counter*.

Decimal count	Binary count 8s 4s 2s 1s D C B A	Decimal count	Binary count 8s 4s 2s 1s D C B A
0	0 0 0 0	8	1 0 0 0
1	0 0 0 1	9	1 0 0 1
2	0 0 1 0	10	1 0 1 0
3	0 0 1 1	11	1 0 1 1
4	0 1 0 0	12	1 1 0 0
5	0 1 0 1	13	1 1 0 1
6	0 1 1 0	14	1 1 1 0
7	0 1 1 1	15	1 1 1 1

Fig. 10-1 Counting sequence for a 4-bit counter

A logic diagram of a mod-16 counter using JK flip-flops is shown in Fig. 10-2. First note that the J and K data inputs of the flip-flops are tied to logical 1. This means that each flip-flop is in its toggle mode. Each clock pulse will then cause the flip-flop to toggle to its opposite state. Note also that the Q output of FF1 (flip-flop 1) is connected directly to the clock (CLK) input to the next unit (FF2), and so forth. Output indicators (lamps or LEDs), shown at the upper right, monitor the binary output of the counter. Indicator A is the LSB (least significant bit), D is the MSB.

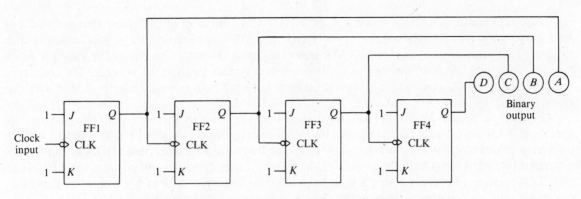

Fig. 10-2 Logic diagram for a 4-bit (mod-16) ripple counter

The mod-16 counter in Fig. 10-2 counts according to the table in Fig. 10-1. It is customary to analyze a counter's operation by using *waveforms* (timing diagrams). Figure 10-3 is a waveform for the mod-16 counter. The top line represents the clock (CLK) input to FF1. The bottom line shows the binary count on the indicators. Note that the binary counter is cleared, or reset to 0000, on the left. Each clock pulse will increase the binary count by 1 as you move to the right on the diagram.

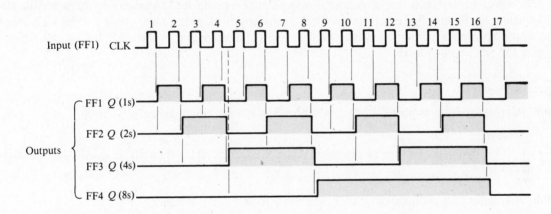

Fig. 10-3 Timing diagram for a mod-16 ripple counter

The bubble on the clock (CLK) input of the *JK* flip-flop shown in Fig. 10-2 means that the unit will toggle on the H-to-L transition (trailing edge) of the clock pulse. Look at clock pulse 1 shown in Fig. 10-3. The H-to-L transition toggles FF1. Output Q of FF1 goes from LOW to HIGH. The binary count is now 0001.

Look at clock pulse 2. The trailing edge of the clock pulse triggers FF1. FF1 toggles, and output Q goes from HIGH to LOW. When output Q of FF1 goes from HIGH to LOW, it in turn toggles FF2 (output Q of FF1 is connected to the CLK input of FF2). FF2 toggles from LOW to HIGH. After clock pulse 2, the binary count has increased to 0010.

Look at clock pulse 3, Fig. 10-3. The trailing edge triggers FF1, which toggles. Output Q of FF1 toggles from LOW to HIGH. The binary count (see bottom line) has increased to 0011.

Look at clock pulse 4, Fig. 10-3. The trailing edge triggers FF1, which toggles, with Q going from HIGH to LOW. This H-to-L transition at Q of FF1 in turn toggles FF2. Output Q of FF2 toggles from HIGH to LOW. This H-to-L transition at Q of FF2 in turn causes FF3 to toggle. Output Q of FF3 toggles from LOW to HIGH. The binary count now stands at 0100.

Look at the dashed line after pulse 4 to the HIGH waveform at Q of FF3. Note that quite a lot of *time* passes before FF3 finally toggles to its HIGH state. That is because FF1 toggles, which in turn toggles FF2, which in turn toggles FF3. All that takes time. This type of counter is called a *ripple counter*. The triggering from flip-flop to flip-flop in effect ripples through the counter. The counter is also referred to as an *asynchronous counter* because not all flip-flops toggle exactly in step with the clock pulse.

Look at the remainder of the waveform shown in Fig. 10-3 to make sure you understand its operation. Note particularly that, on pulse 16, the H-to-L transition toggles FF1. The output of FF1 goes from HIGH to LOW. FF2 is toggled by FF1. The output of FF2 goes from HIGH to LOW. FF3 is toggled by FF2, and so forth. Note that all the flip-flops toggle in turn and go from their HIGH to their LOW states. The binary count is then back to 0000. The counter does not stop at its maximum count; it continues counting as long as the clock pulses are fed into the CLK input of FF1.

Count carefully the number of HIGH pulses under the first 16 clock pulses (in the FF1 output line, Fig. 10-3). You will find eight pulses. Sixteen pulses go into FF1, and only eight pulses come out. This flip-flop is therefore a *frequency divider*. 16 divided by 8 equals 2. FF1 may thus also be considered a *divide-by-2 counter*.

Count the HIGH output pulses at FF2. For 16 clock pulses, only four pulses appear at the output of FF2 (16 divided by 4 equals 4). Output Q of FF2 may be considered a *divide-by-4 counter*. It is found that the output of FF3 is a divide-by-8 counter. The output of FF4 is a divide-by-16 counter. On some devices, such as digital clocks, dividing frequency is a very important job for counters.

The waveform confirms that a counter is a sequential logic device. The memory characteristic also is important; for the flip-flop must "remember" how many clock pulses have arrived at the CLK input. The ripple counter is the simplest type of counter. Its shortcoming is the *time lag* as one flip-flop triggers the next, and so forth.

SOLVED PROBLEMS

10.1 A ripple counter is a(n) _____ (asynchronous, synchronous) device.

Solution:

The ripple counter is an asynchronous device because not all flip-flops trigger exactly in step with the clock pulse.

10.2 A counter that counts from 0 to 7 is called a mod-_____ counter.

Solution:

A counter that counts from 0 to 7 is called a mod-8 counter.

10.3 Draw a logic diagram of a mod-8 ripple counter using three *JK* flip-flops.

Solution:

See Fig. 10-4.

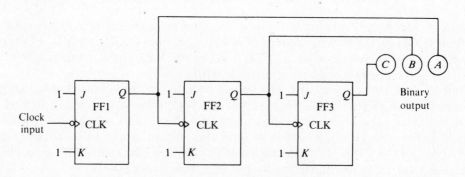

Fig. 10-4 A 3-bit ripple counter

10.4 List the sequence of binary counts that the counter in Prob. 10-3 would go through.

Solution:

The mod-8 counter would count in binary as follows: $000, 001, 010, 011, 100, 101, 110, 111$, and then back to 000, and so forth.

10.5 It is customary to designate FF1 in a counter as the _____ (LSB, MSB) counter.

Solution:

Customarily, FF1 is the LSB counter.

10.6 Refer to Fig. 10-5. What is the binary count after pulse 2?

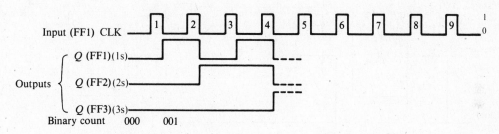

Fig. 10-5 Timing diagram for a mod-8 ripple counter

Solution:

The binary count after pulse 2 is 010.

10.7 Refer to Fig. 10-5. The output of FF1 will go HIGH again on the trailing edge of clock pulse _____.

Solution:

FF1 will go HIGH again on the trailing edge of clock pulse 5.

10.8 Refer to Fig. 10-5. The output of FF2 will go HIGH again on the _____ (leading, trailing) edge of clock pulse 6.

Solution:

FF2 will go HIGH again on the trailing edge of clock pulse 6.

10.9 Refer to Fig. 10-5. The output of FF3 will go LOW again on the H-to-L edge of clock pulse _____.

Solution:

FF3 will go LOW again on the H-to-L edge of clock pulse 8.

10.10 Refer to Fig. 10-5. The binary count after clock pulse 8 will be _____.

Solution:

The binary count after clock pulse 8 will be 000.

10-3 PARALLEL COUNTERS

The asynchronous ripple counter has the limitation of the time lag in triggering all the flip-flops. To cure this problem, *parallel counters* can be used. The logic diagram for a 3-bit parallel counter is shown in Fig. 10-6a. Note that all CLK inputs are tied directly to the input clock. They are wired in *parallel*. Note also that *JK* flip-flops are used. FF1 is the 1s place counter and is always in the toggle mode. FF2 has its *J* and *K* inputs tied to the output of FF1 and will be in the hold or toggle mode. The outputs of FF1 and FF2 are fed into an AND gate. The AND gate controls the mode of operation of FF3. When the AND gate is activated by 1s at *A* and *B*, FF3 will be in its toggle mode. With the AND gate deactivated, FF3 will be in its hold mode. FF2 is the 2s place counter and FF3 the 4s place counter.

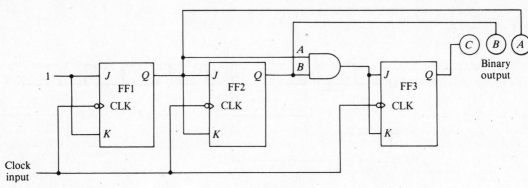

(a) Logic diagram

Decimal count	Binary count		
	4s	2s	1s
	C	*B*	*A*
0	0	0	0
1	0	0	1
2	0	1	0
3	0	1	1
4	1	0	0
5	1	0	1
6	1	1	0
7	1	1	1

(b) Counting sequence

Fig. 10-6 A 3-bit parallel counter

The counting sequence for this 3-bit parallel counter is shown in Fig. 10-6b. Note that this is a modulo-8 (mod-8) counter. The counter will start counting at binary 000 and count up to 111. It will then recycle back to 000 to start the count again.

The waveform (timing diagram) for the parallel mod-8 counter is drawn in Fig. 10-7. The top line represents the clock (CLK) inputs to all three flip-flops. The outputs (at *Q*) of the flip-flops are shown in the middle three lines. The bottom line gives the indicated binary count.

Consider pulse 1, Fig. 10-7. Pulse 1 arrives at each of the three flip-flops. FF1 toggles from LOW to HIGH. FF2 and FF3 do not toggle because they are in the hold mode (*J* and *K* = 0). The binary count is now 001.

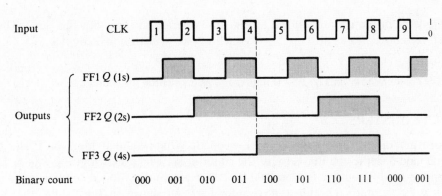

Input   CLK

Outputs { FF1 Q (1s)

FF2 Q (2s)

FF3 Q (4s)

Binary count   000   001   010   011   100   101   110   111   000   001

Fig. 10-7 Timing diagram for a 3-bit parallel counter

Look at pulse 2, Fig. 10-7. Pulse 2 arrives at all the flip-flops, FF1 and FF2 toggle because they are in the toggle mode (J and $K = 1$). FF1 goes from HIGH to LOW while FF2 goes from LOW to HIGH. FF3 is still in the hold mode, and so it does not toggle. The count is now 010.

Pulse 3 arrives at all the flip-flops at the same time. Only FF1 toggles. FF2 and FF3 are in the hold mode because J and $K = 0$. The binary count is now 011.

Consider pulse 4, Fig. 10-7. Note that the AND gate is activated just before the clock pulse goes from HIGH to LOW. The AND gate will put FF3 in the toggle mode (J and $K = 1$). On the H-to-L transition of clock pulse 4, *all* flip-flops toggle. FF1 and FF2 go from HIGH to LOW. FF3 toggles from LOW to HIGH. The binary count is now 100. Note the dashed line below the trailing edge of clock pulse 4. Hardly any time lag is evident from FF1 to FF3 because all the flip-flops are clocked at exactly the same time. That is the advantage of the parallel-type counter. Parallel counters are also called *synchronous counters* because all flip-flops trigger exactly in time with the clock. Parallel counters are more complicated (see the added lines and the AND gate), but they are used when the time lag problem with a ripple counter would cause problems.

Look over the rest of the waveform in Fig. 10-7. Understand that each flip-flop is clocked on each clock pulse. FF1 always toggles. FF2 and FF3 may be in either the toggle or the hold mode.

SOLVED PROBLEMS

10.11 Refer to Fig. 10-7. When the clock pulse 5 is HIGH, FF1 is in its toggle mode, FF2 in its _____ (hold, toggle) mode, and FF3 in its _____ (hold, toggle) mode.

Solution:

When pulse 5 is HIGH, FF1 is in its toggle mode, FF2 in its hold mode, and FF3 in its hold mode.

10.12 Refer to Fig. 10-7. On the trailing edge of clock pulse 6, which flip-flop(s) toggle?

Solution:

On the trailing edge of clock pulse 6 (Fig. 10-7), both FF1 and FF2 toggle.

10.13 Refer to Fig. 10-7. When clock pulse 8 is HIGH, which flip-flops are in the toggle mode?

Solution:

When clock pulse 8 (Fig. 10-7) is HIGH, all three flip-flops are in the toggle mode.

10.14 Refer to Fig. 10-7. What is the binary count after clock pulse 8?

Solution:

The binary count after clock pulse 8 (Fig. 10-7) is 000.

10.15 All flip-flops in the counter shown in Fig. 10-7 operate in step with the clock. The counter is therefore referred to as a(n) _____ (asynchronous, synchronous) counter.

Solution:

The counter shown in Fig. 10-7 is referred to as a synchronous counter.

10-4 OTHER COUNTERS

Suppose a modulo-6 ripple counter were needed. What would it look like? The first step in constructing a mod-6 ripple counter is to list the counting sequence shown in Fig. 10-8a. The counting sequence for the mod-6 counter is from 000 to 101. Note that a 3-bit counter is needed with a 4s counter (C), a 2s counter (B), and a 1s counter (A). As shown in Fig. 10-8a, the 3-bit counter normally counts from 000 to 111. The last two counts on the chart (110 and 111) must be omitted.

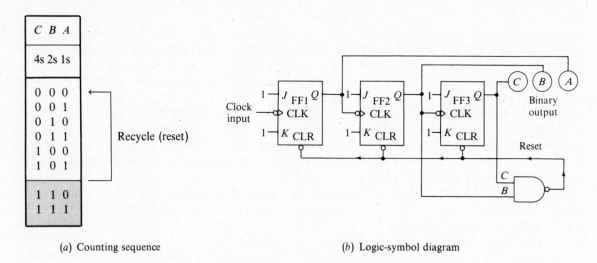

(a) Counting sequence   (b) Logic-symbol diagram

Fig. 10-8 Mod-6 ripple counter

The trick to this mod-6 design problem is to look at the binary count *immediately after the highest count of the counter*. In this case, it is 110. Feed the 110 into a logic circuit that will produce a *clear*, or reset, pulse. The clear pulse goes back to an asynchronous clear input on each JK flip-flop, thus clearing, or resetting, the counter to 000.

The logic circuit needed to clear or reset the JK flip-flops back to 0 is shown in Fig. 10-8b. The 2-input NAND gate will do the job when the outputs of FF2 and FF3 are fed into it. Note from the counting table in Fig. 10-8a that the first time both C and B are 1 is *immediately after* the highest count. Thus when the counter tries to go to 110, it will immediately be cleared or reset to 000.

The mod-6 counter shown in Fig. 10-8b is a ripple counter that is just reset or cleared two counts before its normal maximum count of 111. The NAND gate does the job of resetting the JK flip-flops to 0 by activating the CLR inputs.

Waveforms for the mod-6 ripple counter are diagrammed in Fig. 10-9. The clock (CLK) input to FF1 is shown across the top. The middle three lines show the state of the Q outputs. The bottom line gives the binary count.

The mod-6 counter represented in the diagram in Fig. 10-9 operates as a normal ripple counter until pulse 6. The binary count before pulse 6 is 101, the maximum count for this unit. On the H-to-L transition of clock pulse 6, FF1 toggles from HIGH to LOW. FF1's H-to-L transition triggers FF2, which toggles from LOW to HIGH. At point a, Fig. 10-9, both outputs of FF2 and FF3 are at 1. These two 1s are applied to the NAND gate (see Fig. 10-8b). The NAND gate is activated, producing a 0. The 0 activates the asynchronous CLR input to all the flip-flops, resetting them all to 0. The resetting, or clearing to 000, is shown at point b, Fig. 10-9. The small pulse at point a, Fig. 10-9, is so short that

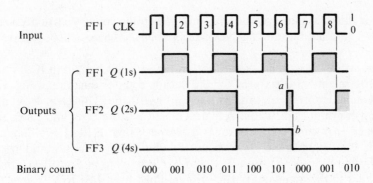

Fig. 10-9 Timing diagram for a mod-6 ripple counter

it does not even light the output indicators. The counter is free to count upward normally again from binary 000.

Look at the trailing edge of pulse 6 (Fig. 10-9) again. Again note the lag between the time pulse 6 goes from HIGH to LOW and the time FF2 and FF3 finally are reset to 0 at point *b*. Engineers refer to this lag time as the propagation time, and it is based on the *propagation delay* of the flip-flop and gate being used. The propagation delay for a typical TTL flip-flop is very short—from 5 to 30 ns (nanoseconds). Some logic families have much longer propagation delays.

A *decade counter* is probably the most widely used counter. It could also be described as a *modulo-10 counter*. Figure 10-10*a* is a diagram of a mod-10 ripple counter. Four *JK* flip-flops plus a NAND gate are used to wire the decade counter. The unit counts just like the mod-16 counter up to

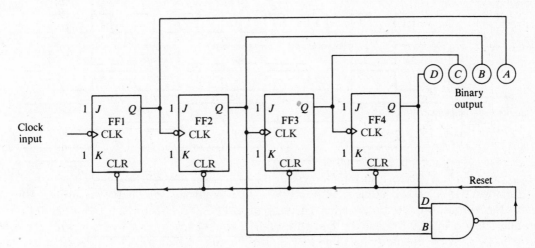

(*a*) Logic diagram for ripple-type decade counter

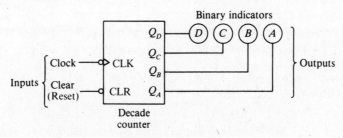

(*b*) Simplified logic symbol for decade counter

Fig. 10-10

1001. Binary 1001 is the maximum count of this unit. When the count tries to advance to 1010, the two 1s ($D = 1$ and $B = 1$) are fed into the NAND gate. The NAND gate is activated, resetting the display to 0000.

A general logic symbol is sometimes used for a counter when bought in IC form. The logic symbol shown in Fig. 10-10b might be substituted for the decade counter diagram in Fig. 10-10a. A *clear* (or reset) input has been added to the decade counter in Fig. 10-10b. This clear input does not appear on the decade counter shown in Fig. 10-10a. A logical 0 activates the reset and clears the output to 0000.

It was mentioned that some counters count downward. Figure 10-11 is a diagram of such a down counter. This unit is a 3-bit ripple down counter. The binary count would be $111, 110, 101, 100, 011, 010, 001, 000$, followed by a recycle to 111, and so forth. Note in Fig. 10-11a that the ripple down counter is very similar to the up counter. The "trigger line" from FF1 to FF2 goes from the $\bar{Q}$ output to the clock input instead of from the Q output to the clock. Otherwise, the up counter and down

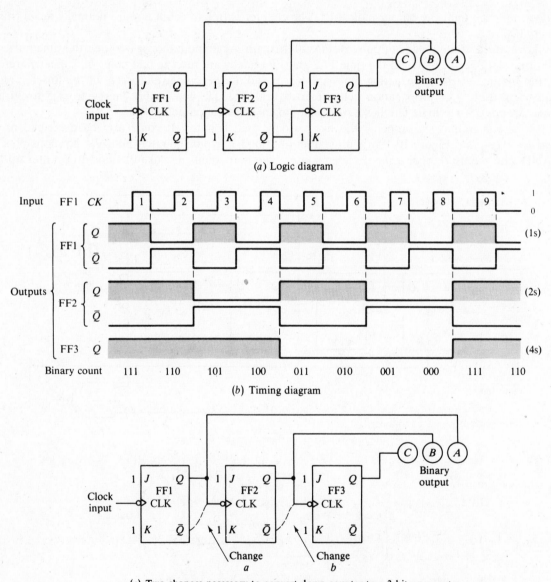

(a) Logic diagram

(b) Timing diagram

(c) Two changes necessary to convert down counter to a 3-bit up counter

Fig. 10-11 A 3-bit ripple down counter

counter are wired in the same way. Note also that each *JK* flip-flop is in its toggle mode (*J* and *K* equal 1).

The waveform in Fig. 10-11*b* aids in the understanding of the operation of the down counter. The top line is the CLK input to FF1. The bottom line is the binary count. Note that the binary count starts at 111 on the left. Two outputs (*Q* and $\overline{Q}$) are shown for both FF1 and FF2. Output *Q* is shown for FF3. The outputs attached to the binary indicators are shown with shading on the timing diagram.

Consider pulse 1 (Fig. 10-11*b*). All flip-flops are set. The binary output on the indicators is 111. On the H-to-L transition of pulse 1, FF1 toggles. Output *Q* goes from HIGH to LOW ($\overline{Q}$ goes from LOW to HIGH). The binary count is now 110.

Look at pulse 2 (Fig. 10-11*b*). On the H-to-L transition of the clock pulse, FF1 toggles. This causes output *Q* to go from LOW to HIGH. Output $\overline{Q}$ goes from HIGH to LOW, thereby causing FF2 to toggle. FF2 toggles, and output *Q* goes from HIGH to LOW ($\overline{Q}$ goes from LOW to HIGH). The binary count is now 101.

Consider pulse 3, Fig. 10-11*b*. Pulse 3 triggers FF1. Output *Q* of FF1 goes LOW while $\overline{Q}$ goes HIGH. The binary output is now 100.

Look at pulse 4 (Fig. 10-11*b*). Pulse 4 triggers FF1. FF1 is set, and output $\overline{Q}$ goes from HIGH to LOW. That causes FF2 to toggle. FF2 is set, and output $\overline{Q}$ goes from HIGH to LOW. That in turn causes FF3 to toggle and reset. The binary output after pulse 4 is then 011.

Look over the rest of the waveform. Particularly note the light vertical lines that show the triggering of the next flip-flop. Remember that the *Q* outputs connect to the output indicators but the $\overline{Q}$ outputs of FF1 and FF2 trigger the next flip-flop.

SOLVED PROBLEMS

10.16 A decade counter has _____ counts and therefore is also called a modulo-_____ counter.

   Solution:

   A decade counter has 10 counts and is called a modulo-10 counter.

10.17 The maximum binary count for a 3-bit counter is _____ (binary number).

   Solution:

   The maximum binary count for a 3-bit counter is binary 111.

10.18 Refer to Fig. 10-8*b*. The job of the NAND gate in this mod-6 counter is to _____ (reset, set) the flip-flops to _____ (binary number) after the counter's maximum number of _____ (binary number).

   Solution:

   The job of the NAND gate shown in Fig. 10-8*b* is to reset the flip-flops to 000 after the counter's maximum number of binary 101.

10.19 Refer to Fig. 10-9. Which flip-flop(s) toggle on the H-to-L transition of clock pulse 4?

   Solution:

   All three flip-flops toggle on the H-to-L transition of clock pulse 4 (Fig. 10-9).

10.20 Refer to Fig. 10-3. The time lag after clock pulse 4 shown with the dashed line is *caused by* the _____ delay of the flip-flops.

   Solution:

   The time lag shown by the dashed line after pulse 4 (Fig. 10-3) is caused by the propagation delay of the flip-flops.

10.21 Refer to Fig. 10-9. Why is the pulse at point *a* very short?

Solution:

    The pulse at point *a*, Fig. 10-9, is very short because, as it goes HIGH, both FF2 and FF3 are set, which causes the NAND gate (see Fig. 10-8*b*) to reset all three flip-flops.

10.22 Refer to Fig. 10-11*a* and *b*. List the next ten binary counts after 010 on this counter.

Solution:

    The next ten binary counts after 010 on the 3-bit down counter of Fig. 10-11 are as follows: $001, 000, 111, 110, 101, 100, 011, 010, 001, 000$.

10.23 Refer to Fig. 10-11*a*. This is a mod-_____ (ripple, synchronous) down counter.

Solution:

    This is a mod-8 ripple down counter.

10.24 List the binary counting sequence of a mod-9 up counter.

Solution:

    The binary counting sequence of a mod-9 up counter is as follows: $0000, 0001, 0010, 0011, 0100, 0101, 0110, 0111, 1000$.

10.25 Refer to Fig. 10-10*a*. If this unit were converted to a mod-9 counter, the two inputs to the NAND gate would be _____ (*D, C, B, A*) and _____ (*D, C, B, A*).

Solution:

    If the unit shown in Fig. 10-10*a* were converted to a mod-9 counter, the two inputs to the NAND gate would be *A* and *D*, so all flip-flops would be reset immediately upon hitting the binary 1001 count.

10.26 Refer to Fig. 10-11*a*. List two wiring changes that will convert this 3-bit down counter to an up counter.

Solution:

    The down counter shown in Fig. 10-11*a* can be converted to an up counter by making the changes shown in Fig. 10-11*c*:
1. Move the wire coming from $\overline{Q}$ of FF1 to output Q of FF1.
2. Move the wire coming from $\overline{Q}$ of FF2 to output Q of FF2.

10.27 Refer to Fig. 10-11*b*. The clock input triggers FF1; the _____ ($Q, \overline{Q}$) output of FF1 triggers FF2; and the _____ ($Q, \overline{Q}$) output of FF2 triggers FF3 in this ripple counter.

Solution:

    The clock triggers FF1; the $\overline{Q}$ output of FF1 triggers FF2; and the $\overline{Q}$ output of FF2 triggers FF3 in the ripple counter shown in Fig. 10-11*b*.

10-5 TTL IC COUNTERS

    Counters can be constructed from individual flip-flops and gates or be purchased from manufacturers in IC form. Several typical general-purpose TTL counter ICs will be detailed in this section.

The 74192 IC is described by the manufacturers as a *TTL synchronous BCD up/down counter*. A block symbol of the 74192 IC decade counter is shown in Fig. 10-12. Note the use of dual clock (CLK) inputs. If the count-up clock input is pulsed, the counter will count upward from 0000 to 1001 (0 to 9 in decimal). If the count-down clock input is pulsed, the counter will count downward from 1001 to 0000 (9 to 0 in decimal). The counter toggles on the L-to-H transition of the clock pulse.

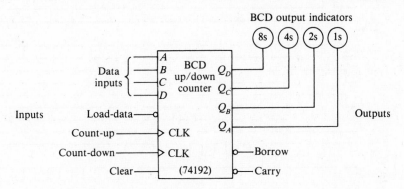

Fig. 10-12   The 74192 synchronous BCD up/down counter IC

The asynchronous clear input to the 74192 counter shown in Fig. 10-12 is activated by a HIGH. When activated, the clear input resets all Q outputs to LOW (0000). The clear input overrides all other inputs. The 74192 counter can be preset to any number by activating the load-data input with a LOW. With a LOW at the load-data input, the data at the data inputs will be asynchronously transferred to the BCD output ($A = Q_A, B = Q_B, C = Q_C, D = Q_D$).

The BCD outputs shown in Fig. 10-12 (Q_D, Q_C, Q_B, Q_A) are the normal outputs from the four flip-flops in the 74192 IC. The carry output is used when *cascading* several counters together. Figure 10-13 shows two 74192 ICs cascaded to form an up counter that will count from BCD 0000 0000 to 1001 1001 (0 to 99 in decimal). Note that the carry output of the 1s up counter is connected directly to the count-up clock input of the 10s up counter. For a cascaded down counter (99 to 0 in decimal), the borrow output of the 1s counter is connected directly into the count-down input of the 10s counter. The count-down input on the 1s counter then becomes the clock input.

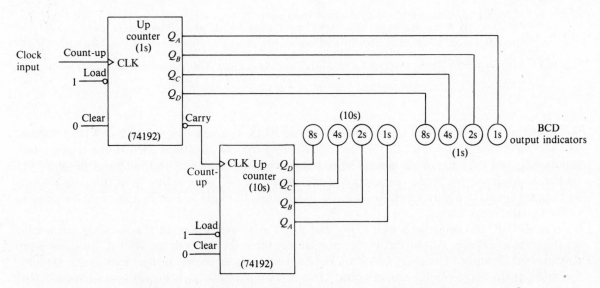

Fig. 10-13   Cascading two 74192 ICs to form a 0–99 BCD up counter

A manufacturer's timing diagram for the 74192 decade counter is shown in Fig. 10-14. Shown from left to right are typical clear, load (preset), count-up, and count-down sequences. The manufacturer's waveforms give much information on the operation of an IC.

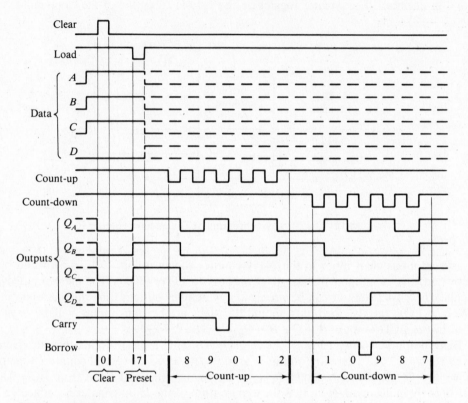

Sequence: (1) Clear outputs to zero.
           (2) Load (preset) to BCD seven.
           (3) Count up to eight, nine, carry, zero, one, and two.
           (4) Count down to one, zero, borrow, nine, eight, and seven.

Notes: (a) Clear overrides load, data, and count inputs.
        (b) When counting up, count-down input must be HIGH; when counting down, count-up input must be HIGH.

Fig. 10-14 Manufacturer's timing diagram for the 74192 decade counter IC (*Courtesy of National Semiconductor Corp.*)

A second counter in IC form is detailed in Fig. 10-15. A block diagram of the TTL 7493 *4-bit binary counter* is shown in Fig. 10-15a. Note the use of four *JK* flip-flops each in the toggle mode. Inputs $\overline{CP_0}$ and $\overline{CP_1}$ are *clock* inputs. Note that normal output Q of the left-hand flip-flop, Fig. 10-15a, is *not connected* to the clock input of the second FF. To form a 4-bit mod-16 ripple counter, an external connection must be made from Q_0 to $\overline{CP_1}$ (pin 12 to pin 1, Fig. 10-15b), with $\overline{CP_0}$ serving as the counter clock input.

The 7493 IC has two reset inputs (MR_1 and MR_2) as shown in Fig. 10-15a. A mode select table for the reset inputs is in Fig. 10-15c. Under normal use, the reset inputs of the 7493 IC *must not be left disconnected* (floating). The reset pins float HIGH, which places the IC in the reset mode. While in the reset mode, the 7493 IC cannot count. The reset inputs are asynchronous and override both clocks.

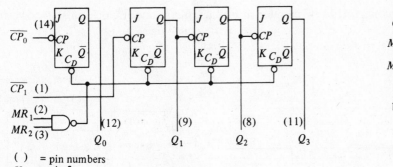

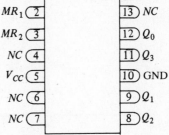

(a) Logic diagram

( ) = pin numbers
V_{CC} = pin 5
GND = pin 10

(b) Pin diagram

Mode of operation	Reset inputs		Outputs			
	MR_1	MR_2	Q_0	Q_1	Q_2	Q_3
Reset	H	H	L	L	L	L
Count	L	H	Count			
Count	H	L	Count			
Count	L	L	Count			

H = HIGH voltage level
L = LOW voltage level

(c) Mode select table

Count	Output				Count	Output			
	Q_3	Q_2	Q_1	Q_0		Q_3	Q_2	Q_1	Q_0
0	L	L	L	L	8	H	L	L	L
1	L	L	L	H	9	H	L	L	H
2	L	L	H	L	10	H	L	H	L
3	L	L	H	H	11	H	L	H	H
4	L	H	L	L	12	H	H	L	L
5	L	H	L	H	13	H	H	L	H
6	L	H	H	L	14	H	H	H	L
7	L	H	H	H	15	H	H	H	H

Output Q_0 is connected to $\overline{CP_1}$.

(d) Counting sequence for 4-bit counter

Fig. 10-15 7493 4-bit binary counter IC

The table in Fig. 10-15d shows the binary counting sequence for the 7493 IC wired as a mod-16 ripple counter. In Fig. 10-15b, note the unusual power connections (V_{CC} = pin 5 and GND = pin 10) on the 7493 counter IC.

SOLVED PROBLEMS

10.28 The 74192 IC is a _____ (binary, decade) counter that will count _____ (down, up, both up and down).

Solution:

The 74192 IC is a decade counter that will count both up and down depending on which clock input is used.

10.29 List the BCD outputs for the 74192 IC counter after each of the input clock pulses shown in Fig. 10-16.

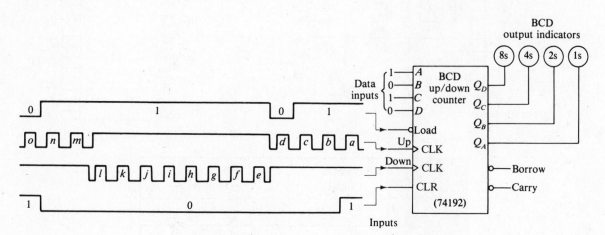

Fig. 10-16 Up/down counter pulse-train problem

Solution:

Refer to the waveforms in Fig. 10-14. The BCD outputs for the 74192 IC counter shown in Fig. 10-16 are:

pulse a = 0000 (clear input overrides all inputs)

pulse b = 0001 (count-up input pulsed)

pulse c = 0010 (count up)

pulse d = 0101 (load input activated with a LOW; the 0101 at the data inputs is loaded into the counter)

pulse e = 0100 (count-down input pulsed)

pulse f = 0011 (count down)

pulse g = 0010 (count down)

pulse h = 0001 (count down)

pulse i = 0000 (count down)

pulse j = 1001 (count-down counter cycles to BCD 1001)

pulse k = 1000 (count-down input pulsed)

pulse l = 0111 (count down)

pulse m = 1000 (count-up input pulsed)

pulse n = 1001 (count up)

pulse o = 0000 (clear input overrides all other inputs)

10.30 The counters are said to be _____ (cascaded, paralleled) when the carry output of one 74192 IC counter is fed into the clock input of the next IC.

Solution:

When the carry output of one 74192 IC is fed into the CLK input of the next IC, the counters are said to be cascaded.

10.31 The 7493 IC is a _____ (ripple-, synchronous-) type counter.

Solution:

The 7493 IC is a ripple-type counter.

10.32 The reset inputs (MR_1 and MR_2) on the 7493 IC are active _____ (HIGH, LOW) inputs.

Solution:

See mode table in Fig. 10-15c. The 7493 IC has active HIGH reset inputs.

10.33 List the binary output from the 7493 counter IC after each input clock pulse shown in Fig. 10-17.

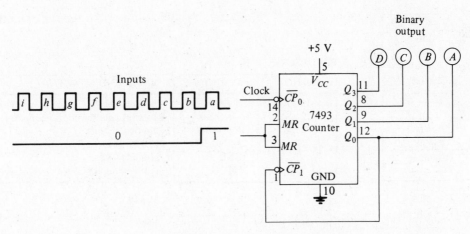

Fig. 10-17 Counter pulse-train problem

Solution:

Refer to Fig. 10-15 for assistance. The binary outputs from the 7493 counter IC shown in Fig. 10-17 are as follows:

pulse a = 0000 (reset)	pulse d = 0011 (count up)	pulse g = 0110 (count up)
pulse b = 0001 (count up)	pulse e = 0100 (count up)	pulse h = 0111 (count up)
pulse c = 0010 (count up)	pulse f = 0101 (count up)	pulse i = 1000 (count up)

10.34 The 7493 counter IC shown in Fig. 10-17 is in the _____ mode of operation during clock pulse a.

Solution:

The 7493 IC shown in Fig. 10-17 is in the reset mode during clock pulse a.

10.35 The 7493 counter IC shown in Fig. 10-17 is in the _____ mode of operation during clock pulse b.

Solution:

The 7493 IC shown in Fig. 10-17 is in the count mode during clock pulse b.

10-6 CMOS IC COUNTERS

A variety of counters are available from IC manufacturers using the CMOS technology. Two representative CMOS counter integrated circuits are featured in this section. The first is a simple ripple counter and the second a more sophisticated presettable synchronous up/down counter.

Manufacturer's data on the first CMOS counter is reproduced in Fig. 10-18. A logic diagram (called a function diagram by the manufacturer) is shown in Fig. 10-18a for the CMOS *74HC393 dual 4-bit binary ripple counter* IC. A more detailed logic diagram of each 4-bit ripple counter is shown in Fig. 10-18c. Note the use of four T flip-flops. The clock inputs ($1\overline{CP}$ and $2\overline{CP}$) are edge-triggered on the HIGH-to-LOW transition of the clock pulse as indicated in the pin description table in Fig. 10-18b. The master reset pins ($1MR$ and $2MR$) on the 74HC393 counter are active HIGH inputs. The flip-flop outputs of the counter are labeled as Q_0 to Q_3 with Q_0 being the LSB while Q_3 holds the MSB of the 4-bit binary number. The 74HC393 dual 4-bit binary counter is packaged in a 14-pin DIP IC which is illustrated in Fig. 10-18d. The 74HC393 counter requires a 5-V dc power supply.

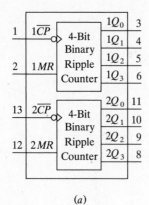

PIN DESCRIPTION

PIN NO.	SYMBOL	NAME AND FUNCTION
1, 13	$1\overline{CP}$, $2\overline{CP}$	clock inputs (HIGH-to-LOW, edge-triggered)
2, 12	$1MR$, $2MR$	asynchronous master reset inputs (active HIGH)
3, 4, 5, 6,	$1Q_0$ to $1Q_3$.	flip-flop outputs
11, 10, 9, 8	$2Q_0$ to $2Q_3$	
7	GND	ground (0 V)
14	V_{CC}	positive supply voltage

(a)   (b)

(c)   (d)

Fig. 10-18 CMOS dual 4-bit binary counter IC (74HC393) (*a*) Function diagram. (*b*) Pin descriptions. (*c*) Detailed logic diagram. (*d*) Pin diagram. (*Courtesy of Signetics Corporation*)

The second featured CMOS counter is the *74HC193 presettable synchronous 4-bit up/down counter IC*. Details of the 74HC193 counter are shown in the manufacturer's data in Fig. 10-19. A function diagram, table of pin descriptions, and pin diagram for the 74HC193 counter are shown in Fig. 10-19*a*, *b*, and *c*. The 74HC193 counter has two edge-triggered clock inputs (CP_U and CP_D) which operate on the LOW-to-HIGH transition of the clock pulse. One clock input is used when counting up (CP_U) while the other is for counting down (CP_D). When using the count up (CP_U) input for implementing an up counter, the count down (CP_D) pin must be tied to HIGH or +5 V.

A truth table detailing the operating modes of the 74HC193 CMOS counter is in Fig. 10-19*d*. The reset mode asynchronously clears the outputs (Q_3, Q_2, Q_1, and Q_0) to binary 0000. The reset pin (MR) is an active HIGH input which overrides all other inputs (such as the load, count, and data inputs). The reset (MR) input is activated with a HIGH briefly on the left side of the waveform diagram in Fig. 10-19*e* when it clears all flip-flops to 0. The parallel load inputs on the 74HC193 counter IC include the four data pins (D_0 to D_3) and the parallel load pin ($\overline{PL}$). The parallel load inputs are for presetting the counter to any given 4-bit count. A preset sequence is shown near the left in the waveform diagram in Fig. 10-19*e* with the $\overline{PL}$ activated with a LOW. During the parallel load operation, binary information from the data inputs (D_3, D_2, D_1, and D_0) is asynchronously transferred to the outputs (Q_3, Q_2, Q_1, and Q_0). In this example (Fig. 10-19*e*), binary 1101 is being loaded into the counter. The reset (MR) input must be LOW during the parallel load operation. Typical count-up and count-down sequences are also shown on the waveform diagram in Fig. 10-19*e*. During the count-up and count-down sequences, the operation of the carry ($\overline{TC_U}$) and borrow ($\overline{TC_D}$)

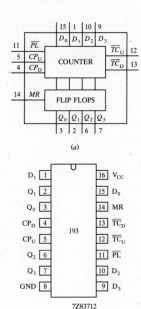

(a)

(c)

7Z83712

PIN DESCRIPTION

PIN NO.	SYMBOL	NAME AND FUNCTION
3, 2, 6, 7	Q_0 to Q_3	flip-flop outputs
4	CP_D	count down clock input*
5	CP_U	count up clock input*
8	GND	ground (0 V)
11	$\overline{PL}$	asynchronous parallel load input (active LOW)
12	$\overline{TC}_U$	terminal count up (carry) output (active LOW)
13	$\overline{TC}_D$	terminal count down (borrow) output (active LOW)
14	MR	asynchronous master reset input (active HIGH)
15, 1, 10, 9	D_0 to D_3	data inputs
16	V_{CC}	positive supply voltage

*LOW-to-HIGH, edge triggered

(b)

OPERATING MODE	INPUTS								OUTPUTS					
	MR	$\overline{PL}$	CP_U	CP_D	D_0	D_1	D_2	D_3	Q_0	Q_1	Q_2	Q_3	$\overline{TC}_U$	$\overline{TC}_D$
reset (clear)	H	X	X	L	X	X	X	X	L	L	L	L	H	L
	H	X	X	H	X	X	X	X	L	L	L	L	H	H
parallel load	L	L	X	L	L	L	L	L	L	L	L	L	H	L
	L	L	X	H	L	L	L	L	L	L	L	L	H	H
	L	L	L	X	H	H	H	H	H	H	H	H	L	H
	L	L	H	X	H	H	H	H	H	H	H	H	H	H
count up	L	H	↑	H	X	X	X	X	count up				H*	H
count down	L	H	H	↑	X	X	X	X	count down				H	H**

*$\overline{TC}_U = CP_U$ at terminal count up (HHHH)
**$\overline{TC}_D = CP_D$ at terminal count down (LLLL)
H = HIGH voltage level
L = LOW voltage level
X = don't care
↑ = LOW-to-HIGH clock transition

(d)

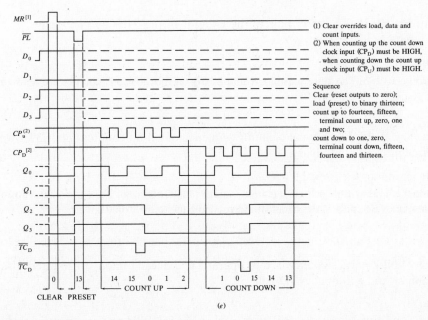

(1) Clear overrides load, data and count inputs.
(2) When counting up the count down clock input (CP_D) must be HIGH, when counting down the count up clock input (CP_U) must be HIGH.

Sequence
Clear (reset outputs to zero); load (preset) to binary thirteen; count up to fourteen, fifteen, terminal count up, zero, one and two; count down to one, zero, terminal count down, fifteen, fourteen and thirteen.

(e)

Fig. 10-19 CMOS presettable 4-bit synchronous up/down counter IC (74HC193) (*a*) Function diagram. (*b*) Pin descriptions. (*c*) Pin diagram. (*d*) Truth table. (*e*) Typical clear, preset, and count sequence. (*Courtesy of Signetics Corporation*)

outputs can be noted. These carry and borrow outputs are used when cascading counters (producing 8-, 12-, or 16-bit devices) as either up or down counters. Note that the carry and borrow outputs ($\overline{TC}_U$ and $\overline{TC}_D$) generate a negative pulse for either a carry or a borrow. The 74HC193 counter is housed in a 16-pin DIP and operates on a 5-V dc power supply.

SOLVED PROBLEMS

10.36 Refer to Fig. 10-18. The 74HC393 IC contains _____ (one, two, four) 4-bit binary _____ (ripple, synchronous) counters in a single DIP package.

Solution:

The 74HC393 IC contains two 4-bit binary ripple counters.

10.37 Refer to Fig. 10-18. The clock inputs to the 74HC393 counters are _____ (edge, level)-triggered on the _____ (H-to-L, L-to-H) edge of the clock pulse.

Solution:

See Fig. 10-18b. The clock inputs to the 74HC393 counters are edge-triggered on the H-to-L edge of the clock pulse.

10.38 Refer to Fig. 10-18. Each counter in the 74HC393 IC contains _____ (three D, four T) flip-flops.

Solution:

See Fig. 10-18c. Each counter in the 74HC393 IC contains four T flip-flops.

10.39 Refer to Fig. 10-18. The reset pins to the counter in the 74HC393 are active _____ (HIGH, LOW) inputs.

Solution:

See Fig. 10-18b. The reset pins ($1MR$ and $2MR$) on the 74HC393 counter are active HIGH inputs.

10.40 Refer to Fig. 10-18. The normal counting sequence of a 4-bit counter (74HC393) would be from 0000 through _____ in binary.

Solution:

Normal counting sequence for the 4-bit counter (74HC393) would be from 0000_2 through 1111_2, and with continued clock pulses, the count would recycle back to 0000, 0001, etc.

10.41 Draw the 4-bit binary counter wired to operate as a decade (mod-10) counter. Use a 74HC393 4-bit counter and a 2-input AND gate.

Solution:

One method of converting a 4-bit binary counter into a decade counter using the 74HC393 IC is shown in Fig. 10-20.

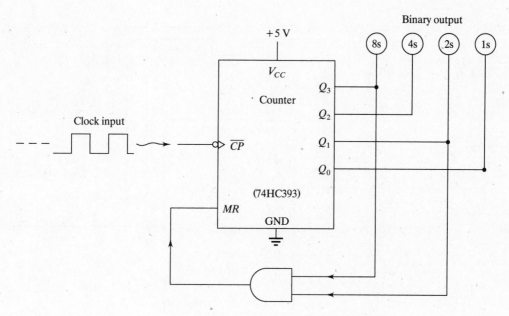

Fig. 10-20 A decade (mod-10) counter circuit

10.42 The 74HC193 IC is described as a(n) _____ (4, 8)-bit presettable _____ (ripple, synchronous) up/down counter manufactured using _____ (CMOS, TTL) technology.

Solution:

The 74HC193 IC is described as a CMOS presettable 4-bit synchronous up/down counter.

10.43 Refer to Fig. 10-19. Why does the 74HC193 counter have two clock inputs?

Solution:

The 74HC193 counter has CP_U (count up) and CP_D (count down) clock inputs. The CP_U pin is used if the design calls for an up counter or the CP_D input is used when developing a down counter. The two clock inputs make the 74HC193 counter a more versatile IC.

10.44 Refer to Fig. 10-19. The _____ (parallel load, reset) pin is an asynchronous active HIGH input that causes the output of the 74HC193 counter to be cleared to 0000 when activated.

Solution:

The reset (*MR*) pin is an asynchronous active HIGH input that causes the output of the 74HC193 counter to be cleared to 0000 when activated.

10.45 Draw a mod-6 counter that has a counting sequence of 001, 010, 011, 100, 101, 110, 001, 010, etc. This is the type of counter that might be used to simulate the roll of a die in a dice game. Use the 74HC193 IC and a 3-input NAND.

Solution:

See Fig. 10-21.

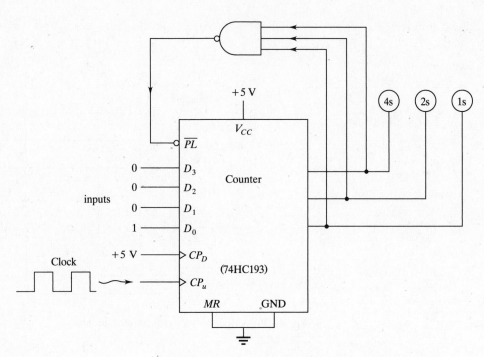

Fig. 10-21 A mod-6 counter circuit (counts 1 to 6)

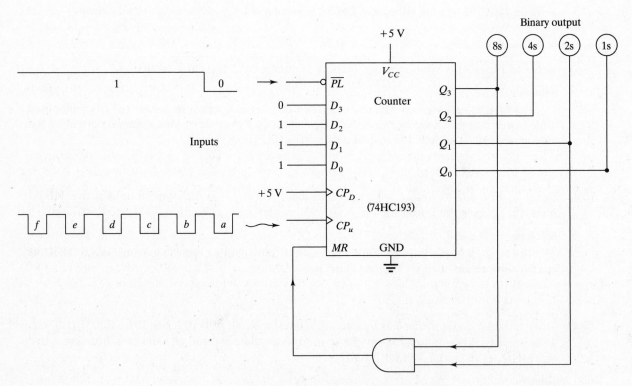

Fig. 10-22 Counter pulse-train problem

10.46 Refer to Fig. 10-22. The 74HC193 IC is wired as a mod-_____ (number) counter in this circuit.

Solution:

The 74HC193 IC is wired as a mod-10 (decade) counter in the circuit in Fig. 10-22.

10.47 Refer to Fig. 10-22. List the mode of operation during each pulse a to f. (Use answers parallel load, count up, or count down.)

Solution:

pulse a = parallel load
pulse b to f = count up

10.48 Refer to Fig. 10-22. List the binary output for the 74HC193 counter after each pulse a to f.

Solution:

The binary output of the decade counter in Fig. 10-22 after each pulse is as follows:
pulse a = 0111 (parallel load to 0111)
pulse b = 1000
pulse c = 1001
pulse d = 0000 (reset to 0000)
pulse e = 0001
pulse f = 0010

10-7 FREQUENCY DIVISION: THE DIGITAL CLOCK

The idea of using a counter for frequency division was introduced in Sec. 10-2. It was mentioned that, for the 4-bit counter shown in Fig. 10-2, the A output could be considered a divide-by-2 output because it divides the clock input frequency in half. Likewise, B (Fig. 10-2) can serve as a divide-by-4 output, C is a divide-by-8 output, and D is a divide-by-16 frequency division output.

One digital system that makes extensive use of counters is diagrammed in Fig. 10-23. The digital clock uses counters as *frequency dividers* in the lower section. The counters shown in Fig. 10-23 are also used as *count accumulators*. The count accumulator's job is to count the input pulses and serve as a temporary memory while passing the current time through the decoders onto the time displays. The block diagram in Fig. 10-23 represents a 6-digit 24-h digital clock.

The input to the frequency dividers shown in Fig. 10-23 is a 60-Hz square wave. The divide-by-60 blocks could be constructed by using a divide-by-6 counter feeding a divide-by-10 counter. A block diagram of such an arrangement is in Fig. 10-24a. The divide-by-6 counter on the left divides the 60 Hz to 10 Hz. The divide-by-10 counter on the right divides the 10 Hz to 1 Hz, or 1 pulse per second. The divide-by-60 block is shown implemented by using 7493 ICs in Fig. 10-24b.

The divide-by-10 counter shown in Fig. 10-24b is implemented by first making an external connection from Q_0 to $\overline{CP_1}$. This makes the 7493 IC a 4-bit binary counter. Second, the IC must be converted to a decade or mod-10 counter. This is accomplished by resetting the counter outputs to 0 when binary 1010 first appears. The resetting is done by using the HIGH outputs from Q_3 and Q_1 tied back to the two reset inputs on the 7493 IC.

The divide-by-6 counter shown in Fig. 10-24b is wired like the unit diagrammed earlier in Fig. 10-8. The first flip-flop inside the 7493 IC package is not used, so $\overline{CP_1}$ becomes the clock input to the divide-by-6 counter.

The 0 to 59 count accumulators in the block diagram of the digital clock shown in Fig. 10-23 are actually two counters. A block diagram showing more detail of the seconds count accumulator/decoder/display section is sketched in Fig. 10-25. A decade (mod-10) counter is needed to accumulate

OUTPUT

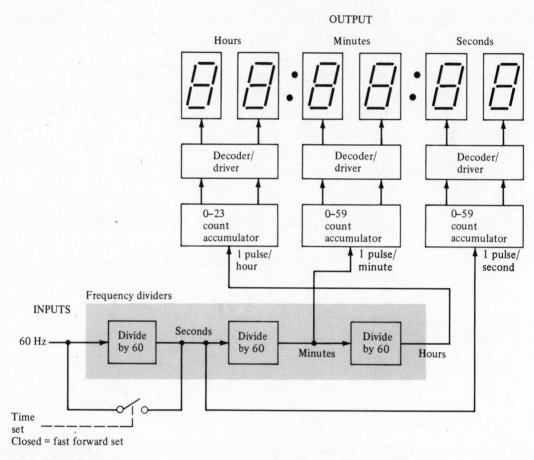

Fig. 10-23 Block diagram of a digital clock (*Roger L. Tokheim*, Digital Electronics, *3d ed., McGraw-Hill, New York, 1990*)

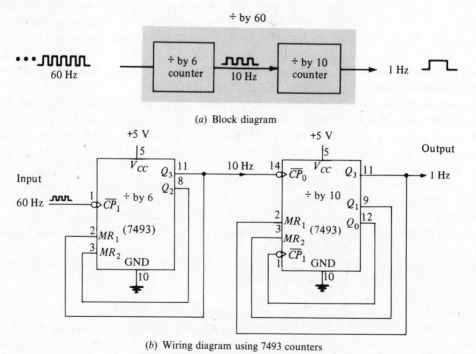

(a) Block diagram

(b) Wiring diagram using 7493 counters

Fig. 10-24 Divide-by-60 counter

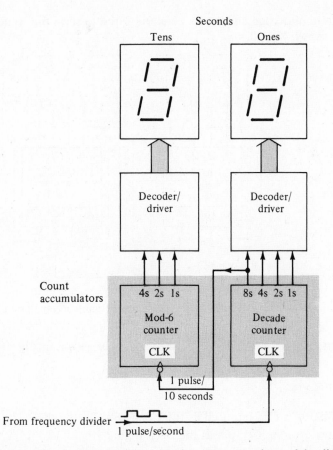

Fig. 10-25 Detailed block diagram of the seconds count accumulator of the digital clock

the 1s of seconds. This decade counter is driven directly from the output of the first divide-by-60 frequency divider. As the decade counter sequences from 9 back to 0, it generates a "carry pulse" which is sent to the mod-6 10s of seconds counter. The decoder/drivers serve to decode the BCD to seven-segment display output.

The minutes and hours count accumulators shown in Fig. 10-23 are connected similarly to the seconds accumulator. The minutes count accumulator would consist of a decade and mod-6 counter (like the seconds accumulator). However, the hours count accumulator would consist of a decade and a mod-3 counter (or mod-2 counter for a 12-h clock).

SOLVED PROBLEMS

10.49 The divide-by-60 block of the digital clock shown in Fig. 10-23 could be constructed by using two _____ .

Solution:

See Fig. 10-24. The divide-by-60 block shown in Fig. 10-23 could be wired by using two counters.

10.50 The 0 to 59 count accumulator of the digital clock shown in Fig. 10-23 could be constructed by using _____ .

Solution:

See Fig. 10-25. The 0 to 59 count accumulator shown in Fig. 10-23 could be wired by using two counters.

10.51 Draw a diagram of decade and mod-6 counters wired to form the count accumulator shown in Fig. 10-25. Use two 7493 ICs.

Solution:

See Fig. 10-26.

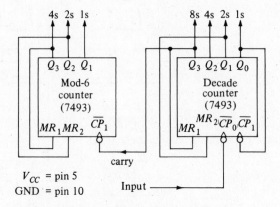

Fig. 10-26 Wiring of a 0 to 59 count accumulator circuit

10.52 Draw a diagram of the divide-by-60 frequency divider shown in Fig. 10-23. Use two 7493 ICs.

Solution:

See Fig. 10-24*b*.

10.53 Why is the 7493 IC's $\overline{CP}_1$ pin used as the clock input of the divide-by-6 counter, whereas the decade counter uses the $\overline{CP}_0$ pin as the clock input?

Solution:

Refer to Fig. 10-15*a*. The divide-by-6 counter uses only the three *JK* flip-flops shown on the right in Fig. 10-15*a* and uses the $\overline{CP}_1$ pin as the clock input. The decade counter uses all four flip-flops in the 7493 IC and uses the $\overline{CP}_0$ pin as the clock input.

Supplementary Problems

10.54 A counter that counts from 0 to 4 is called a mod-_____ counter.   *Ans.*   5

10.55 Draw a logic diagram of a 5-bit ripple up counter using five *JK* flip-flops.   *Ans.*   See Fig. 10-27.

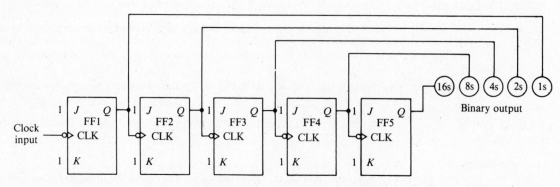

Fig. 10-27 A 5-bit ripple up counter

10.56 The maximum binary count of a 5-bit counter is ___(a)___ (binary number), which equals ___(b)___ in decimals.   *Ans.* (*a*) 11111 (*b*) 31

10.57 In a 4-bit counter, FF4 is usually designated as the _____ (LSB, MSB) counter.
Ans. MSB (most significant bit)

10.58 Refer to Fig. 10-3. On the H-to-L transition of clock pulse 8, how many flip-flops toggle?
Ans. All four

10.59 Refer to Fig. 10-3. On the trailing edge of clock pulse 15, which flip-flop(s) toggle?
Ans. Only FF1 toggles.

10.60 Refer to Fig. 10-3. With clock pulse 16 HIGH, what is the state of each flip-flop?
Ans. All four flip-flops are set (Q outputs are HIGH).

10.61 Refer to Fig. 10-3. After the trailing edge of clock pulse 16, the binary count is ___(a)___ (binary number) and all four flip-flops are ___(b)___ (reset, set).   *Ans.* (*a*) 0000 (*b*) reset

10.62 Refer to Fig. 10-3. Which flip-flop affects FF4 and makes it toggle?
   Ans. Output Q of FF3 is connected to the CLK input of FF4 and makes it toggle when the pulse goes from HIGH to LOW.

10.63 Refer to Fig. 10-5. What is the binary count after pulse 4?   *Ans.* 100

10.64 Refer to Fig. 10-5. The output Q of FF2 will go HIGH again on the trailing edge of clock pulse _____ (5, 6).   *Ans.* 6

10.65 Refer to Fig. 10-5. The output of FF1 will go HIGH on the _____ (leading, trailing) edge of clock pulse 5.   *Ans.* trailing

10.66 Refer to Fig. 10-5. After clock pulse 7, FF1 is _____ (reset, set), FF2 is _____ (reset, set), and FF3 is _____ (reset, set).   *Ans.* All the flip-flops are set ($Q = 1$).

10.67 Refer to Fig. 10-5. Which flip-flop(s) toggle on the H-to-L transition of clock pulse 7?
Ans. Only FF1 toggles.

10.68 Refer to Fig. 10-5. The binary count after clock pulse 9 will be _____.   *Ans.* 001

10.69 The _____ (parallel, ripple) counter is an example of a synchronous device.   *Ans.* parallel

10.70 Refer to Fig. 10-7. The fact that all flip-flops toggle at exactly the same time (see dashed line) means this timing diagram is for a(n) _____ (asynchronous, synchronous) counter.   *Ans.* synchronous

10.71 Refer to Fig. 10-7. When clock pulse 6 is HIGH, FF1 is in its toggle mode, FF2 in its ___(a)___ (hold, toggle) mode, and FF3 in its ___(b)___ (hold, toggle) mode.   *Ans.* (*a*) toggle (*b*) hold

10.72 The _____ (parallel, ripple) counter is the more complicated device.   *Ans.* parallel

10.73 The basic building block for combinational logic circuits is the gate. The basic building block for sequential logic circuits is the _____.   *Ans.* flip-flop

10.74   Refer to Fig. 10-28. The clear (or reset) input on the counter is activated by a _____ (HIGH, LOW).
       Ans.   The clear input on the counter shown in Fig. 10-28 is activated by a LOW, or logical 0. This is symbolized by the small bubble at the clear input.

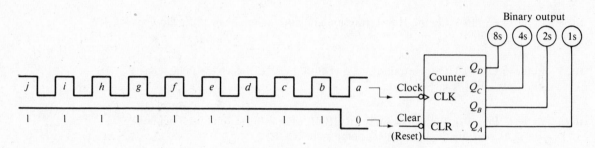

Fig. 10-28   Counter pulse-train problem

10.75   List the binary output after each of the clock pulses for the decade up counter shown in Fig. 10-28.
       Ans.   pulse $a = 0000$ pulse $c = 0010$ pulse $e = 0100$ pulse $g = 0110$ pulse $i = 1000$
                 pulse $b = 0001$ pulse $d = 0011$ pulse $f = 0101$ pulse $h = 0111$ pulse $j = 1001$.

10.76   Assume the counter shown in Fig. 10-28 is a mod-16 down counter. List the binary output after each clock pulse.
       Ans.   pulse $a = 0000$ pulse $c = 1110$ pulse $e = 1100$ pulse $g = 1010$ pulse $i = 1000$
                 pulse $b = 1111$ pulse $d = 1101$ pulse $f = 1011$ pulse $h = 1001$ pulse $j = 0111$

10.77   Draw a logic diagram of a mod-12 ripple up counter by using four *JK* flip-flops (with clear inputs) and a 2-input NAND gate.   *Ans.*   See Fig. 10-29.

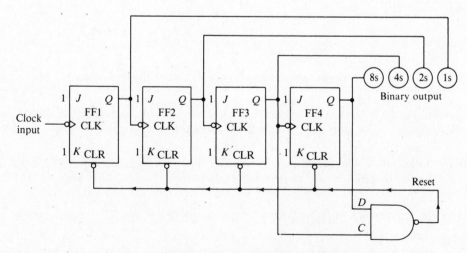

Fig. 10-29   Mod-12 ripple up counter

10.78 Draw a logic diagram for a divide-by-5 ripple counter by using three *JK* flip-flops (with clear inputs) and a 2-input NAND gate. Show clock input and only the divide-by-5 output.   *Ans.*   See Fig. 10-30.

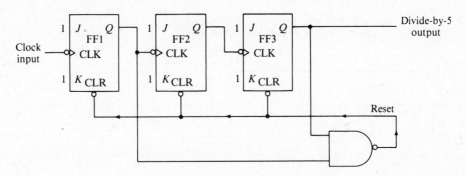

Fig. 10-30 Divide-by-5 ripple counter

10.79 Refer to Fig. 10-31. The clear (CLR) input to the 74192 counter IC is an active _____ (HIGH, LOW) input.   *Ans.*   HIGH

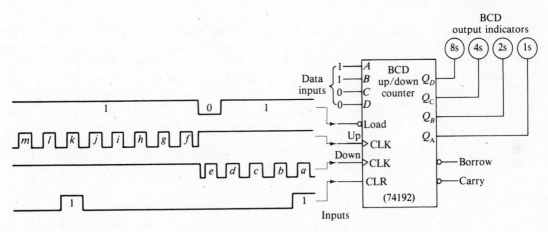

Fig. 10-31 Counter pulse-train problem

10.80 List the BCD outputs of the 74192 IC counter after each of the input clock pulses shown in Fig. 10-31.
   Ans.   pulse *a* = 0000 (clear)   pulse *f* = 0100   pulse *j* = 1000
            pulse *b* = 1001   pulse *g* = 0101   pulse *k* = 0000 (clear)
            pulse *c* = 1000   pulse *h* = 0110   pulse *l* = 0001
            pulse *d* = 0111   pulse *i* = 0111   pulse *m* = 0010
            pulse *e* = 0011 (load)

10.81 Refer to Fig. 10-32. List the binary outputs of the 7493 counter IC after each clock pulse.
   Ans.   pulse *a* = 000 (reset)   pulse *d* = 011 (count up)   pulse *g* = 110 (count up)
            pulse *b* = 001 (count up)   pulse *e* = 100 (count up)   pulse *h* = 111 (count up)
            pulse *c* = 010 (count up)   pulse *f* = 101 (count up)   pulse *i* = 000 (count up)

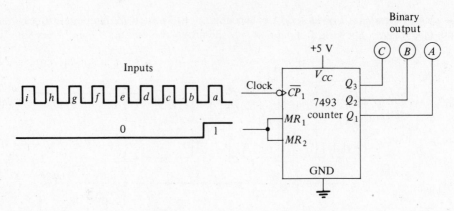

Fig. 10-32 Counter pulse-train problem

10.82   The 7493 IC detailed in Fig. 10-32 is wired as a mod-_____ up counter.   *Ans.*   mod-8

10.83   The 7493 IC shown in Fig. 10-32 is in the _____ mode of operation during clock pulse *a*.
Ans.   reset (or clear)

10.84   Refer to Fig. 10-18. The 74HC393 IC is described by the manufacturer as a CMOS dual _____ (decade,
4-bit binary) counter.   *Ans.*   4-bit binary

10.85   Refer to Fig. 10-18. The 74HC393 IC is a _____ (ripple, synchronous) counter.   *Ans.*   ripple

10.86   Refer to Fig. 10-33. The 74HC393 IC is wired as a mod-_____ (decimal number) counter in this circuit.
Ans.   8

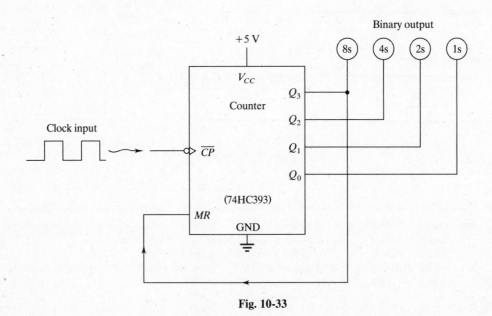

Fig. 10-33

10.87   Refer to Fig. 10-33. The circuit counts from a low of binary 0000 to a high of _____.   *Ans.*   0111

10.88   Refer to Fig. 10-19. On the 74HC193 IC, if both the reset (MR) and parallel load ($\overline{PL}$) pins are activated
at the same time, the _____ input will override all others.   *Ans.*   reset (MR)

10.89 Refer to Fig. 10-34. What is the mode of operation for the 74HC193 counter during clock pulse a?
    Ans.   parallel load

10.90 Refer to Fig. 10-34. What is the mode of operation for the 74HC193 counter during clock pulse b?
    Ans.   count down

10.91 Refer to Fig. 10-34. What is the mode of operation for the 74HC193 counter during clock pulse f?
    Ans.   reset (or clear)

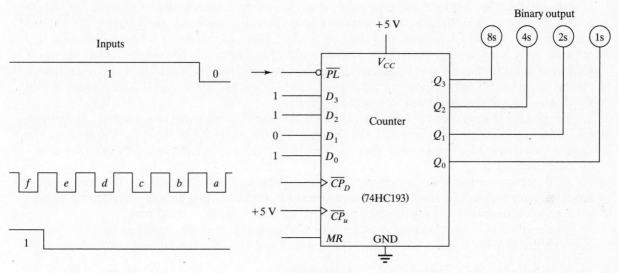

Fig. 10-34

10.92 Refer to Fig. 10-34. List the binary output after each of the clock pulses for the 74HC193 counter circuit.
    Ans.   pulse $a = 1101$
             pulse $b = 1100$
             pulse $c = 1011$
             pulse $d = 1010$
             pulse $e = 1001$
             pulse $f = 0000$

10.93 Refer to Fig. 10-2. If the frequency of the clock input were 1 MHz, the frequency at output A of FF1 would be _____.   *Ans.*   500 kHz, or 0.5 MHz

10.94 Refer to Fig. 10-2. If the frequency of the clock input were 1 MHz, the frequency at output C of FF3 would be _____.   *Ans.*   125 kHz

10.95 Refer to Fig. 10-23. Digital devices called _____ are used to implement the divide-by-60 circuits in this digital clock.   *Ans.*   counters, or counter ICs

10.96 Refer to Fig. 10-23. Digital devices called _____ are used to implement the count accumulators in this digital clock.   *Ans.*   counters, or counter ICs

10.97 Refer to Fig. 10-24a. If the input frequency on the left were 600 kHz, the output frequency would be _____.   *Ans.*   10 kHz.

Chapter 11

Shift Registers

11-1 INTRODUCTION

The *shift register* is one of the most widely used functional devices in digital systems. The simple pocket calculator illustrates the shift register's characteristics. To enter the number 246 on the calculator, the 2 key is depressed and released. A 2 is displayed. Next the 4 key is depressed and released. A 24 is displayed. Finally, the 6 key is depressed and released. The number 246 is displayed. On a typical calculator, the 2 is first shown on the right of the display. When the 4 key is depressed, the 2 is shifted to the left to make room for the 4. The numbers are progressively shifted to the left on the display. This register operates as a shift-left register.

Besides the *shifting characteristic*, the calculator also exhibits a *memory characteristic*. The proper calculator key (such as 2) is depressed and *released*, but the number still shows on the display. The register "remembers" which key was pressed. This temporary memory characteristic is vital in many digital circuits.

Shift registers are classed as sequential logic circuits, and as such they are constructed from flip-flops. Shift registers are used as temporary memories and for shifting data to the left or right. Shift registers are also used for changing serial to parallel data or parallel to serial data.

One method of identifying shift registers is by how data is *loaded into* and *read from* the storage units. Figure 11-1 is a register 8 bits wide. The registers in Fig. 11-1 are classified as:

1. Serial-in serial-out (Fig. 11-1a)
2. Serial-in parallel-out (Fig. 11-1b)
3. Parallel-in serial-out (Fig. 11-1c)
4. Parallel-in parallel-out (Fig. 11-1d)

The diagrams in Fig. 11-1 illustrate the idea of each type of register.

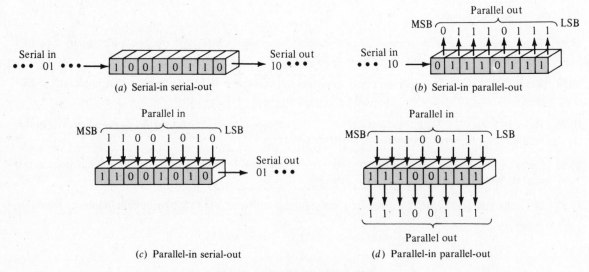

Fig. 11-1 Types of shift registers

11-2   SERIAL-LOAD SHIFT REGISTER

A simple 4-bit shift register is illustrated in Fig. 11-2. Note the use of four D flip-flops. Data bits (0s and 1s) are fed into the D input of FF1. This input is labeled as the serial-data input. The clear input will reset all four flip-flops to 0 when activated by a LOW. A pulse at the clock input will shift the data from the serial-data input to position A (Q of FF1). The indicators (A, B, C, D) across the top of Fig. 11-2 show the contents of each flip-flop or the contents of the register. This register can be classified as a *serial-in parallel-out* unit if data is read from the parallel outputs (A, B, C, D) across the top (Fig. 11-2).

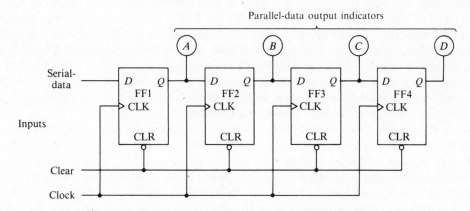

Fig. 11-2   Logic diagram of a 4-bit serial-load shift-right register

Assume all the flip-flops shown in Fig. 11-2 are reset ($Q = 0$). The output is then 0000. Place the clear input at 1. Place a 1 at the data input. Pulse the clock input *once*. The outputs will then read 1000 ($A = 1, B = 0, C = 0, D = 0$). Place a 0 at the data input. Pulse the clock input a second time. The output now reads 0100. After a third pulse, the output reads 0010. After a fourth pulse, the output reads 0001. The binary word 0001 has been loaded into the register one bit at a time. This is called *serial loading*. Note that, on each clock pulse, the register shifts data to the right. This register could then be called a *serial-load shift-right register*.

As with other sequential logic circuits, waveforms (timing diagrams) are an aid to understanding circuit operation. Figure 11-3 illustrates the operation of the 4-bit serial-load shift-right register. The

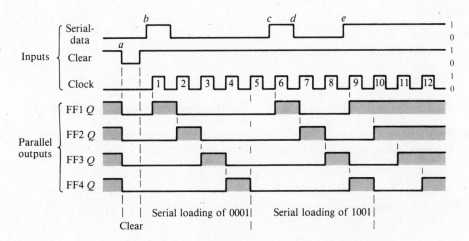

Fig. 11-3   Timing diagram for a 4-bit serial-load shift-right register

three inputs (serial-data, clear, and clock) to the register are shown across the top. The parallel outputs are shown in the middle-four lines. Note that the outputs are taken from the normal output (Q) of each flip-flop. The bottom line describes several functions of the shift register.

Consider the initial conditions of all the flip-flops shown in Fig. 11-3. All are set. At point a on the clear-input waveform, all the flip-flops are reset to 0000. The clear input operates asynchronously and overrides all other inputs. Note that the clear input is an active-LOW input.

At point b on the serial-data input, a HIGH is placed on the D input to FF1. On the leading edge of clock pulse 1, the HIGH is transferred to output Q of FF1. The output now reads 1000. Clock pulse 2 transfers a 0 to output Q of FF1. At the same time, the 1 at input D to FF2 is transferred to output Q of this flip-flop. The output is now 0100. Clock pulse 3 transfers a 0 to the output of FF1. The 1 at input D of FF3 is transferred to the output of this flip-flop. The output of the register is now 0010. Clock pulse 4 transfers a 0 to the output of FF1. The 1 at input D of FF4 is transferred to the output of this flip-flop. The output of the register is now 0001. It took four clock pulses (pulses 1 through 4, Fig. 11-3) to serially load the 4-bit word 0001 into the register.

Consider clock pulse 5 (Fig. 11-3). Just before pulse 5, the register contents are 0001. Clock pulse 5 adds a new 0 at the left (at Q of FF1), and the 1 on the right is shifted out of the register and is lost. The result is that the register contents are 0000 after clock pulse 5.

Consider clock pulses 6 through 9 (Fig. 11-3). These four pulses are used to serially load the binary word 1001 into the register. At point c the serial-data input is placed at 1. On the L-to-H transition of clock pulse 6, this 1 is transferred from the D input of FF1 to its Q output. After pulse 6 the register reads 1000. The serial-data input is returned to 0 at point d. Clock pulses 7 and 8 shift the 1 to the right. After clock pulse 8, the register reads 0010. The serial data input is placed at 1 at point e. On the leading edge of clock pulse 9, this 1 is placed at output Q of FF1 and the other data is shifted one place to the right. The register contents after clock pulse 9 are 1001. It took four clock pulses (6 through 9) to serially load 1001 into the register.

Consider clock pulses 10 through 12 (Fig. 11-3). The serial-data input remains at 1 during these pulses. Before pulse 10 the register contents are 1001. On each pulse, a 1 is added to output Q of FF1 and the other 1s are shifted to the right. After clock pulse 12, the register contents are 1111.

If output D of FF4 in Fig. 11-2 is considered the only output, this storage unit could be classified as a serial-in serial-out shift register.

SOLVED PROBLEMS

11.1 The 4-bit shift register described in this section uses _____ (decimal number) _____ (D, T) flip-flops.

Solution:

The 4-bit register uses four D flip-flops.

11.2 The flip-flops shown in Fig. 11-2 are _____ (leading-, trailing-) edge triggered.

Solution:

The flip-flops shown in Fig. 11-2 are leading-edge triggered.

11.3 In Fig. 11-2, the shift-right operation means to shift data from _____ (FF1, FF4) to _____ (FF1, FF4).

Solution:

By definition, shift right means to shift data from FF1 to FF4 in Fig. 11-2.

11.4 Refer to Fig. 11-2. Clear is an active-_____ (HIGH, LOW) input.

Solution:

    Clear is an active-LOW input in Fig. 11-2, as shown by the bubbles on the CLR inputs of each *D* flip-flop.

11.5 Refer to Fig. 11-3. The clear is a(n) _____ (asynchronous, synchronous) input.

Solution:

    The clear is an asynchronous input to the register (Fig. 11-3).

11.6 Refer to Fig. 11-4. List the states of the output indicators of the shift register *after* each clock pulse (bit *A* on left, bit *D* on right).

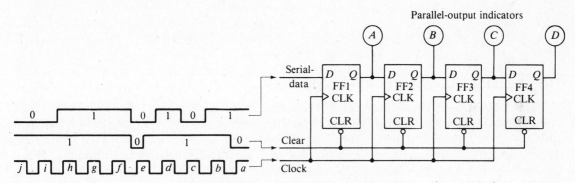

Fig. 11-4 Shift-register pulse-train problem

Solution:

    The output states of the register shown in Fig. 11-4 are as follows:

pulse *a* = 0000	Clear mode resets all FFs to 0.	pulse *f* = 1000	Shift-right mode moves bits one position to the right.
pulse *b* = 1000	Shift-right mode moves bits one position to the right on leading edge of clock pulse. Note that the 1 at the *D* input to FF1 is shifted to the *Q* output of FF1.	pulse *g* = 1100	Shift-right mode moves bits one position to the right. Note that a 1 is being shifted into the leftmost position from input *D* of FF1.
pulse *c* = 0100	Shift-right mode moves bits one position to the right. Note that the 0 at the *D* input to FF1 is shifted to the *Q* output of FF1.	pulse *h* = 1110	Shift-right mode moves bits one position to the right.
		pulse *i* = 0111	Shift-right mode. Note the 0 being loaded into the left position from input *D* of FF1.
pulse *d* = 1010	Shift-right mode moves bits one position to the right. Note that the 1 at *D* input to FF1 is shifted to the *Q* output of FF1.	pulse *j* = 0011	Shift-right mode. Note the 0 being loaded into the left position.
pulse *e* = 0000	Temporarily the output goes to 0101 on leading edge of the clock pulse. Then the clear input is activated, resetting all FFs to 0.		

11.7 Refer to Fig. 11-4. This is a _____ (parallel-, serial-) load shift-_____ (left, right) register.

Solution:

The device shown in Fig. 11-4 is a serial-load shift-right register.

11.8 Refer to Fig. 11-4. After clearing, it takes _____ clock pulse(s) to load a 4-bit word into this register.

Solution:

It takes four clock pulses to serially load the register shown in Fig. 11-4.

11.9 Refer to Fig. 11-4. The CLK inputs to the flip-flops are wired in _____ (parallel, series), and therefore all shifts take place at the same time.

Solution:

The CLK inputs to the flip-flops shown in Fig. 11-4 are wired in parallel.

11-3 PARALLEL-LOAD SHIFT REGISTER

The disadvantage of the serial-load shift register is that it takes many clock pulses to load the unit. A parallel-load shift register loads all bits of information immediately. One simple 4-bit parallel-load shift register is diagrammed in Fig. 11-5. Note the use of *JK* flip-flops with both CLR and *PS* inputs. The inputs at the left are the clear, clock, and four parallel-data (parallel-load) inputs. The clock connects each CLK input in parallel. The clear connects each CLR input in parallel. The *PS* input for each flip-flop is brought out for parallel-data loading. The output indicators across the top of Fig. 11-5 show the state of output *Q* of each flip-flop. Note the wiring of the *JK* flip-flops. Especially note the two *feedback* lines running from the *Q* of FF4 back to the *J* of FF1 and from the $\overline{Q}$ of FF4 back to the *K* of FF1. These are recirculating lines, and they save the data that would normally be lost out the right end of the register. It is said that the data will *recirculate* through the register.

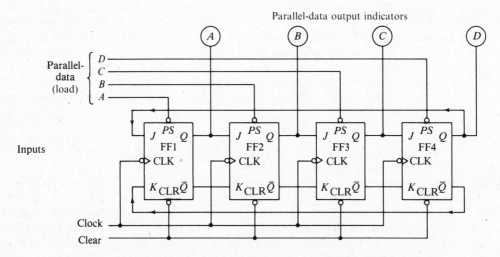

Fig. 11-5 Logic diagram of a 4-bit parallel-load recirculating shift-right register

Note from the *JK* flip-flop logic symbols in Fig. 11-5 that the *PS* and CLR inputs are active-LOW inputs. They are also asynchronous and override all other inputs. Assume that these *JK* flip-flops are pulse-triggered units.

A waveform diagram for the *parallel-load recirculating shift-right register* is shown in Fig. 11-6. The top four lines on the diagram are the parallel-data inputs, or load inputs. These are normally HIGH and are placed LOW only when loading. The clear and clock inputs are shown near the center of the diagram.

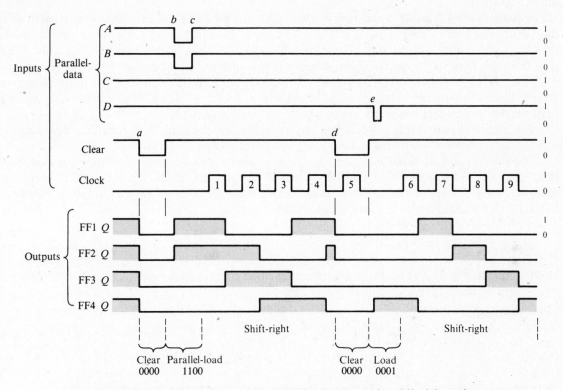

Fig. 11-6 Timing diagram for a 4-bit parallel-load recirculating shift-right register

The four shaded waveforms in Fig. 11-6 are the outputs at Q of each *JK* flip-flop. Across the bottom of the diagram are functions being performed by the register.

Consider the outputs on the left side of Fig. 11-6. The outputs are 1111 before point a on the clear waveform. At point a, the outputs are immediately reset to 0000. The clear input is asynchronous and therefore needs no clock pulse to reset the register.

At point b the A and B parallel-data inputs are activated. Being asynchronous inputs, the outputs of FF1 and FF2 go HIGH immediately. At point c, the A and B parallel-data inputs are deactivated. The register is now loaded with 1100.

On the trailing edge of clock pulse 1, the two 1s shift one position to the right. The result is 0110 after clock pulse 1. Another right shift takes place on the trailing edge of clock pulse 2. The result after pulse 2 at the outputs of the *JK* flip-flops is 0011.

Consider clock pulse 3, shown in Fig. 11-6. The output was 0011 before pulse 3. On the trailing edge of pulse 3, a right shift takes place. The 1 at Q of FF4 would normally be lost, but because of the recirculating lines (see Fig. 11-5) it is shifted back around to Q of FF1. The result is that the register contents are 1001 after clock pulse 3. Likewise, clock pulse 4 shifts the register one place to the right. The 1 at Q of FF4 is shifted to Q of FF1. The results are that, after pulse 4, the register contains 1100. This is the same data that was loaded in the register prior to clock pulse 1. It took four pulses to recirculate the data to its original position.

Consider point d on the clear waveform in Fig. 11-6. It is an asynchronous input; therefore, as soon as it goes LOW, all flip-flops are reset. Clock pulse 5 has no effect because the clear input overrides the clock.

Consider point *e* on the parallel-load waveform in Fig. 11-6. For a very short time, parallel-data input *D* is activated and then deactivated. It loads 0001 into the register. Clock pulse 6 recirculates the 1 at output *Q* of FF4 to FF1. After pulse 6, the register contains 1000. Pulses 7, 8, and 9 shift the single 1 to the right three places. After the four pulses (6 through 9), the data is the same as the original: 0001.

A careful look at Figs. 11-5 and 11-6 will show that the *JK* flip-flops are always operating in either the *set* or *reset* modes. Before pulse 6 (Fig. 11-6), the *Q* outputs are 0001. However, remember that the complementary $\overline{Q}$ outputs are at the same instant 1110. On the trailing edge of clock pulse 6, FF1 goes from the reset to the set condition because it has inputs of $J = 1$ and $K = 0$. FF2 has inputs of $J = 0$ and $K = 1$ and therefore stays in the reset condition. FF3 has inputs of $J = 0$ and $K = 1$ and therefore stays in the reset condition. FF4 has inputs of $J = 0$ and $K = 1$. FF4 changes state and goes from the set to the reset condition.

The circuit shown in Fig. 11-5 is but one of many parallel-load shift registers. Because these registers are somewhat complicated, they are often purchased in IC form.

The shift register shown in Fig. 11-5 might also be called a *ring counter* if a single 1 is loaded into the register. If a continuous series of pulses arrives at the clock input, the lone HIGH output will sequence in a "ring" in the register. Each output (*A*, *B*, *C*, and *D*) can then be turned on (HIGH = on) in sequence as the ring counter shifts.

SOLVED PROBLEMS

11.10 Refer to Fig. 11-5. The parallel-load recirculating register uses four _____ (*D*, *JK*) flip-flops with asynchronous _____ and _____ inputs.

Solution:

The register shown in Fig. 11-5 uses four *JK* flip-flops with asynchronous clear (CLR) and preset (*PS*) inputs.

11.11 Refer to Fig. 11-5. The asynchronous inputs (*PS* and CLR) to the *JK* flip-flops have active-_____ (HIGH, LOW) inputs.

Solution:

The asynchronous inputs to the flip-flops shown in Fig. 11-5 have active-LOW inputs.

11.12 Refer to Fig. 11-5. This register is a shift-_____ (left, right) device because it shifts data from output *Q* of _____ (FF1, FF4) to _____ (FF1, FF4).

Solution:

The register shown in Fig. 11-5 is a shift-right device because it shifts data from FF1 to FF4.

11.13 Refer to Fig. 11-5. It takes _____ clock pulse(s) to load a 4-bit number in this shift register.

Solution:

It takes zero clock pulses to load the register shown in Fig. 11-5. The *PS* (parallel-load) inputs are asynchronous and therefore do not need a clock pulse to load the register.

11.14 Refer to Fig. 11-5. The *JK* flip-flops are always in either the _____ or the _____ mode in this register.

Solution:

The *JK* flip-flops are always in either the reset or set mode in the register shown in Fig. 11-5.

11.15 The *JK* flip-flop is in its set mode when input $J =$ _____ (0, 1) and input $K =$ _____ (0, 1).

Solution:

   The *JK* flip-flop is in its set mode when $J = 1$ and $K = 0$.

11.16 The *JK* flip-flop is in its reset mode when input $J =$ _____ (0, 1) and input $K =$ _____ (0, 1).

Solution:

   The *JK* flip-flop is in its reset mode when $J = 0$ and $K = 1$.

11.17 List the state of the output indicators *after* each clock pulse on the shift-right register shown in Fig. 11-7.

Solution:

   The outputs of the register shown in Fig. 11-7 after each clock pulse are as follows:

pulse $a = 000$	Clear mode resets all FFs to 0.	pulse $f = 000$	Clear mode resets all FFs to 0.
pulse $b = 010$	Parallel-load mode sets outputs to 100. On trailing edge of pulse, register shifts right one position to 010.	pulse $g = 101$	Temporarily the parallel-load inputs *B* and *C* load 011 into the register. On the trailing edge of the clock pulse, the shift-right mode causes the bits to shift right one position. The 1 at *C* is recirculated to position *A*.
pulse $c = 001$	Shift-right mode shifts bits one position to the right. The 0 at *C* is recirculated back to *A*.		
pulse $d = 100$	Shift-right mode shifts bits one position to the right. The 1 at *C* is recirculated back to *A*.	pulse $h = 110$	Shift-right mode. The 1 at *C* is recirculated back to *A*.
		pulse $i = 011$	Shift-right mode.
pulse $e = 010$	Shift-right mode shifts bits one position to the right.	pulse $j = 111$	Parallel-load mode loads all FFs with a 1.

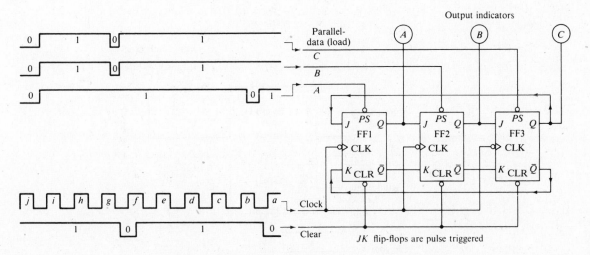

Fig. 11-7 Parallel-load shift-register pulse-train problem

11.18 Refer to Fig. 11-7. What is the mode of operation of each flip-flop while clock pulse c is HIGH?

Solution:

The modes of operation of the flip-flops while clock pulse c is HIGH (Fig. 11-7) are as follows:

$$\text{FF1 mode} = \text{reset } (J = 0, K = 1)$$
$$\text{FF2 mode} = \text{reset } (J = 0, K = 1)$$
$$\text{FF3 mode} = \text{set } (J = 1, K = 0)$$

11.19 Refer to Fig. 11-7. What is the mode of operation of each flip-flop when pulse j is HIGH?

Solution:

All flip-flops are being asynchronously preset by the active parallel-data inputs. All flip-flops are in the set mode ($J = 1, K = 0$).

11.20 Refer to Fig. 11-7. This digital device is a _____-bit _____ (nonrecirculating, recirculating) shift-_____ (left, right) register.

Solution:

The digital device shown in Fig. 11-7 is a 3-bit, recirculating shift-right register.

11-4   TTL SHIFT REGISTERS

Integrated-circuit manufacturers market many shift registers. The one that has been selected for description is a universal shift register. A block logic symbol for the commercial TTL 74194 4-bit universal shift register is shown in Fig. 11-8. The 74194 register has 10 inputs and 4 outputs. The outputs are connected to the normal (Q) outputs of each flip-flop inside the IC.

Consider the inputs on the 74194 register shown in Fig. 11-8. The parallel-load inputs (A, B, C, D) are the top four inputs. The next two inputs are for feeding data into the register *serially* (one bit at a time). The shift-right serial input (D_{SR}) feeds bits into position A (Q_A) as the register is shifted to the right. The shift-left serial input (D_{SL}) feeds bits into position D (Q_D) as the register is shifted to the left. The clock input (CLK) triggers the four flip-flops on the L-to-H transition of the clock pulse. When activated with a LOW, the clear (CLR) input resets each flip-flop to 0. The mode controls

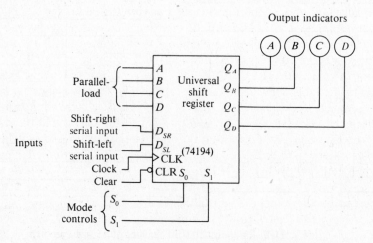

Fig. 11-8 Logic symbol for the 74194 universal shift-register IC

instruct the register through a gating network to shift right, shift left, parallel-load, or hold (do nothing). Of course, the 74194, which is a TTL IC, has a +5 V and GND power-supply connection. The power-supply connections are not usually shown on the logic symbol.

A mode-select–function table for the 74194 shift register is shown in Fig. 11-9. The operating modes for the shift register are listed in the left section of the table. The operating modes are reset, hold, shift-left, shift-right, and parallel-load.

Operating mode	Inputs							Outputs			
	CLK	CLR	S_1	S_0	D_{SR}	D_{SL}	D_n	Q_0	Q_1	Q_2	Q_3
Reset (clear)	X	L	X	X	X	X	X	L	L	L	L
Hold (do nothing)	X	H	l*	l*	X	X	X	q_0	q_1	q_2	q_3
Shift-left	↑	H	h	l*	X	l	X	q_1	q_2	q_3	L
	↑	H	h	l*	X	h	X	q_1	q_2	q_3	H
Shift-right	↑	H	l*	h	l	X	X	L	q_0	q_1	q_2
	↑	H	l*	h	h	X	X	H	q_0	q_1	q_2
Parallel-load	↑	H	h	h	X	X	d_n	d_0	d_1	d_2	d_3

H = HIGH voltage level
h = HIGH voltage level one setup time prior to the L-to-H clock transition
L = LOW voltage level
l = LOW voltage level one setup time prior to the L-to-H clock transition
$d_n(q_n)$ = (Lowercase letters indicate the state of the referenced input [or output] one setup time prior to the L-to-H clock transition.)
X = don't care
↑ = L-to-H clock transition

* The H-to-L transition of the S_0 and S_1 inputs on the 74194 should only take place while CK is HIGH for conventional operation.

Fig. 11-9 Mode select/function table for the 74194 universal shift-register IC

Consider the reset (clear) mode of the shift register shown in Fig. 11-9. When the CLR input is LOW, it overrides all other inputs (all other inputs are X on the table) and clears the outputs to 0000 (shown as LLLL in table). Note that the outputs are identified with a Q_0 instead of Q_A, Q_1 for Q_B, and so forth. Identification of inputs and outputs does vary from manufacturer to manufacturer.

The remaining four modes of operation shown in Fig. 11-9 are controlled by the mode controls (S_0 and S_1). When both mode controls are LOW ($S_0 = 0$, $S_1 = 0$), the shift register is in the hold mode and will do nothing. The table shows that the outputs (at Q_0 through Q_3) are displayed, however.

Consider the shift-left line in Fig. 11-9. The two mode controls are set properly ($S_0 = 0$, $S_1 = 1$), and data is fed into the shift-left serial input (D_{SL}). Note that the 1s and 0s at the shift-left serial input are transferred to the Q_3 (D) position as the register shifts one place to the left. The shift takes place on the L-to-H transition of the clock pulse, as shown by the arrow pointing upward on the table.

Look at the shift-right line in Fig. 11-9. The mode controls are set ($S_0 = 1$, $S_1 = 0$). Data is placed at the shift-right serial input (D_{SR}). On the L-to-H transition of the clock pulse, the bit at input D_{SR} is transferred to the Q_0 (A) output as the register shifts one place to the right.

The final operating mode for the universal shift register is shown on the bottom line in Fig. 11-9. To parallel load (also called *broadside load*) the 74194 register, set the two mode controls ($S_0 = 1$, $S_1 = 1$). On the L-to-H transition of the clock pulse, the data at the parallel-load inputs will be transferred to the appropriate outputs. Note that the parallel-load inputs are *not* asynchronous as they were on the previous parallel-load register. The parallel-load operation takes place in step with a single clock pulse.

The 74194 register is indeed universal. Data can be loaded either serially or in parallel. Data can be read out in parallel or in serial form [output from one point such as Q_D (Q_3)]. The register can hold (or do nothing) on command. The register can shift right or shift left. This 4-bit register is but one of many units manufactured in IC form.

A few other TTL shift registers include the 7494 4-bit and 7496 5-bit shift registers. Also listed in data manuals are the 74164 8-bit serial-in, parallel-out and 74165 8-bit serial/parallel-in, serial-out shift registers. Other shift registers are available from chip manufacturers in the various TTL subfamilies such as the 74LS395A 4-bit cascadeable shift register with 3-state outputs.

SOLVED PROBLEMS

11.21 List the five modes of operation of the 74194 shift register.

   Solution:

   The five modes of operation of the 74194 register are as follows:
   (*a*) reset (clear)   (*c*) shift-left   (*e*) parallel-load
   (*b*) hold   (*d*) shift-right

11.22 Refer to Fig. 11-9. The single asynchronous input on the 74194 register that overrides all other inputs is the _____ input.

   Solution:

   The clear is the only asynchronous input on the 74194 register.

11.23 Refer to Fig. 11-9. What effect does a clock pulse have when the 74194 register is in the hold mode?

   Solution:

   The 74194 register does nothing on a clock pulse when it is in the hold mode.

11.24 Refer to Fig. 11-9. It takes _____ clock pulse(s) to parallel load four bits in the 74194 register.

   Solution:

   It takes one clock pulse to parallel load the 74194 shift register.

11.25 The 74194 input labeled D_{SR} would be used when $S_0 =$ _____ (0, 1) and $S_1 =$ _____ (0, 1).

   Solution:

   The D_{SR} input (shift-right serial input) would be used during the shift-right mode, and therefore $S_0 = 1$ and $S_1 = 0$.

11.26 The 74194 register uses _____ (positive-edge, pulse) triggering.

   Solution:

   The 74194 register uses positive-edge triggering.

11.27 List the operating mode of the shift register (74194) for each of the pulses shown in Fig. 11-10.

   Solution:

   Refer to the S_1 and S_0 columns of the mode-select table in Fig. 11-9. The operating mode of the register for each pulse shown in Fig. 11-10 is as follows:
   pulse a = reset (clear)   pulse d = shift-left   pulse g = shift-right   pulse j = shift-left
   pulse b = parallel-load   pulse e = shift-left   pulse h = reset (clear)
   pulse c = shift-left   pulse f = shift-right   pulse i = hold

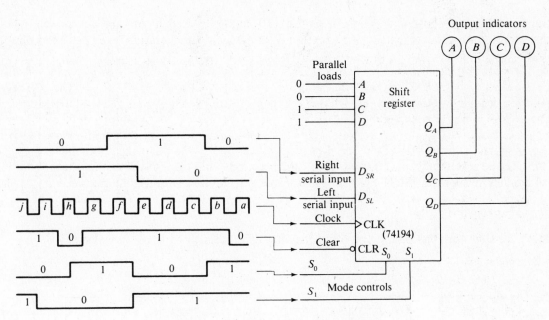

Fig. 11-10 Shift-register pulse-train problem

11.28 List the state of the output indicators after each pulse for the 74194 shift register shown in Fig. 11-10.

Solution:

The output indicators read as follows for the register shown in Fig. 11-10 (A on left, D on right):

pulse $a = 0000$ Reset mode which clears all outputs to 0.

pulse $b = 0011$ Parallel-load mode which loads four parallel-load inputs into register.

pulse $c = 0110$ Shift-left mode moves bits one position to the left. Note that a 0 is being serially loaded into position D from the left serial input.

pulse $d = 1100$ Shift-left mode moves bits one position to the left. Note that a 0 is being serially loaded into position D from the left serial input.

pulse $e = 1000$ Shift-left mode moves bits one position to the left. Note that a 0 is being serially loaded into position D from the left serial input.

pulse $f = 1100$ Shift-right mode moves bits one position to the right. Note that a 1 is being serially loaded into position A from the right serial input.

pulse $g = 0110$ Shift-right mode moves bits one position to the right. Note that a 0 is being serially loaded into position A from the right serial input.

pulse $h = 0000$ The clear input overrides all other inputs and resets all outputs to 0.

pulse $i = 0000$ Hold mode commands the register to do nothing.

pulse $j = 0001$ Shift-left mode moves bits one position to the left. Note that a 1 is being serially loaded into position D from the left serial input.

11-5 CMOS SHIFT REGISTERS

A wide variety of CMOS shift registers are available from chip manufacturers. The *74HC164 8-bit serial-in parallel-out shift-register IC* is featured in this section. Information from the manufacturer's data manual is reproduced in Fig. 11-11.

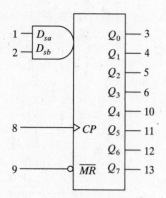

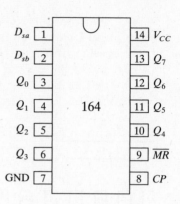

(a) Simplified logic symbol

(b) Pin diagram

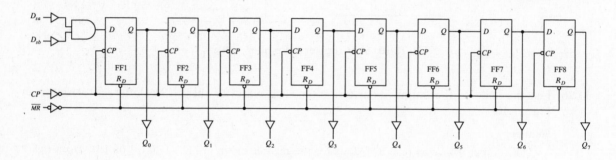

(c) Detailed logic diagram

Operating modes	Inputs				Outputs	
	$\overline{MR}$	CP	D_{sa}	D_{sb}	Q_0	Q_1–Q_7
reset (clear)	L	X	X	X	L	L–L
shift	H	↑	l	l	L	q_0–q_6
	H	↑	l	h	L	q_0–q_6
	H	↑	h	l	L	q_0–q_6
	H	↑	h	h	H	q_0–q_6

H = HIGH voltage level
h = HIGH voltage level one setup time prior to the LOW-to-HIGH
   clock transition
L = LOW voltage level
l = LOW voltage level one setup time prior to the LOW-to-HIGH
   clock transition
q = lowercase letters indicate the state of the referenced input
   one setup time prior to the LOW-to-HIGH clock transiton
↑ = LOW-to-HIGH clock transition

(d) Truth table

Fig. 11-11 The 74HC164 shift register IC

The 74HC164 CMOS IC is an 8-bit edge-triggered shift register that permits only serial data input. Eight parallel outputs are available (Q_0 to Q_7) from each of the eight internal flip-flops (Fig. 11-11c). The clock input (CP) to the 74HC164 is edge-triggered and shifts data on the LOW-to-HIGH transition of the clock pulse. Data is entered one bit at a time (serially) through one of two data inputs (D_{sa} or D_{sb}). The simplified logic diagram in Fig. 11-11a shows that the data inputs (D_{sa} and D_{sb}) are ANDed together. This means that one input can be used as an active-HIGH data enable input while serial data is fed into the second data input. If no data enable input is required, both data inputs (D_{sa} and D_{sb}) are tied together and used as a single serial data input. In Fig. 11-11c, each clock pulse will shift data one position to the right (from Q_0 to Q_7) in the shift register. The master reset ($\overline{MR}$) pin on the 74HC164 IC is an active-LOW input which resets all eight flip-flops and clears the outputs to 0. The master reset ($\overline{MR}$) is an asynchronous input that overrides all other inputs. The 74HC164 shift register is packaged in a 14-pin DIP IC, shown in Fig. 11-11b. A truth table detailing the operating modes of the 74HC164 IC is reproduced in Fig. 11-11d. The 74HC164 IC operates on a 5-V dc power supply.

Manufacturers produce a variety of CMOS shift registers. If you are wiring shift registers using D flip-flops, the 4076 and 40174 ICs are available. The 4014 8-stage static shift-register IC is a serial-in parallel-out device. The 4031 64-stage static shift register is a serial-in serial-out unit. The 4035 4-bit shift register is a parallel-in parallel-out storage unit. The 4034 8-bit static shift register is a universal 3-state bidirectional parallel/serial-input/output unit which can input from and output to bus lines. Many other shift registers are also available in the 74HC and 74HCT series of CMOS ICs.

SOLVED PROBLEMS

11.29 Refer to Fig. 11-11. The _____ input to the 74HC164 shift register overrides all others when activated with a LOW.

Solution:
The active-LOW master reset ($\overline{MR}$) input to the 74HC164 shift register overrides all others.

11.30 The master reset ($\overline{MR}$) pin on the 74HC164 IC is a(n) _____ (asynchronous, synchronous) input.

Solution:
The reset ($\overline{MR}$) pin on the 74HC164 IC is an asynchronous input.

11.31 Refer to Fig. 11-11. The clock input (CP) to the 74HC164 IC is _____ (edge, pulse)-triggered and shifts data on the _____ (H-to-L, L-to-H) transition of the clock pulse.

Solution:
The clock input (CP) to the 74HC164 IC is edge-triggered and shifts data on the L-to-H transition of the clock pulse.

11.32 The 74HC164 is an 8-bit _____ (parallel, serial)-in parallel-out shift register.

Solution:
The 74HC164 is an 8-bit serial-in parallel-out shift register.

11.33 Refer to Fig. 11-11. Why does the 74HC164 IC have two serial data inputs (see D_{sa} and D_{sb})?

Solution:

The 74HC164 IC has two ANDed serial data inputs (D_{sa} and D_{sb}). Two serial data inputs allow one to be used as an active-HIGH serial data enable input to turn the data input on and off.

11.34 Refer to Fig. 11-12. The 74HC164 shift register is in the _____ (reset, shift) mode of operation during clock pulse *a*.

Solution:

The master reset ($\overline{MR}$) input is activated with a LOW during pulse *a* so the shift register is in the reset mode. Remember that the reset (MR) input overrides all others.

11.35 Refer to Fig. 11-12. The 74HC164 IC is in the _____ (reset, shift) mode of operation during clock pulse *b*.

Solution:

Reset (MR) is deactivated so the 74HC164 IC shifts right one position loading a 1 bit from the data input (D_{sb}) into the Q_0 position. The results after pulse *b* are 10000000.

11.36 Refer to Fig. 11-12. The serial-in data inputs are _____ (activated, deactivated) during clock pulse *c*.

Solution:

See Fig. 11-12. The top input to the serial data AND gate is LOW during clock pulse *c*, which deactivates the entire serial data input.

11.37 List the state of the output indicators after each clock pulse for the 74HC164 shift register shown in Fig. 11-12.

Solution:

The output indicators read as follows for the shift register shown in Fig. 11-12 (Q_0 on left, Q_7 on right):

pulse *a* = 0000 0000   Reset mode
pulse *b* = 1000 0000   Shift right—serial load a 1 into Q_0
pulse *c* = 0100 0000   Shift right—serial load a 0 into Q_0
pulse *d* = 0010 0000   Shift right—serial input disabled
pulse *e* = 0001 0000   Shift right—serial input disabled
pulse *f* = 0000 1000   Shift right—serial input disabled
pulse *g* = 0000 0100   Shift right—serial input disabled
pulse *h* = 0000 0010   Shift right—serial input disabled
pulse *i* = 0000 0001   Shift right—serial input disabled
pulse *j* = 0000 0000   Shift right—serial input disabled

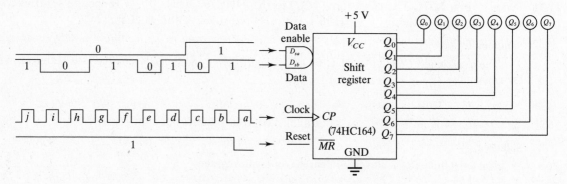

Fig. 11-12 Shift-register pulse-train problem

Supplementary Problems

11.38 Draw the logic diagram for a 5-bit serial-load shift-right register. Use five *D* flip-flops. Label inputs as clock, clear, and serial-data. Label outputs as *A*, *B*, *C*, *D*, and *E*.   *Ans.*   See Fig. 11-13.

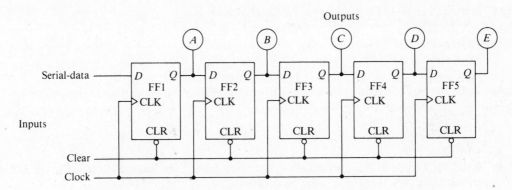

Fig. 11-13 Logic diagram for a 5-bit serial-load shift-right register

11.39 It takes _____ clock pulse(s) to load a 5-bit serial-load shift register.   *Ans.*   five

11.40 Refer to Fig. 11-14. The *data* input is a   *(a)*   (parallel, serial) data input to this   *(b)*   -bit shift- _____   *(c)*   (left, right) register.   *Ans.*   *(a)* serial   *(b)* 3   *(c)* right

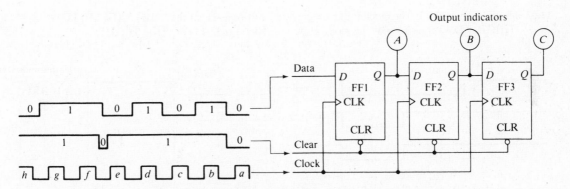

Fig. 11-14 Serial-load shift-register pulse-train problem

11.41 List the output after each clock pulse for the shift register shown in Fig. 11-14 (*A* on left, *C* on right).
     Ans.   pulse *a* = 000   pulse *c* = 010   pulse *e* = 010   pulse *g* = 110
          pulse *b* = 100   pulse *d* = 101   pulse *f* = 100   pulse *h* = 011

11.42 Refer to Fig. 11-14. List the two synchronous inputs on this register.   *Ans.*   data (serial), clock

11.43 Refer to Fig. 11-14. It takes _____ clock pulse(s) to load this register with 011.   *Ans.*   three

11.44 Refer to Fig. 11-7. This 3-bit parallel-load shift register uses   *(a)*   (*D*, *JK*) flip-flops and is a   *(b)*   (nonrecirculating, recirculating) unit.   *Ans.*   *(a)* *JK*   *(b)* recirculating

11.45 Refer to Fig. 11-7. What is the mode of operation of each *JK* flip-flop while clock pulse *d* is HIGH?
       Ans. FF1 mode = set ($J = 1, K = 0$)
            FF2 mode = reset ($J = 0, K = 1$)
            FF3 mode = reset ($J = 0, K = 1$)

11.46 Refer to Fig. 11-7. The active state for the clear input is _____ (0, 1).   *Ans.* 0

11.47 Refer to Fig. 11-7. The output indicators on this register read $A =$ _(a)_ , $B =$ _(b)_ , and $C =$ _(c)_
       when pulse *g* is HIGH.   *Ans.* (*a*) 0 (*b*) 1 (*c*) 1

11.48 Refer to Fig. 11-7. What is the mode of operation of each *JK* flip-flop while clock pulse *h* is HIGH?
       Ans. FF1 mode = set ($J = 1, K = 0$)
            FF2 mode = set ($J = 1, K = 0$)
            FF3 mode = reset ($J = 0, K = 1$)

11.49 Refer to Fig. 11-7. The two lines with arrows going back from FF3 to FF1 are called _____
       (recirculating, reset) lines.   *Ans.* recirculating

11.50 Refer to Fig. 11-7. The *JK* flip-flops are triggered on the _(a)_ (HIGH, LOW)-to- _(b)_ (HIGH,
       LOW) transition of the clock pulse.   *Ans.* (*a*) HIGH (*b*) LOW

11.51 Refer to Fig. 11-7. List the outputs of the register while each clock pulse is HIGH (just before the H-to-L
       transition of the clock).
       Ans. pulse $a = 000$ pulse $d = 001$ pulse $g = 011$ pulse $j = 111$
            pulse $b = 100$ pulse $e = 100$ pulse $h = 101$
            pulse $c = 010$ pulse $f = 000$ pulse $i = 110$

11.52 Refer to Fig. 11-15. The clock input triggers the shift register on the _(a)_ (HIGH, LOW)-to- _(b)_
       (HIGH, LOW) transition of the clock pulse.   *Ans.* (*a*) LOW (*b*) HIGH

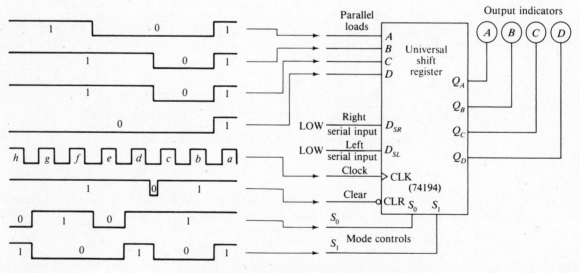

Fig. 11-15 Universal shift-register pulse-train problem

11.53 Refer to Fig. 11-15. The clear connection to the 74194 register is an active _____ (HIGH, LOW) input
       and overrides all others.   *Ans.* LOW

11.54   List the operating mode of the 74194 shift register for each clock pulse shown in Fig. 11-15.
       Ans.   pulse *a* = parallel-load   pulse *d* = parallel-load   pulse *g* = shift-right
                pulse *b* = shift-right   pulse *e* = hold   pulse *h* = shift-left
                pulse *c* = shift-right   pulse *f* = shift-right

11.55   List the states of the output indicators of the register shown in Fig. 11-15 while each clock pulse is HIGH.
       Ans.   pulse *a* = 1111   pulse *c* = 0011   pulse *e* = 0110   pulse *g* = 0001
                pulse *b* = 0111   pulse *d* = 0110   pulse *f* = 0011   pulse *h* = 0010

11.56   Refer to Fig. 11-15. Just *before* clock pulse *d*, the output indicators read _____. Why?
       Ans.   Because of the clear pulse, the output indicators read 0000 just before clock pulse *d* (Fig. 11-15).

11.57   Refer to Fig. 11-15. During pulse *a*, this register is set up for _____ (parallel, serial) loading.
       Ans.   parallel

11.58   Another term for parallel loading is _____ loading.   *Ans.*   broadside

11.59   A shift register is classified as a _____ (combinational, sequential) logic circuit.   *Ans.*   sequential

11.60   The storage unit shown in Fig. 11-4 might be classified as a serial-in _____ -out register.
       Ans.   parallel

11.61   Refer to Fig. 11-5. The recirculating shift register might also be called a _____ counter.   *Ans.*   ring

11.62   Refer to Fig. 11-16. Which part of the figure illustrates the idea of a serial-in parallel-out register?
       Ans.   part *b*

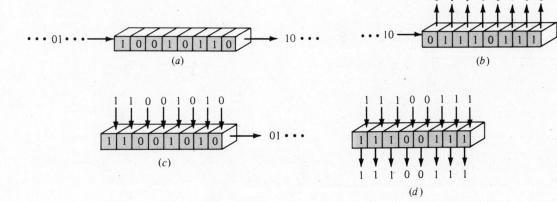

Fig. 11-16 Types of registers

11.63   Refer to Fig. 11-16. Which part of the figure illustrates the idea of a parallel-in serial-out register?
       Ans.   part *c*

11.64   The 74HC164 IC is best represented by which type of register pictured in Fig. 11-16?
       Ans.   *b* (serial-in parallel-out)

11.65   Refer to Fig. 11-12. This shift register is an example of a _____ (CMOS, TTL) IC.   *Ans.*   CMOS

11.66   Refer to Fig. 11-12. The master reset ($\overline{MR}$) pin on the 74HC164 is an active-_____ (HIGH, LOW) input and overrides all others.   *Ans.*   LOW

11.67   Refer to Fig. 11-12. The data enable (D_{sa}) input is an active-_____ (HIGH, LOW) input in this example.   *Ans.*   HIGH

11.68   Refer to Fig. 11-12. Assuming the data enable (D_{sa}) input is *HIGH the entire time* (pulses *a* to *i*), list the states of the output indicators for the shift register after each clock pulse (Q_0 on left, Q_7 on right).
   Ans.   Assuming D_{sa} input = HIGH, then
            pulse *a* = 0000 0000
            pulse *b* = 1000 0000
            pulse *c* = 0100 0000
            pulse *d* = 1010 0000
            pulse *e* = 0101 0000
            pulse *f* = 1010 1000
            pulse *g* = 1101 0100
            pulse *h* = 0110 1010
            pulse *i* = 0011 0101
            pulse *j* = 1001 1010

Chapter 12

Microcomputer Memory

12-1 INTRODUCTION

In a new electronic product, designers must choose to use either analog or digital devices. If the unit must input, process, or output alphanumeric data, the choice is clearly digital. Also, if the unit has any type of memory or stored program, the choice is clearly digital. Digital circuitry is becoming more popular, but most complex electronic systems contain both analog and digital devices.

Microcomputer memory is one example of the application of data storage devices called memory. A simplified microcomputer system is shown in Fig. 12-1. The keyboard is the input device, while the output device is the video monitor. The central processing unit (CPU) controls the operation of the microcomputer system and processes data. The internal memory of a typical microcomputer system is composed of three types of *semiconductor memory*. The *nonvolatile* semiconductor memory is shown in Fig. 12-1 as *ROM* (*read-only memory*) and *NVRAM* (*nonvolatile RAM*). The *volatile* semiconductor memory is shown as *RAM* (*random-access memory*).

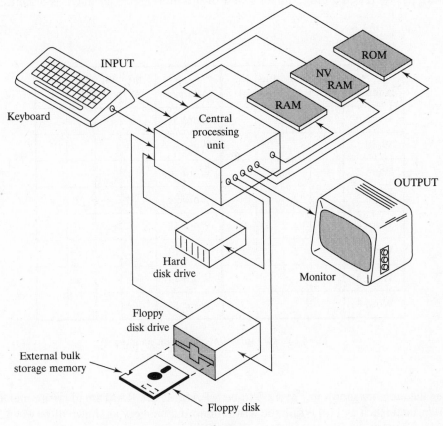

Fig. 12-1

Data and most programs are commonly stored on *magnetic bulk storage* devices such as *floppy disks* or *hard disks*. The disk drive is the unit that reads from or writes to either the floppy disk or the hard disk. Strictly speaking, each device, such as the keyboard, video monitor, disk drive, and CPU, has smaller memory devices. These memory devices usually take the form of registers and latches, but they can contain smaller ROM and RAM semiconductor memory devices.

279

The RAM and ROM storage devices come as ICs and are typically mounted on printed-circuit boards as depicted in Fig. 12-1. It is usual to have at least one ROM and many RAM ICs in a microcomputer.

12-2   RANDOM-ACCESS MEMORY (RAM)

Semiconductor memories are classified as volatile and nonvolatile. A volatile memory is one that loses its data when power is turned off. The *RAM* (*random-access memory*) is a volatile semiconductor memory widely used in modern microcomputers to hold data and programs temporarily. The RAM is also described as a *read/write memory*. Storing data in a RAM is called the *write operation* or writing. Detecting or recalling data from RAM is called the *read operation* or reading. When data is read from memory, the contents of RAM are not destroyed.

Consider the table in Fig. 12-2. This is a representation of the inside of a 64-bit memory. The 64 squares (mostly blank) represent the 64 memory cells inside the 64-bit memory. The memory is organized into 16 groups of 4 bits each. Each 4-bit group is called a *word*. This memory is said to be organized as a 16×4 memory. It contains 16 words of 4 bits each. The representation of the 64-bit memory shown in Fig. 12-2 is a programmer's view of this unit. Electronically it is organized somewhat differently.

Address	Bit D	Bit C	Bit B	Bit A	Address	Bit D	Bit C	Bit B	Bit A
Word 0					Word 8				
Word 1					Word 9				
Word 2					Word 10				
Word 3					Word 11				
Word 4					Word 12				
Word 5	1	1	0	1	Word 13				
Word 6					Word 14				
Word 7					Word 15				

Fig. 12-2 Organization of a 64-bit memory

Consider the memory shown in Fig. 12-2 to be a RAM. If the RAM (read/write memory) were in the *write mode*, data (such as 1101) could be written into the memory as shown after word 5. The write process is similar to writing on a scratch pad. If the RAM were in the *read mode*, data (such as 1101) could be read from the memory. The process is similar to reading the 1101 from word location 5 in Fig. 12-2. A RAM memory of this type is sometimes called a *scratch-pad memory*. Reading word 5 (1101) does not destroy the contents of the memory; it is said that the read process is *nondestructive*. The memory in Fig. 12-2 is a *random-access memory* because one can skip down to word 5 or any other word with ease.

A logic diagram of a simple RAM IC is drawn in Fig. 12-3a. The 74F189 TTL RAM IC is a 64-bit read/write random-access memory. The 74F189 IC is from the newer Fairchild advanced Schottky TTL, FAST, a subfamily that exhibits a combination of performance and efficiency unapproached by any other TTL family. Its internal organization is similar to that shown in Fig. 12-2. It has 16 words, each of which is 4 bits long for a total of 64 memory locations.

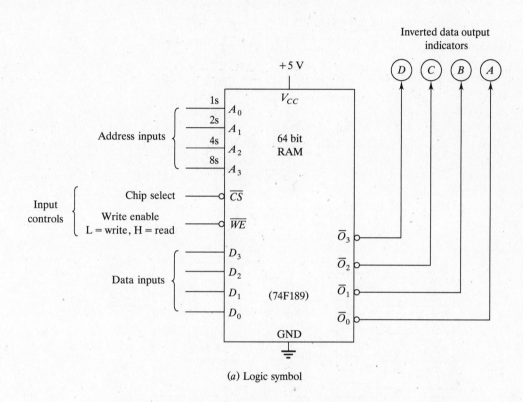

(a) Logic symbol

	Inputs		
Operating mode	$\overline{CS}$	$\overline{WE}$	Condition of outputs
Write	L	L	High impedance
Read	L	H	Complement of stored data
Store (inhibit)	H	X	High impedance

H = HIGH voltage level
L = LOW voltage level
X = Irrelevant

(b) Operating modes

Fig. 12-3 74F189 64-bit static RAM

The modes of operation for the 74F189 RAM IC are detailed in Fig. 12-3b. The first line of the chart illustrates the *write mode*. Note from Fig. 12-3b that both control inputs ($\overline{CS}$ and $\overline{WE}$) are LOW. During the write operation, the 4 input data bits (D_3, D_2, D_1, D_0) are written into the memory word location specified by the address inputs. For instance, to write 1101 into word location 5 as shown in Fig. 12-2, the data inputs must be $D_3 = 1$, $D_2 = 1$, $D_1 = 0$, and $D_0 = 1$ and the address inputs must be $A_3 = 0$, $A_2 = 1$, $A_1 = 0$, $A_0 = 1$. Next, the write enable input ($\overline{WE}$) must be LOW. Finally, the chip

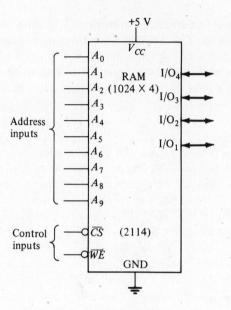

(a) Logic diagram

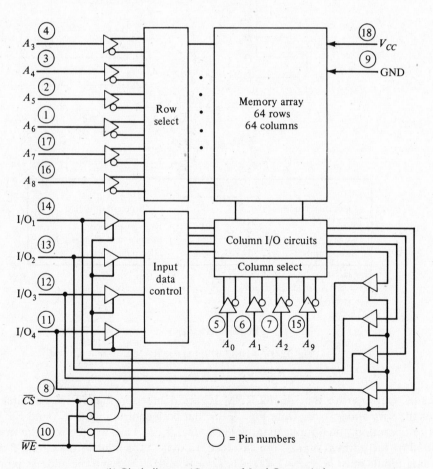

(b) Block diagram (*Courtesy of Intel Corporation*)

Fig. 12-4 MOS static RAM (1024 × 4)

select ($\overline{CS}$) input must go LOW. Memory address location 5 (word 5) now contains the data 1101. It will be noted that the outputs of the 74F189 RAM remain in their high impedance state during the write operation.

The second line of the table in Fig. 12-3b shows the *read mode* for the 74F189 RAM. The input controls must be set so that $\overline{CS}$ = LOW and $\overline{WE}$ = HIGH. The contents of the memory location addressed will appear at the outputs ($\overline{O}_3, \overline{O}_2, \overline{O}_1, \overline{O}_0$) in *complementary* form. Note the invert bubbles at the outputs on the logic symbol in Fig. 12-3a. For example, to read the contents of word 5 (address location 5) in the memory in Fig. 12-2, the chip select ($\overline{CS}$) must be activated with a LOW and the write enable ($\overline{WE}$) must be deactivated with a HIGH. The address inputs must be $A_3 = 0$, $A_2 = 1$, $A_1 = 0$, $A_0 = 1$. The outputs indicate 0010, which is the complement of the true data 1101 located at address location 5 in the RAM. It should be understood that the read operation does not destroy the stored data in memory location 5 but outputs an inverted *copy* of that data.

The bottom line in the table in Fig. 12-3b illustrates the *store* or *inhibit mode*. When the chip select ($\overline{CS}$) input is deactivated with a HIGH, the outputs go to a high impedance state (they float) and the read and write operations are inhibited. It can be said that the memory is "storing" data.

The 74F189 IC is an example of a static RAM. A static RAM can be fabricated using either bipolar or MOS technology. The static RAM uses flip-flops (or similar circuits) as memory cells and holds its data as long as power is supplied to the chip. Large-capacity temporary storage units are usually *dynamic RAMs*. A dynamic RAM's memory cell is based on an MOS device that stores a charge (much like a small capacitor). The difficulty with dynamic RAMs (DRAMs) is that all memory cells must be refreshed every few milliseconds to retain data.

The 2114 static RAM is a popular MOS memory IC. It will store 4096 bits, which are organized into 1024 words of 4 bits each. A logic diagram of the 2114 RAM is in Fig. 12-4a. Note that the 2114 RAM has 10 address lines which can access 1024 (2^{10}) words. It has the familiar chip select ($\overline{CS}$) and write enable ($\overline{WE}$) control inputs, I/O_1, I/O_2, I/O_3, and I/O_4 are inputs when the RAM is in the write mode and outputs when the IC is in the read mode. The 2114 RAM is powered by a +5-V power supply.

A block diagram of the 2114 RAM is shown in Fig. 12-4b. Note especially the three-state buffers used to isolate the data bus from the input/output (I/O) pins. Note that the address lines also are buffered. The 2114 RAM comes in 18-pin DIP IC form.

Microprocessor-based systems (such as microcomputers) typically store and transfer data in 8-bit groups called words. Two 2114 RAMs are connected in Fig. 12-5 to form a RAM memory of 1024 words each 8 bits wide. This is referred to as 1K of memory in most microcomputers, that is, 1024 *bytes* (8-bit groups) of memory. Note that the left 2114 RAM furnishes the least significant 4 bits of a

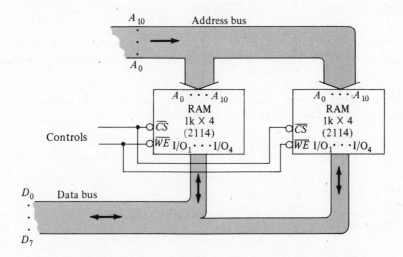

Fig. 12-5 Combining two 1K × 4 RAMs to form a 1K × 8 RAM

word and the right 2114 RAM furnishes the 4 most significant bits. The 2114 RAM has the proper buffering to interface with the system address bus and data bus.

Often-mentioned characteristics of RAMs are size (in bits) and organization (words × bits per word). For the 2114 RAM this would be 4096 bits, or 1024 × 4. For the 74F189 RAM this would be 64 bits, or 16 × 4. A second characteristic might be the technology used to fabricate the chip. This would be *NMOS* (*N-channel metal oxide semiconductor*) for the 2114 RAM. The 74F189 uses the new Fairchild advanced Schottky TTL technology. A third characteristic might be the type of output. Both the 2114 and 74F189 RAMs have three-state outputs.

A fourth characteristic might be the access time (speed) of the memory chip. The *access time* is the time it takes to locate and read data from a RAM. The access time of the 2114 RAM may be from 50 to 450 ns, depending on which version you specify. The access time of the 74F189 RAM is only about 10 ns. The 74F189 is said to be a faster memory. Faster memories are more expensive than their slower counterparts.

A fifth characteristic might be the type of memory: either static (SRAM) or dynamic (DRAM). Both the 2114 and 74F189 ICs are static RAMs. The packaging and power supply voltage are two other common specifications for RAMs. The 2114 RAM is packaged in an 18-pin DIP. The 74F189 IC is housed in either a 16-pin DIP or in a 16-pin LCC (leadless chip carrier) package. Both the 2114 and 74F189 RAMs operate on a 5-V dc power supply.

SOLVED PROBLEMS

12.1   The letters ROM stand for _____-_____ _____ in digital electronics.

   Solution:

   The letters ROM stand for read-only memory.

12.2   The letters RAM literally stand for _____-_____ memory, but in practice they designate a _____ _____ memory.

   Solution:

   The letters RAM literally stand for random-access memory, but in practice they designate a read/write memory.

12.3   The ROM is _____ (nonvolatile, volatile), whereas the RAM is a _____ memory.

   Solution:

   The ROM is nonvolatile, whereas the RAM is a volatile memory.

12.4   The _____ (read, write) process in a RAM consists of putting data into the memory, whereas the _____ (read, write) process consists of revealing the stored contents of a memory location.

   Solution:

   The write process in a RAM consists of putting data into the memory, whereas the read process consists of revealing the stored contents of a memory location.

12.5   The _____ (RAM, ROM) is easily erased.

   Solution:

   The RAM is easily erased.

12.6   A 32 × 8 memory would contain _____ words, each _____ bits long, for a total capacity of _____ bits.

   Solution:

   A 32 × 8 memory would contain 32 words, each 8 bits long, for a total capacity of 256 bits.

12.7 List the mode of operation of the 74F189 RAM for each input pulse shown in Fig. 12-6.

Solution:

The $\overline{WE}$ and $\overline{CS}$ inputs in the memory control the operation of the RAM. The modes of operation for the RAM in Fig. 12-6 are as follows:

pulses a to i = write
pulse j = stored (inhibit read and write)
pulse k = read
pulse l = read

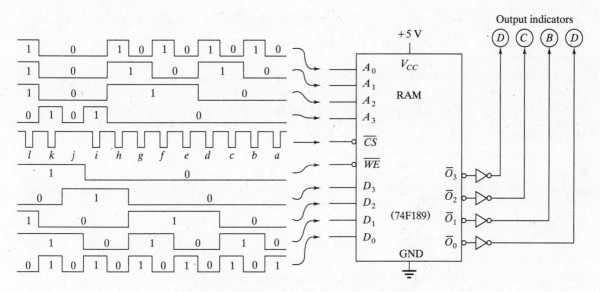

Fig. 12-6 Static RAM pulse-train problem

12.8 List the memory contents of word locations 0 through 8 after pulse l, Fig. 12-6.

Solution:

The memory contents of the 74F189 RAM (Fig. 12-6) after pulse l are as follows:

word 0 = 0001	Written into memory location 0 during pulse a
word 1 = 0010	Written into memory location 1 during pulse b
word 2 = 0011	Written into memory location 2 during pulse c
word 3 = 0100	Written into memory location 3 during pulse d
word 4 = 0101	Written into memory location 4 during pulse e
word 5 = 0110	Written into memory location 5 during pulse f
word 6 = 0111	Written into memory location 6 during pulse g
word 7 = 1000	Written into memory location 7 during pulse h
word 8 = 1001	Written into memory location 8 during pulse i

12.9   What is the state of the output indicators during input pulse k in Fig. 12-6?

   Solution:

   Output indicators read 1001 during pulse k which is a copy of the contents of memory location 8.

12.10   What is the state of the output indicators during input pulse l in Fig. 12-6?

   Solution:

   Output indicators read 1000 during pulse l which is a copy of the contents of memory location 7.

12.11   The access time for the _____ (2114, 74F189) RAM is shorter, and therefore it is considered a faster chip.

   Solution:

   The access time for the 74F189 RAM is shorter.

12.12   Each memory cell of this static RAM is similar to a _____ (capacitor, flip-flop).

   Solution:

   Each memory cell in a static RAM is similar to a flip-flop.

12.13   See Fig. 12-4. The 10 address lines entering the 2114 RAM can address _____ different words.

   Solution:

   The 10 address lines on the 2114 shown in Fig. 12-4 can address a total of 1024 (2^{10}) words in RAM.

12.14   What is meant when a microcomputer is said to have 16K of memory?

   Solution:

   A computer said to have 16K of memory has a 16 384-byte memory. Such a memory would have a total capacity of 131 072 bits ($16 384 \times 8 = 131 072$).

12.15   Refer to Fig. 12-5. Why can the I/O pins of the RAMs be connected directly to the data bus?

   Solution:

   The 2114 RAMs shown in Fig. 12-5 have three-state buffered I/O terminals.

12.16   The time it takes to locate and read data from a RAM is called the _____ (access, interface) time.

   Solution:

   The time it takes to locate and read data from a RAM is called the access time.

12.17   The 2114 IC shown in Fig. 12-4 is a _____ (dynamic, static) RAM.

   Solution:

   The 2114 IC shown in Fig. 12-4 is a static RAM.

12-3   READ-ONLY MEMORY (ROM)

   Microcomputers must store permanent information in the form of a system or monitor program, typically in a *read-only memory* (*ROM*). The ROM is programmed by the manufacturer to the user's specifications. Smaller ROMs can be used to solve combinational logic problems (to implement truth tables).

ROMs are classified as nonvolatile storage devices because they do not lose their data when power is turned off. The read-only memory is also referred to as the *mask-programmed ROM*. The ROM is used only in high-volume production applications because the initial set-up costs are high. Varieties of programmable read-only memories (PROMs) might be used for low-volume applications.

Consider the problem of converting from decimal to Gray code (see Sec. 2-3). The truth table for this problem is in Fig. 12-7a. This conversion could be made by using a simple diode ROM circuit such as the one shown in Fig. 12-7b. If the rotary switch has selected the decimal 2 position, what will the ROM output indicators display? The outputs (D, C, B, A) will indicate LLHH or 0011. The D and C outputs are connected directly to ground through the resistor and read LOW. The B and A outputs are connected to $+5$ V through two forward-biased diodes, and the output voltage will read about $+2$–3 V, which is a logical HIGH.

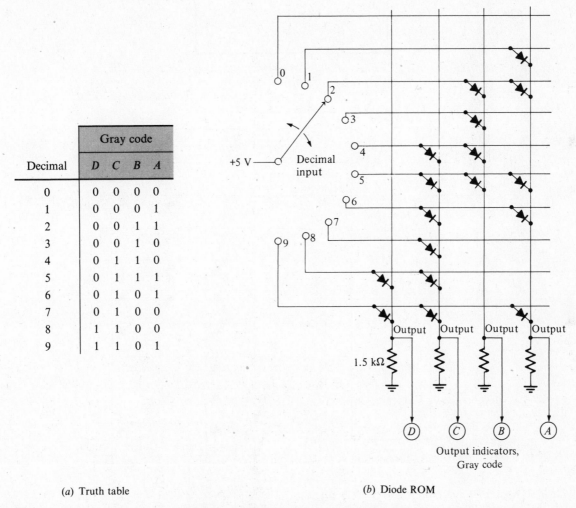

Decimal	Gray code			
	D	C	B	A
0	0	0	0	0
1	0	0	0	1
2	0	0	1	1
3	0	0	1	0
4	0	1	1	0
5	0	1	1	1
6	0	1	0	1
7	0	1	0	0
8	1	1	0	0
9	1	1	0	1

(a) Truth table

(b) Diode ROM

Fig. 12-7 Decimal–to–Gray code conversion

Note that the pattern of diodes in the diode ROM matrix (Fig. 12-7b) is similar to the pattern of 1s in the truth table (Fig. 12-7a). The circuit in Fig. 12-7b is considered a ROM that is permanently programmed as a decimal–to–Gray code decoder. Each new position of the rotary switch will give the correct Gray code output as defined in the truth table. In a memory, such as the one shown in Fig. 12-7, each position of the rotary switch is referred to as an *address*.

A slight refinement in the diode ROM is shown in Fig. 12-8. Figure 12-8a is the truth table for a binary–to–Gray code converter. The diode ROM circuit shown in Fig. 12-8b has an added 1-of-10

Binary				Gray code			
8s	4s	2s	1s	D	C	B	A
0	0	0	0	0	0	0	0
0	0	0	1	0	0	0	1
0	0	1	0	0	0	1	1
0	0	1	1	0	0	1	0
0	1	0	0	0	1	1	0
0	1	0	1	0	1	1	1
0	1	1	0	0	1	0	1
0	1	1	1	0	1	0	0
1	0	0	0	1	1	0	0
1	0	0	1	1	1	0	1

(*a*) Truth table

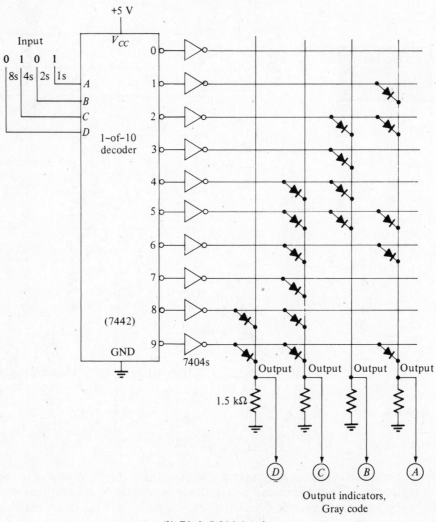

(*b*) Diode ROM decoder

Fig. 12-8 Binary–to–Gray code conversion

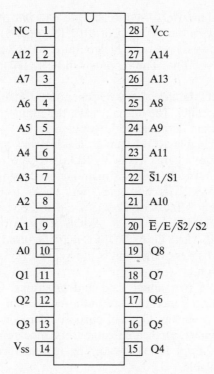

(a) Pin diagram (*Reprinted by permission of Texas Instruments*)

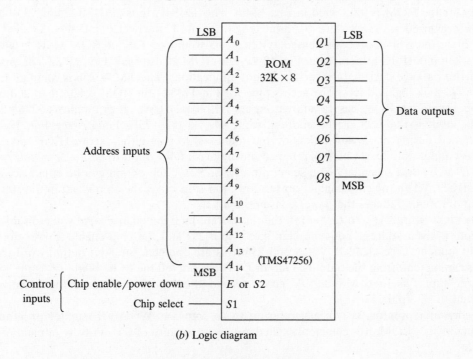

(b) Logic diagram

Fig. 12-9 TMS47256 32K × 8 ROM

decoder (TTL 7442 IC) and inverters used to activate *only one of the 10 rows* in the diode ROM. The example in Fig. 12-8b shows a binary input of 0101 (decimal 5). This activates output 5 of the 7442 with a LOW; and that drives the inverter, which outputs a HIGH. The HIGH forward-biases the three diodes connected to the row 5 line. The outputs will be LHHH or 0111. This is also specified in the truth table.

The primitive diode ROMs have many disadvantages. Their logic levels are marginal, and they have very limited drive capability. They do not have the input and output buffering that is needed to work with systems that contain data and address busses.

Practical mask-programmable ROMs are available from many manufacturers. They can range from very small units to quite large capacity ROMs. Some of these commercial ROMs can be purchased in familiar DIP form. ROMs are manufactured using TTL, CMOS, NMOS, PMOS, and GaAs (gallium arsenide) process technologies. The GaAs technology yields very fast digital ICs. Currently ROMs using either NMOS or CMOS seem to be very popular. As an example, one very small unit is the Harris 82HM141C 512×8 NMOS ROM with an access time of under 70 ns. A similar unit, the very fast 14GM048 GaAs ROM is manufactured by Tri Quint Semiconductor and has access time of less than 1.5 ns. One large unit is the Sharp LH5316000 $2M \times 16$ CMOS ROM, with an access time of less than 200 ns.

ROMs are used to hold permanent data and programs. Computer system programs, look-up tables, decoders, and character generators are but a few uses of the ROM. They can also be used for solving combinational logic problems. General-purpose microcomputers contain a larger proportion of RAM for their internal memory. Dedicated computers allocate more addresses to ROM and usually contain only smaller amounts of RAM. According to one recent listing, more than 300 different commercial ROMs are available.

As an example, a pin diagram of a commercial TMS47256 ROM is shown in Fig. 12-9a. A logic diagram for the ROM is illustrated in Fig. 12-9b. The TMS47256 is an NMOS 262 144-bit read-only memory organized as 32768 words of 8-bit length. From a practical point, this is called a $32K \times 8$ ROM, or in most microprocessor-based systems this would be 32K of ROM (32K bytes of ROM). While a 28-pin DIP IC is pictured in Fig. 12-9a, the ROM is also available in a 32-lead plastic leaded chip carrier package designed for surface mount applications. The TMS47256 is compatible with most TTL and CMOS logic devices. The access time for the TMS47256 ROM is less than 200 ns.

The TMS47256 ROM has 15 address inputs (A_0 to A_{14}), which can address 32768 (2^{15}) words. The A_0 input is the LSB of the address, while A_{14} is the MSB. Pin 22 (see Fig. 12-9a) can be programmed during mask fabrication by the manufacturer as either active-HIGH or active-LOW *chip-select* input. Pin 20 can be programmed during mask fabrication to be a *chip-enable/power-down* input (E or $\bar{E}$) or a secondary chip-select pin ($S2$ or $\overline{S2}$). Each option can be either active-LOW or active-HIGH. When the chip-enable/power-down pin is inactive, the chip is put in the standby mode. The standby mode reduces the power consumption.

The eight outputs (Q_1 to Q_8) are in the three-state high-impedance state when disabled. To *read* data from a given address, both the chip select (pin 22) and the chip-enable/power-down (pin 20) control inputs must be enabled. When both controls are enabled, the 8-bit output word from a given address can be read from the outputs. Output Q_1 is considered the LSB, while Q_8 is the MSB. A 5-V dc power supply is used with $+5$ V connected to the V_{CC} (pin 28) and the negative (ground) connected to V_{SS} (pin 14).

A computer program is typically referred to as *software*. When a computer program is stored permanently in a ROM, it is commonly called *firmware* because of the difficulty of making changes in the code.

SOLVED PROBLEMS

12.18 A _____ (RAM, ROM) is classified as a nonvolatile storage device.

Solution:

A ROM is classified as a nonvolatile storage device because it does not lose its data when power is turned off.

12.19 A _____ (RAM, ROM) is programmed by the computer operator.

Solution:

A RAM is programmed by the computer operator.

12.20 Refer to Fig. 12-8b. What is the function of the diode ROM?

Solution:

The function of the ROM shown in Fig. 12-8b is as a simple binary–to–Gray code converter.

12.21 Refer to Fig. 12-8b. List the state of the output for each binary input from 0000 to 1001.

Solution:

See the Gray code output for each binary count in Fig. 12-8a.

12.22 Refer to Fig. 12-8b. When the binary input is 0001, output 1 is activated and the output of its inverter goes _____ (HIGH, LOW), thereby forward-biasing a diode in the _____ (A, B, C, D) column.

Solution:

Binary 0001 activates output 1, Fig. 12-8b. This causes the inverter output to go HIGH, which forward-biases the single diode in the A column. The ROM output reads 0001.

12.23 Refer to Fig. 12-10. List the state of the ROM outputs during each pulse.

Solution:

The ROM outputs during each pulse are as follows:

pulse $a = 0011$ (address = 0)   pulse $e = 1001$ (address = 6)   pulse $h = 0101$ (address = 2)
pulse $b = 0110$ (address = 3)   pulse $f = 0100$ (address = 1)   pulse $i = 0111$ (address = 4)
pulse $c = 1100$ (address = 9)   pulse $g = 1011$ (address = 8)   pulse $j = 1000$ (address = 5)
pulse $d = 1010$ (address = 7)

12.24 The TMS47256 RAM shown in Fig. 12-9 can address _____ words. Each word is _____-bits wide.

Solution:

The TMS47256 RAM can address 32768 (32K) words each 8-bits wide.

12.25 The TMS47256 IC is a _____ (field, mask)-programmable read-only memory.

Solution:

The TMS47256 IC is a mask-programmable ROM which is programmed by the manufacturer to the user's specifications.

12.26 The TMS47256 ROM has _____ (number) address lines which can address _____ (16, 32)K bytes of memory.

Solution:

See Fig. 12-9b. The TMS47256 ROM has 15 address lines (A_0 to A_{14}) which can address 32K bytes of memory.

12.27 With the chip-enable/power-down pin of the TMS47256 ROM _____ (disabled, enabled), the chip goes into the standby mode, which reduces power consumption.

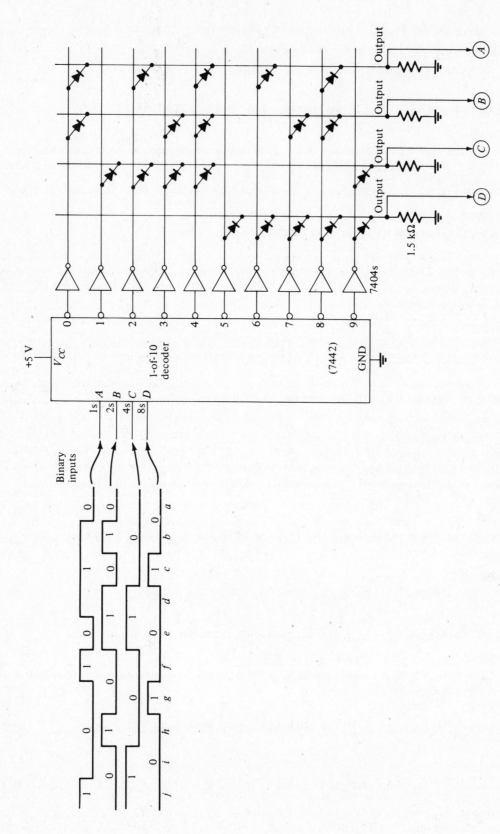

Fig. 12-10 ROM pulse-train problem

Solution:

> With the chip-enable/power-down input (pin 20) disabled, the chip goes into the standby mode, which reduces power consumption.

12.28   Refer to Fig. 12-9. Which two control inputs to the TMS47256 ROM must be enabled for stored data to be read from the outputs?

Solution:

> Both the chip-select (pin 22) and chip-enable/power-down (pin 20) control inputs must be enabled for stored date to be read from the outputs.

12-4 PROGRAMMABLE READ-ONLY MEMORY

Mask-programmable ROMs are programmed by the manufacturer by using photographic masks to expose the silicon die. *Mask-programmable ROMs* have long development times, and their initial costs are high. They are usually referred to simply as ROMs.

Field *programmable ROMs* (*PROMs*) also are available. They shorten development time and lower costs. It is also much easier to correct program errors and update products when PROMs can be programmed (burned) by the local developer. The regular PROM can be programmed only once like a ROM, but its advantage is that it can be made in limited quantities and can be programmed in the local lab or shop.

A variation of the PROM is an *erasable PROM* (*EPROM*). The EPROM is programmed or burned at the local level by using a *PROM burner*. If the EPROM must be reused or reprogrammed, a special quartz window on the top of the IC is used. Ultraviolet (UV) light is directed at the EPROM chip under the window for about an hour. The UV light erases the EPROM by setting all the memory cells to a logical 1. The EPROM can then be reprogrammed. Figure 12-11 illustrates a typical 24-pin EPROM DIP IC. Note the rectangular EPROM chip visible through the quartz window on top of the IC. These units may sometimes be called *UV-erasable PROMs*.

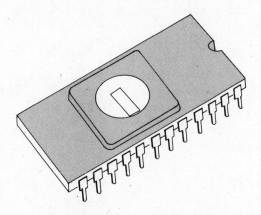

Fig. 12-11 UV erasable PROM

A third variation of the PROM is an *electrically erasable PROM*, also referred to as an EEPROM or E^2PROM. Because an EEPROM can be erased electrically, it is possible to erase and reprogram it while it remains on the circuit board. This is not possible with the PROM or the UV-erasable PROM. The EEPROM can also reprogram parts of the code on the chip 1 byte at a time.

A fourth variation of the PROM is the *flash EPROM*. The flash EPROM is very similar to the EEPROM in that it can be reprogrammed while on the circuit board. The flash EPROM is different

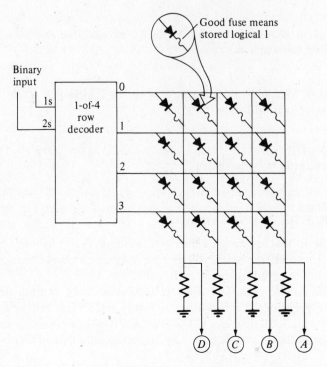

(a) Before programming (all logical 1s)

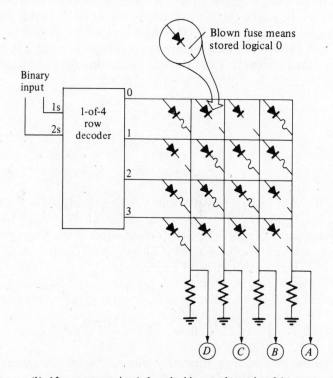

(b) After programming (selected addresses changed to 0s)

Fig. 12-12 Diode PROM

from the EEPROM in that the entire chip is erased and then reprogrammed. The advantages of the flash EPROM over the older EEPROM is it has a simpler storage unit so more bits can be stored on a single chip. Also flash EPROMs can be erased and reprogrammed much faster than EEPROMs. The disadvantages of the flash EPROM are that 12 or 12.75 V are required for reprogramming and that a single byte cannot be reprogrammed as on an EEPROM.

The basic idea of a programmable ROM (PROM) is illustrated in Fig. 12-12a. This is a simple 16-bit (4×4) PROM. It is similar to the diode ROM studied in the preceding section. Note that each of the memory cells contains a diode and a good fuse. That means that each of the memory cells in Fig. 12-12a contains a logical 1, which is how the PROM might look *before programming*.

The PROM shown in Fig. 12-12b has been programmed with seven 0s. To program or *burn the PROM*, tiny fuses must be blown as shown in Fig. 12-12b. A blown fuse in this case disconnects the diode and means that a logical 0 is permanently stored in the memory cell. Because of the permanent nature of burning a PROM, the unit cannot be reprogrammed. A PROM of the type shown in Fig. 12-12 can be programmed only once.

A popular EPROM family is the 27XX series. It is available from many manufacturers such as Intel and Advanced Micro Devices. A short summary of some models in the 27XX series is shown in Fig. 12-13. Note that all models are organized with byte-wide (8-bit-wide) outputs. Many versions of each of these basic numbers are available. Examples are low-power CMOS units, EPROMs with different access times, and even pin-compatible PROMs, EEPROMs, and ROMs.

EPROM 27XX	Organization	Number of bits
2708	1 024 × 8	8 192
2716	2 048 × 8	16 384
2732	4 096 × 8	32 768
2764	8 192 × 8	65 536
27128	16 384 × 8	131 072
27256	32 768 × 8	262 144
27512	65 536 × 8	524 288

Fig. 12-13 Selected members of the 277XX series EPROM family

A sample IC from the 27XX series EPROM family is shown in Fig. 12-14. The pin diagram in Fig. 12-14a is for the *2732A 32K (4K $\times$ 8) Ultraviolet Erasable PROM*. The 2732A EPROM has 12 address pins ($A_0 - A_{11}$) which can access the 4096 (2^{12}) byte-wide words in the memory. The 2732A EPROM uses a 5-V power supply and can be erased by using ultraviolet (UV) light. The chip enable ($\overline{CE}$) input is like the chip select ($\overline{CS}$) inputs seen on other memory chips. It is activated with a LOW.

The $\overline{OE}/V_{pp}$ pin serves a dual purpose; it has one purpose during reading and another during writing. Under normal use the EPROM is being read. A LOW at the output enable ($\overline{OE}$) pin during a memory read activates the three-state output buffers driving the data bus of the computer system. The eight output pins on the 2732 EPROM are labeled $O_0 - O_7$. The block diagram in Fig. 12-14b shows the organization of the 2732A EPROM IC.

When the 2732A EPROM is erased, all memory cells are returned to logical 1. Data is introduced by changing selected memory cells to 0s. The 2732A is in the *programming mode* (writing into EPROM) when the dual-purpose $\overline{OE}/V_{pp}$ input is at 21 V. During programming (writing), the input data is applied to the data output pins ($O_0 - O_7$). The word to be programmed into the EPROM is addressed by using the 12 address lines. A very short (less than 55 ms) TTL level LOW pulse is then applied to the $\overline{CE}$ input.

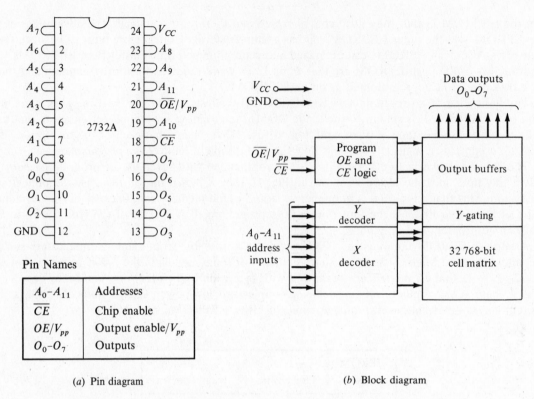

(a) Pin diagram

(b) Block diagram

Fig. 12-14 The 2732A 32K UV erasable PROM (*Courtesy of Intel Corporation*)

EPROM erasing and programming is handled by special equipment called PROM burners. After erasing and reprogramming, it is common to cover the EPROM window (see Fig. 12-11) with an opaque sticker. The sticker protects the chip from UV light from fluorescent lights and sunlight. The EPROM can be erased by direct sunlight in about one week or room level fluorescent lighting in about three years.

One of the disadvantages of the typical RAM is that it is volatile. When power is turned off, all data is lost. To solve this problem, nonvolatile static RAMs have been developed. Currently nonvolatile read/write memories are implemented by (1) using a CMOS SRAM with battery backup or (2) using a newer semiconductor NVSRAM (nonvolatile static RAM).

Static RAMs have both read and write capabilities but are volatile memories. One straightforward solution to the volatility problem is to furnish a *battery backup* for the SRAM. CMOS RAMs are used with battery backup because they consume little power. A long-life battery (such as a lithium battery) is used to back up data on the normally volatile CMOS SRAM when the power fails. During normal operation the regular dc power supply provides power for the SRAM. When power is turned off, a special circuit senses the drop in voltage and switches the SRAM to its standby battery power. Backup batteries have life expectancies of about 10 years.

A newer product called a *nonvolatile RAM* has become available. The nonvolatile RAM is commonly referred to as a *NVRAM, NOVRAM,* or *NVSRAM.* The NVRAM has the advantage of having both read and write capabilities but does not have the disadvantage of being a volatile memory or having a battery backup.

A logic diagram of a commercial NVSRAM is shown in Fig. 12-15*a*. The names of the pins are given in the chart in Fig. 12-15*b*. Notice from the logic diagram that the NVSRAM has two parallel memory arrays. The front memory array is a normal static RAM, while the back is an EEPROM. Each SRAM storage location has a parallel EEPROM memory cell. During normal operation the SRAM array is written into and read from just like with any SRAM. When the dc power supply voltage drops, a circuit automatically senses the drop in dc supply voltage and performs the *store* operation, and all

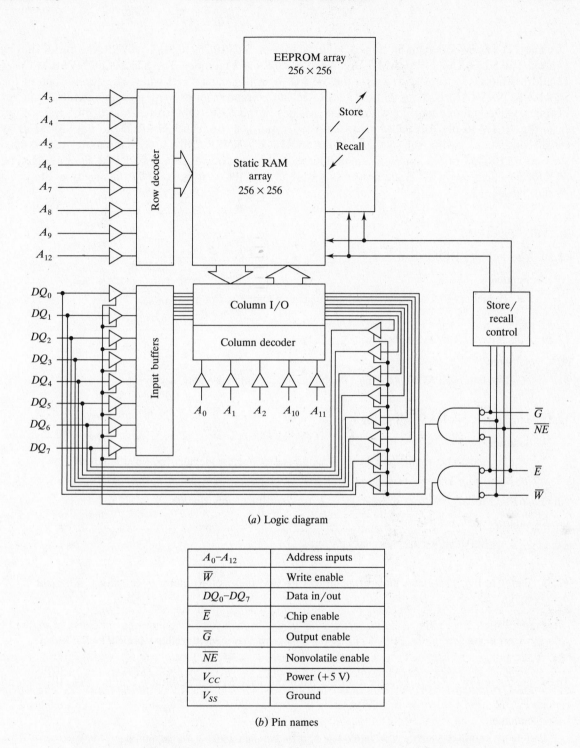

(a) Logic diagram

A_0–A_{12}	Address inputs
$\overline{W}$	Write enable
DQ_0–DQ_7	Data in/out
$\overline{E}$	Chip enable
$\overline{G}$	Output enable
$\overline{NE}$	Nonvolatile enable
V_{CC}	Power (+5 V)
V_{SS}	Ground

(b) Pin names

Fig. 12-15 STK10C68 CMOS nonvolatile SRAM (*Courtesy of Simtek Corporation*)

data on the volatile SRAM array is stored in the nonvolatile EEPROM array. This store operation is shown with an arrow pointing from the SRAM to the EEPROM on the logic diagram in Fig. 12-15a. With the power off, the EEPROM array within the NVSRAM holds a duplicate of the last data in the SRAM array. When power to the chip is turned on, the NVSRAM automatically performs the *recall* operation shown with an arrow pointing from the EEPROM to the SRAM in Fig. 12-15a. The recall operation copies all the data from the EEPROM array in the NVSRAM to the SRAM array.

The NVSRAM detailed in Fig. 12-15a is for an STK10C68 CMOS NVSRAM produced by Simtek. The STK10C68 NVSRAM is organized as an $8K \times 8$ memory. The STK10C68 NVSRAM uses 13 address lines (A_0 to A_{12}) to access the 8192 (2^{13}) words, each 8 bits wide. The access time of the STK10C68 NVSRAM is about 25 ns. The SRAM can be read from or written to an unlimited number of times, while independent nonvolatile data resides in the EEPROM array. Data can be transferred from the SRAM to the EEPROM array (store operation), or from the EEPROM to the SRAM array (recall operation), using the $\overline{NE}$ pin. The STK10C68 NVSRAM can handle more than 10000 store-to-EEPROM operations and an unlimited number of recall-from-EEPROM operations. The STK10C68 operates on a 5-V dc power supply. The STK10C68 is packaged in a variety of standard 28-pin packages.

SOLVED PROBLEMS

12.29 The letters PROM stand for _____.

   Solution:

   The letters PROM stand for programmable read-only memory.

12.30 The letters EPROM stand for _____.

   Solution:

   The letters EPROM stand for erasable programmable read-only memory.

12.31 The letters NVRAM stand for _____.

   Solution:

   The letters NVRAM stand for nonvolatile random-access memory (nonvolatile RAM).

12.32 A PROM can be programmed _____ (many times, only once).

   Solution:

   A PROM can be programmed only once.

12.33 Refer to Fig. 12-12b. A blown fuse in this PROM means that memory cell stores a logical _____ (0, 1).

   Solution:

   A blown fuse in the PROM in Fig. 12-12b means that memory cell is storing a logical 0.

12.34 Refer to Fig. 12-12b. List the outputs from the PROM for the binary inputs of 00, 01, 10, and 11.

   Solution:

   The outputs from the PROM in Fig. 12-12b for each address are as follows:
   address 00 output = 1001 (row 0)   address 10 output = 1110 (row 2)
   address 01 output = 0111 (row 1)   address 11 output = 1000 (row 3)

12.35 Refer to Fig. 12-11. What is the purpose of the window in the EPROM?

   Solution:

   A strong ultraviolet (UV) light directed through the window of the IC in Fig. 12-11 will erase the EPROM chip.

12.36 What is the advantage of an EPROM over a PROM?

Solution:

The EPROM can be erased and used over, whereas the PROM can be programmed only once.

12.37 The 2732A IC shown in Fig. 12-14 is a(n) _____ (EPROM, RAM) memory unit.

Solution:

The 2732A IC shown in Fig. 12-14 is an EPROM memory unit.

12.38 Refer to Fig. 12-11. Why would an opaque sticker be placed over the window of the EPROM after programming?

Solution:

An opaque sticker is commonly placed over the window of an EPROM (see Fig. 12-11) to keep sunlight and fluorescent light from erasing the memory unit.

12.39 Refer to Fig. 12-14. What is the purpose of the $\overline{OE}/V_{pp}$ input pin on the 2732A EPROM?

Solution:

The $\overline{OE}/V_{pp}$ pin on the 2732A EPROM shown in Fig. 12-14 has a dual purpose. In the read mode, the $\overline{OE}$ pin is the output enable to turn on the three-state buffers so they can drive the data bus. In the program mode, the V_{pp} pin is held at 21 V, which allows writing into the EPROM through the O_0–O_7 pins.

12.40 The letters SRAM stand for _____.

Solution:

The letters SRAM stand for static RAM, or static random-access memory.

12.41 The letters NVSRAM stand for _____.

Solution:

The letters NVSRAM stand for nonvolatile static random-access memory.

12.42 What two methods are currently used to form nonvolatile static RAMs?

Solution:

Currently nonvolatile SRAM memories are produced by (1) using a CMOS SRAM with battery backup and (2) using a NVSRAM (see Fig. 12-15a).

12.43 SRAMs with battery backup generally use a long-life battery such as a _____ (carbon-zinc, lithium) battery to supply standby power when the dc power supply is turned off.

Solution:

SRAMs with battery backup commonly use lithium batteries to supply standby power when the dc power supply is off.

12.44 The NVSRAM in Fig. 12-15 might also be called a NVRAM or _____ (DRAM, NOVRAM).

Solution:

The NVSRAM in Fig. 12-15 might also be called a NVRAM or NOVRAM.

12.45 The NVSRAM contains both a static RAM and a nonvolatile _____ (EEPROM, ROM) of the same size.

Solution:

See Fig. 12-15*a*. The NVSRAM contains both a static RAM and a nonvolatile EEPROM of the same size.

12.46 Refer to Fig. 12-15. As power is turned off, the STK10C68 NVSRAM automatically _____ (recalls, stores) the data on the SRAM to the EEPROM.

Solution:

As power is turned off, the STK10C68 NVSRAM automatically stores (copies) the data on the SRAM to the EEPROM.

12.47 Refer to Fig. 12-15. As power is first turned on, the STK10C68 NVSRAM automatically _____ (recalls, stores) the data from the EEPROM to the SRAM.

Solution:

As power is first turned on, the STK10C68 NVSRAM automatically recalls (copies) data from the EEPROM to the SRAM.

12.48 Refer to Fig. 12-15. What is the purpose of the eight *DQ* pins on the STK10C68 NVSRAM?

Solution:

The *DQ* pins serve as eight parallel data outputs during memory read operations or data inputs during a memory write operation.

12-5 MICROCOMPUTER BULK STORAGE

Program and data storage in a computer system is sometimes classified as either *internal* or *external*. In a microcomputer, the internal storage devices are semiconductor RAM, ROM (or EPROM), and various registers. Currently the most common form of external storage for microcomputers is the magnetic disk. Magnetic disks are subdivided into hard or floppy disks. The most common form of magnetic disk used with microcomputers is the *floppy disk*. Typical types of memory devices used in microcomputers are summarized in Fig. 12-1. External storage is also referred to as *bulk storage*.

Data is stored on floppy disks in much the same way it is on magnetic tapes. The disk drive unit reads and writes on the floppy disk. This is like the play and record functions of a tape recorder. Reading from a disk has an advantage over reading from a tape because the disk is a *random-access* instead of a sequential-access device. The disk drive can access any point on the floppy disk in a very short time. In contrast, access to information on a tape is very slow.

Floppy disks, or diskettes, come in several sizes. Those most commonly used with microcomputers are the 5.25-in and the newer 3.5-in sizes. There is also an 8-in version of the floppy disk available. A diagram of a typical 5.25-in floppy disk is shown in Fig. 12-16*a*. The thin, circular, plastic floppy is permanently enclosed in a plastic jacket. The plastic disk is coated with a magnetic material, iron oxide or barium ferrite. Several holes are cut in both sides of the jacket. These are illustrated in Fig. 12-16*a*.

The round center hole in the jacket provides access to the center area of the disk. The hub of the disk drive clamps on this area to spin the disk at a constant speed (300 or 360 rpm). The larger hole in the jacket near the bottom of the disk shown in Fig. 12-16*a* exposes part of the disk to the read/write head of the disk drive. The read/write head touches the spinning floppy disk to store data on the disk (to write) or retrieve data from it (to read). The small round hole cut in the jacket and disk is used as

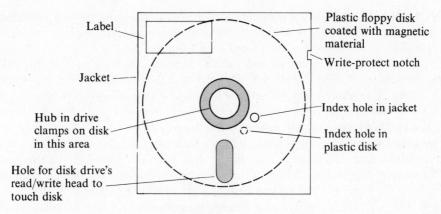

Label

Jacket

Hub in drive
clamps on disk
in this area

Hole for disk drive's
read/write head to
touch disk

Plastic floppy disk
coated with magnetic
material

Write-protect notch

Index hole in jacket

Index hole in
plastic disk

(a) Features of the disk

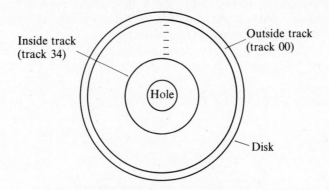

Inside track
(track 34)

Hole

Outside track
(track 00)

Disk

(b) Location of invisible tracks on disk

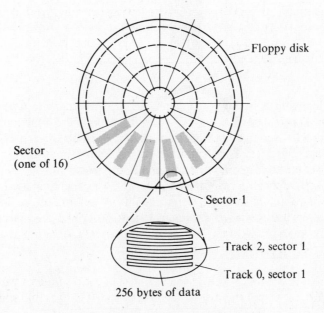

Floppy disk

Sector
(one of 16)

Sector 1

Track 2, sector 1

Track 0, sector 1

256 bytes of data

(c) Location of invisible sectors on disk

Fig. 12-16 A 5.25-in floppy disk

an index hole by disk drives on a few computers. If covered, the write-protect notch on the 5.25-in floppy disk prevents data from being written to the disk. When the write-protect notch is open, as in Fig. 12-16a, the disk drive can both write to and read from the disk.

Floppy disks are organized in *tracks* and *sectors*. Figure 12-16b shows how one microcomputer manufacturer formats the 5.25-in floppy disk. The disk is organized into 35 circular tracks numbered from 00 to 34 (00 to 22 in hexadecimal). Each track is divided into 16 sectors, which are shown in Fig. 12-16c. Each sector has 35 short tracks, as shown near the bottom of Fig. 12-16c. By using this format, each short track can hold 256 eight-bit words, or 256 *bytes*.

When formatted as shown in Fig. 12-16c, a floppy disk can hold about 140K bytes of data. That is about 1 million bits of data on a single 5.25-in floppy disk. It should be noted that there is no standard method of formatting floppy disks. Many microcomputer manufacturers format their disks to hold much more data. That includes reading and writing on both sides of the disk.

The floppy disk is a random-access bulk storage memory device which is widely used with home, school, and office microcomputers. Care must be taken when handling floppy disks. Do not touch the magnetic disk itself, and do not press hard when writing on the plastic jacket (5.25-in and 8-in). A felt-tip pen is recommended for labeling floppy disks. Magnetic fields and high temperatures also can harm data stored on floppy disks. Because of the danger of surface abrasion, keep disks in a clean area and protect the thin magnetic coating from scratches.

A diagram of the 3.5-in floppy disk is shown in Fig. 12-17. The case is made of rigid plastic for maximum protection of the floppy disk housed inside. The drawing of the 3.5-in disk in Fig. 12-17 is a

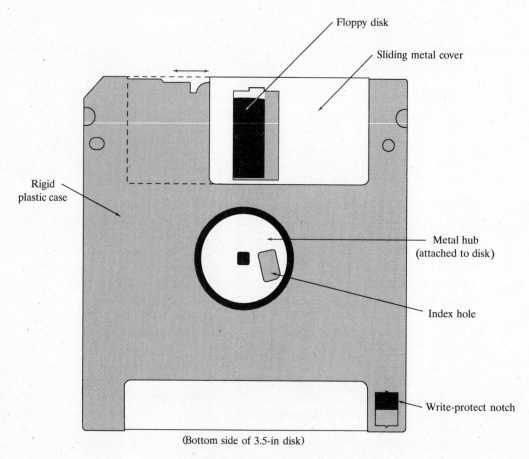

(Bottom side of 3.5-in disk)

Fig. 12-17 A 3.5-in floppy disk

view from the *underside* of the storage unit. The center of the plastic case is cut out (on the bottom only), revealing a metal drive hub which is connected to the floppy disk. A sliding metal cover is shown in Fig. 12-17 moved to the right, revealing a rectangular cutout in the rigid plastic case exposing the floppy disk. The floppy disk is accessible from both the bottom and top sides of the disk so the disk drive read/write heads can retrieve/store data on both sides. When released, the sliding metal cover, which is spring-loaded, snaps back to the left (in Fig. 12-17) to protect the surface of the floppy disk. A write-protect notch is shown at the lower right of the 3.5-in disk in Fig. 12-17. If the write-protect hole is closed by sliding the built-in cover upward (as shown in Fig. 12-17), the disk drive can both write to and read from the disk. This is sometimes called the unlocked position. If the hole is open (slide cover downward in Fig. 12-17), the disk drive can *only* read from the disk. This is sometimes called the locked position. An index hole is cut in the metal hub for timing purposes.

The 3.5-in disk shown in Fig. 12-17 is a newer development, compared to the 5.25-in and 8-in floppy disks. Precision disk drives commonly access 80 tracks on both sides of the disk. Common formats on the 3.5-in disk allow it to store either 400K, 720K, or 800K bytes. Available with suitable disk drives, high-density 3.5-in disks (FDHD—floppy disk high density) have storage capacity of 1.44M bytes. Most modern microcomputers come with at least one disk drive used to read from and write to 3.5-in floppy disks.

Another bulk storage method that is very popular on microcomputers, as well as large computer systems, is the *hard disk*, a rigid metal disk coated with magnetic material. These disks may be arranged as shown in Fig. 12-18. Notice that read/write heads float just above the surface of the spinning hard disks. The motor spins the hard disk at about 3000 rpm, which is about 10 times faster than the rotation of a floppy disk. The drive units are very precise, and the hard disk may be permanently mounted with the air filtered to keep out unwanted dust and smoke which can hamper operation. Removable hard disks, such as the 5.25-in cartridge drive, are also available. Currently 20M-, 40M-, and 80M-byte hard drives are common on home, school, and small business microcomputers. Larger-capacity units are also widely used in business. Two advantages of hard disks over floppy disks are (1) they store many times more information and (2) they can access information faster.

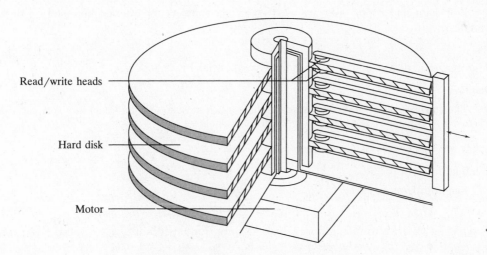

Fig. 12-18 Hard disk drive mechanism

Read/write heads

Hard disk

Motor

Hard disk drives are sometimes called *Winchester* drives. Microcomputers with hard drives are very common and typically also have a floppy disk drive attached to the system so the data and programs on the hard drive can be backed up for use in the event of hard disk failure.

Still another bulk storage method that shows great promise is the *optical disk*. The optical disk is a relative of the laser videodisk. Optical disks are available in three types: (1) read-only, (2) write-once read-many (WORM), and (3) read/write. The read-only disk (optical ROM) is good for prerecorded information like an encyclopedia. The WORM optical disk can be written to once and then read from many times.

Read/write optical disks have large storage capacities and are similar in function to a hard disk. The technology used for writing and reading with the optical disk is different from the magnetic hard disk. The *magneto-optical* disk drive uses a laser in conjunction with a coil of wire to erase, write to, and read from the metal-coated disk. A popular magneto-optical disk has a storage capacity of 128M bytes on a removable 3.5-in optical disk. These optical disks look much like the 3.5-in floppy disk except they are thicker and contain an optical disk. These removable disks are sometimes called *rewritable magneto-optical disks*. A 5.25-in magneto-optical disk drive is also available with removable cartridges with a storage capacity of 650M bytes. Because the magneto-optical disk can be removed from the disk drive, it is a suitable medium for backup storage or for transferring large amounts of data or programs from one machine to another.

One of the least expensive methods for storing vast amounts of data for backup is to use magnetic tape. Some drives are available that use inexpensive digital audio tape (DAT); however, access to data on tape is very slow.

SOLVED PROBLEMS

12.49 Refer to Fig. 12-1. Which device(s) on the microcomputer shown could be classified as internal storage?

Solution:

Both the semiconductor RAM, ROM, and NVRAM on the microcomputer shown in Fig. 12-1 could both be classified as internal storage. The floppy disk is external storage.

12.50 What two types of magnetic disks are used on microcomputers?

Solution:

Both hard and floppy disks are used on microcomputers for external bulk storage of data and programs.

12.51 The magnetic disk is a _____ (random-, sequential-) access device.

Solution:

The magnetic disk is a random-access device, which means it can find data in a very short time.

12.52 What are three sizes of floppy disks?

Solution:

Floppy disks come in the 3.5-, 5.25-, and 8-in sizes.

12.53 A typical disk drive spins the floppy disk at a constant speed of _____ (300, 3000) rpm.

Solution:

A disk drive spins the floppy disk at a constant speed of 300 rpm (one manufacturer's specification). Hard disks might spin at 3000 rpm.

12.54 To store data on a floppy disk is called _____.

Solution:

To store data on a floppy disk is called writing (write operation).

12.55 Briefly, how is data organized on a floppy disk?

Solution:

Data is organized in tracks and sectors. See Fig. 12-16b and c for more detail on the format used by one microcomputer manufacturer.

12.56 Refer to Fig. 12-16c. By using this format, a floppy disk can hold about _____ bytes of information.

Solution:

By using the format shown in Fig. 12-16c, a floppy disk can hold about 140K ($16 \times 256 \times 35 = 143\,360$ bytes) of information.

12.57 List some precautions that must be observed when 5.25-in floppy disks are handled.

Solution:

The following are some precautions when floppy disks are handled:
1. Do not touch the magnetic disk itself.
2. Mark the disk lightly or with felt-tip pens when labeling.
3. Keep the disk away from strong magnetic fields.
4. Keep the disk away from high temperatures.
5. Keep the disk clean.
6. Protect the disk from scratches or surface abrasion.
7. Do not bend or fold the disk.

12.58 What advantage does a hard disk drive have over a floppy disk?

Solution:

The hard drive has a much greater storage capacity and a quicker access time.

12.59 The WORM optical disk can be written to _____ (once, about 10000 times) and read from many times.

Solution:

The WORM (write-once read-many) optical disk can be written to once and read from many times.

12.60 The _____ (magnetic hard disk, magneto-optical disk) drive uses a laser in conjunction with a coil of wire to erase, write to, and read from the disk.

Solution:

The magneto-optical disk drive uses a laser in conjunction with a coil of wire to erase, write to, and read from the optical disk.

12.61 The popular removable 3.5-in rewritable magneto-optical disk has a capacity of about _____ (400K, 128M) bytes and is commonly used for backup storage or for transferring large amounts of data from one machine to another.

Solution:

The popular removable 3.5-in rewritable magneto-optical disk has a capacity of about 128M bytes.

Supplementary Problems

12.62   Refer to Fig. 12-1. List the five types of memory used by this microcomputer system.
Ans.   RAM, ROM, NVRAM, floppy disk, and hard disk

12.63   Refer to Fig. 12-1. Which type of memory in this system is volatile?
Ans.   RAM (read/write memory)

12.64   Refer to Fig. 12-1. What three types of storage devices are semiconductor memories in this system?
Ans.   RAM, ROM, and NVRAM

12.65   Refer to Fig. 12-1. The medium for storing data on the _____ (floppy disk, RAM) is magnetic.
Ans.   floppy disk

12.66   A read/write memory could be a _____ (RAM, ROM).   *Ans.*   RAM

12.67   Refer to Fig. 12-1. This semiconductor memory has the read/write capabilities of a RAM with the nonvolatile characteristics of a ROM.   *Ans.*   NVRAM

12.68   The RAM is a _(a)_ (nonvolatile, volatile) memory that _(b)_ (can, cannot) be erased by turning off the power.   *Ans.* (a) volatile (b) can

12.69   The _____ (dynamic, static) RAM uses a memory cell similar to a flip-flop.   *Ans.*   static

12.70   Refer to Fig. 12-5. The system is said to have _____ (1K, 8K) of memory.   *Ans.*   1K (1024 bytes)

12.71   A 256×4 memory would contain _(a)_ words, each _(b)_ bits long, for a total capacity of _(c)_ bits.   *Ans.* (a) 256 (b) 4 (c) 1024

12.72   Entering data in a RAM is the _____ (read, write) operation.   *Ans.*   write

12.73   The _____ (read, write) mode of operation of a RAM means revealing the contents of a memory location.   *Ans.*   read

12.74   A _____ (RAM, ROM) can be repeatedly programmed by the user.   *Ans.*   RAM

12.75   A ROM is a _____ (permanent, temporary) memory.   *Ans.*   permanent

12.76   A _____ (RAM, ROM) is programmed by the manufacturer to the user's specifications.
Ans.   ROM

12.77   Refer to Fig. 12-7b. What is the function of this simple diode ROM?
Ans.   decimal–to–Gray code decoder

12.78   Refer to Fig. 12-7b. List the state of the outputs for each decimal input (0–9).
Ans.   See the table in Fig. 12-7a.

12.79   Refer to Fig. 12-7b. Which diode(s) are forward-biased when the input switch is at decimal 2?
Ans.   two diodes in columns *A* and *B* in row 2 in Fig. 12-7b.

12.80   Larger-capacity ROMs (such as $512K \times 8$ ROMs) use _____ (bipolar, CMOS) technology in their manufacture.   *Ans.*   CMOS

12.81   A $131\,072 \times 8$ ROM would have a total capacity of _____ bits   *Ans.*   1 048 576

12.82   A $65\,536 \times 8$ ROM would need _____ (8, 16) address line pins on the IC.   *Ans.*   16 ($2^{16} = 65\,536$)

12.83   What is a computer program called when it is permanently stored in a ROM?   *Ans.*   firmware

12.84   In a general-purpose microcomputer, a greater proportion of internal memory is probably allocated to _____ (RAM, ROM).   *Ans.*   RAM

12.85   In a dedicated computer, a greater proportion of internal memory is probably allocated to _____ (RAM, ROM).   *Ans.*   ROM

12.86   The letters EEPROM stand for _____.
Ans.   electrically erasable programmable read-only memory

12.87   A mask-programmable read-only memory is commonly called a(n) _____.   *Ans.*   ROM

12.88   Refer to Fig. 12-12. This is an example of a(n) _____ (EPROM, PROM).   *Ans.*   PROM

12.89   Refer to Fig. 12-12*b*. A good fuse (see row 0, column *D*) in the PROM means that the memory cell stores a logical _____ (0, 1).   *Ans.*   1

12.90   Refer to Fig. 12-11. This IC is a(n) _____ (EPROM, ROM).
Ans.   EPROM (ultraviolet-erasable PROM)

12.91   An EPROM is considered a _____ (nonvolatile, volatile) memory device.   *Ans.*   nonvolatile

12.92   The abbreviation $E^2 PROM$ stands for _____.
Ans.   electrically erasable programmable read-only memory (same as EEPROM)

12.93   EPROMs are programmed in the _____ (factory, local lab).   *Ans.*   local lab

12.94   What is the equipment that is used to program EPROMs called?   *Ans.*   PROM burner

12.95   The letters SRAM stand for _____.   *Ans.*   static random-access memory (static RAM)

12.96   The letters RWM stand for _____ when dealing with semiconductor memories.
Ans.   read/write memory (same as RAM)

12.97   A RWM is more commonly referred to as a(n)_____.   *Ans.*   RAM

12.98   Magnetic _____ (disks, tapes) are random-access devices and have a short access time.   *Ans.*   disks

12.99   To retrieve data from a floppy disk is called _____.   *Ans.*   reading

12.100   "Winchester" is another name for what magnetic storage device?   *Ans.*   hard disk drive

12.101   The 74F189 RAM IC is from the newer _____ subfamily that exhibits an outstanding combination of performance and efficiency.   *Ans.*   Fairchild advanced Schottky TTL, FAST

12.102   A short access time for a RAM, ROM, or PROM means it is _____ (faster, slower).
Ans.   faster (A faster chip can be used in higher-frequency circuits.)

12.103   Semiconductor memory ICs manufactured using the _____ (CMOS, GaAs) process technology are the fastest chips.   *Ans.*   GaAs (gallium arsenide)

12.104   Refer to Fig. 12-9. The TMS47256 $32K \times 8$ ROM IC would have _____ address inputs and _____ data outputs.   *Ans.*   15 (A_0 to A_{14}), 8 (Q_1 to Q_8)

12.105 The flash EPROM is very similar to the _____ (EEPROM, NOVRAM).   *Ans.*   EEPROM

12.106 The letters NVSRAM stands for _____ .
Ans.   nonvolatile static random-access memory, or nonvolatile static RAM

12.107 Refer to Fig. 12-15. The STK10C68 NVSRAM has 64K bits of memory organized with _____ words each _____-bits wide.   *Ans.*   8K (8192 words), 8

12.108 Refer to Fig. 12-15. The STK10C68 IC is considered a _____ (nonvolatile, volatile) memory unit.
Ans.   nonvolatile (does not lose data on loss of power)

12.109 Refer to Fig. 12-17. Disk drives that use the 3.5-in floppy disk most often read from and write to _____ (both sides, one side) of the memory disk.   *Ans.*   both

12.110 Refer to Fig. 12-17. The 3.5-in floppy disk is *write-protected* and can only be read from when the hole in the write-protect slot is _____ (closed, open).
Ans.   open (This is the opposite of the 5.25-in disk in Fig. 12-16a.)

12.111 The _____ (floppy, hard) disk has the advantage over the other in that it can store more data and can access the information faster.   *Ans.*   hard

12.112 The 3.5-in _____ (magnetic floppy, rewritable magneto-optical) disk has a storage capacity of about 128M bytes and uses a laser diode and coil of wire for erasing, reading, and writing.
Ans.   rewritable magneto-optical

Chapter 13

Other Devices and Techniques

13-1 INTRODUCTION

In examining manufacturers' TTL, CMOS, and memory data manuals, you would find several types of ICs that have not been investigated in the first 12 chapters of this book. This will be a "catchall" chapter to include devices and techniques that do not fit neatly into other chapters but are topics that are included in many of the standard textbooks in the field. Included will be multiplexers/data selectors and multiplexing, demultiplexers, an introduction to digital data transmission, latches and three-state buffers, programmable logic devices, magnitude comparators, and Schmitt trigger devices.

13-2 DATA SELECTOR / MULTIPLEXERS

A *data selector* is the electronic version of a one-way rotary switch. Figure 13-1 shows a single-pole, eight-position rotary switch on the left. The eight inputs (0–7) are shown on the left, and a single output on the right is labeled Y. A data selector is shown on the right. The data at input 2 (a logical 1) is being transferred through the contacts of the rotary switch. Similarly, the data at input 2 (a logical 1) is being transferred through the circuitry of the data selector on the right. The data position is selected by mechanically turning the rotor on the rotary switch. The data position is selected in the data selector by placing the proper binary number on the data-select inputs (C, B, A). The data selector permits data to flow only from input to output, whereas the rotary switch allows data to flow in both directions. A data selector can be thought of as being similar to a one-way rotary switch.

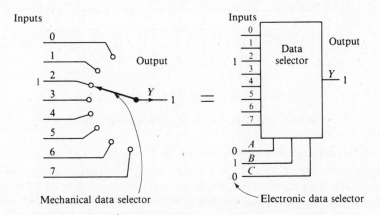

Fig. 13-1 Comparison of a rotary switch and a data selector

A commercial data selector is shown in block-diagram form in Fig. 13-2a. This TTL IC is identified as a 74150 *16-input data selector/multiplexer* by the manufacturers. Note the 16 data inputs at the top left. The 74150 has a single inverted output labeled W. Four data-select inputs (D, C, B, A) are identified at the lower left in Fig. 13-2a. A LOW at the strobe input will enable the data selector and can be thought of as a main on-off switch.

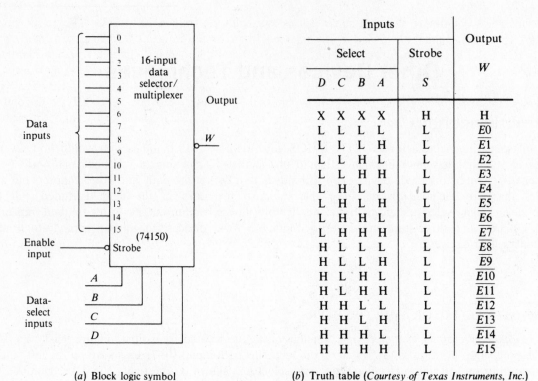

(a) Block logic symbol

| Inputs | | | | | Output |
| Select | | | | Strobe | W |
D	C	B	A	S	
X	X	X	X	H	H
L	L	L	L	L	$\overline{E0}$
L	L	L	H	L	$\overline{E1}$
L	L	H	L	L	$\overline{E2}$
L	L	H	H	L	$\overline{E3}$
L	H	L	L	L	$\overline{E4}$
L	H	L	H	L	$\overline{E5}$
L	H	H	L	L	$\overline{E6}$
L	H	H	H	L	$\overline{E7}$
H	L	L	L	L	$\overline{E8}$
H	L	L	H	L	$\overline{E9}$
H	L	H	L	L	$\overline{E10}$
H	L	H	H	L	$\overline{E11}$
H	H	L	L	L	$\overline{E12}$
H	H	L	H	L	$\overline{E13}$
H	H	H	L	L	$\overline{E14}$
H	H	H	H	L	$\overline{E15}$

(b) Truth table (*Courtesy of Texas Instruments, Inc.*)

(c) Pin diagram (*Courtesy of Texas Instruments, Inc.*)

Fig. 13-2 The TTL 74150 data selector/multiplexer IC

Consider the 74150 data selector truth table in Fig. 13-2b. Line 1 shows the strobe (enable) input HIGH, which disables the entire unit. Line 2 shows all the data-select inputs LOW as well as the strobe input being LOW. This enables the information at data input 0 to be transferred to output W. The data at output W will appear in its inverted form, as symbolized by the $\overline{E0}$ in the output column of the truth table. As the binary count increases (0001, 0010, 0011, and so forth) down the truth table, each data input in turn is connected to output W of the data selector.

The 74150 IC is packaged in a 24-pin package. The pin diagram for this IC is shown in Fig. 13-2c. Besides the 21 inputs and one output shown on the block diagram, the pin diagram also identifies the power connections (V_{CC} and GND). Being a TTL IC, the 74150 requires a 5-V power supply.

Note the use of the term "data selector/*multiplexer*" to identify the 74150 IC. A 74150 digital multiplexer can be used to transmit a 16-bit parallel word into serial form. This is accomplished by connecting a counter to the data-select inputs and counting from 0000 to 1111. The 16-bit parallel word at the data inputs (0–15) is then transferred to the output in *serial form* (one at a time).

The 74150 data selector/multiplexer can also be used to solve difficult combinational logic problems. Consider the truth table on the left in Fig. 13-3. The *simplified* Boolean expression for this truth table is $\overline{A}\,\overline{B}\,\overline{C}\,\overline{D} + AB\overline{C}\,\overline{D} + \overline{A}BC\overline{D} + A\overline{B}\,\overline{C}D + \overline{A}B\overline{C}D + \overline{A}\,\overline{B}CD + ABCD = Y$. Many ICs would be needed to implement this complicated expression by using AND-OR logic or NAND combinational logic circuits. The data selector is an easy method of solving this otherwise difficult problem.

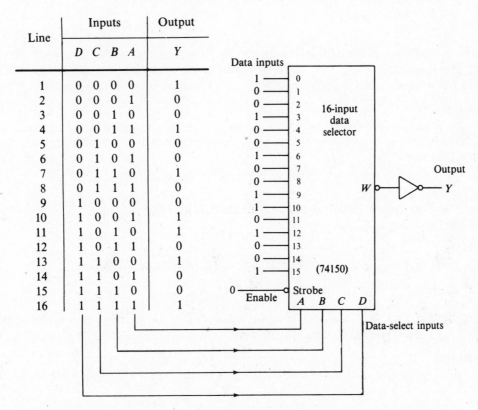

Fig. 13-3 Using the 74150 data selector to solve a combinational logic problem

A combinational logic problem is posed by the truth table in Fig. 13-3. A 16-input data selector is being used to solve the problem. The 16 data inputs (0–15) to the 74150 IC have logic levels corresponding to the output column of the truth table. Line 1 in the truth table has an input of binary 0000 (decimal 0) and an output of 1. The 1 is then applied to the 0 data input of the data selector. Line 2 in the truth table has an input of binary 0001 (decimal 1) and an output of 0. The 0 is then applied to the 1 input of the data selector. The input logic levels (D, C, B, A) from the truth table are applied to the data-select inputs of the 74150 data selector. The enable input of the 74150 IC is placed at 0, and the unit solves the logic problem in the truth table. Note that, because of the inverted output of the 74150 data selector, an inverter is shown added at the right in Fig. 13-3. The data selector solution to this combinational logic problem was a quick and easy one-package solution.

SOLVED PROBLEMS

13.1 A data selector is also called a _____.

Solution:

A data selector is also called a multiplexer.

13.2 A data selector is comparable to a mechanical _____ switch.

Solution:

A data selector is comparable to a mechanical one-way rotary switch.

13.3 Refer to Fig. 13-2. If the data selects on the 74150 IC equal $D = 1$, $C = 0$, $B = 1$, $A = 1$ and if the chip is enabled by a _____ (HIGH, LOW) at the strobe input, _____ (inverted, normal) data will be transferred from data input _____ (decimal number) to output W.

Solution:

Based on the truth table in Fig. 13-2, if the data selects on the 74150 IC equal 1011 (HLHH in truth table) and if the chip is enabled by a LOW at the strobe input, inverted data will be transferred from data input 11 to output W.

13.4 Refer to Fig. 13-2. A HIGH at the strobe input of the 74150 IC will _____ (disable, enable) the data selector.

Solution:

A HIGH at the strobe input of the 74150 IC will disable the data selector.

13.5 Refer to Fig. 13-3. If the data-select inputs equal $D = 1$, $C = 0$, $B = 1$, $A = 0$, the output Y will be _____ (HIGH, LOW).

Solution:

If the data-select inputs equal 1010, output Y of the data selector shown in Fig. 13-3 will be HIGH.

13.6 Often the single-package method of solving a combinational logic problem involves using _____ (a data selector, NAND logic).

Solution:

Often the single-package method of solving a combinational logic problem involves using a data selector.

13.7 Draw a block diagram of a 74150 data selector being used to solve the logic problem described by the Boolean expression $\overline{A}\,\overline{B}\,\overline{C}D + \overline{A}BCD + ABC\overline{D} + A\overline{B}\,\overline{C}\,\overline{D} + ABCD = Y$.

Solution:

See Fig. 13-4. The procedure is to first prepare from the Boolean expression a truth table similar to that in Fig. 13-3. Each 0 and 1 in the output column of the truth table will be placed on the corresponding data input of the data selector. An inverter is placed at output W of the 74150 data selector to read out noninverted data at Y.

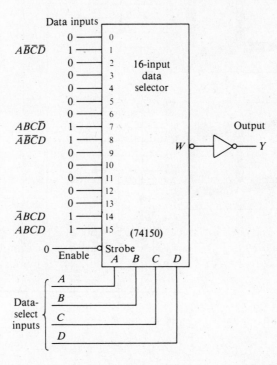

Fig. 13-4 Solution of a combinational logic problem by using a 74150 data selector

13-3 MULTIPLEXING DISPLAYS

Many electronic systems use alphanumeric displays. In fact, alphanumeric displays are a first clue that an electronic system contains at least some digital circuitry.

A simple 0 to 99 counter system with a digital readout is diagrammed in Fig. 13-5. The 0 to 99 counter system is used to illustrate the idea of *display multiplexing*. The counters are driven by a low-frequency clock (1 Hz). The outputs from the two decade counters are alternately fed through the

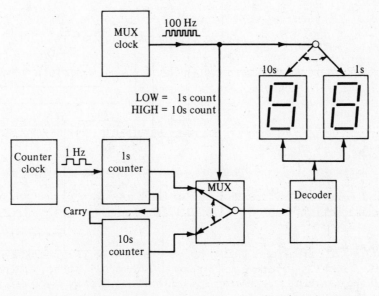

Fig. 13-5 Block diagram of 0 to 99 counter using multiplexed displays

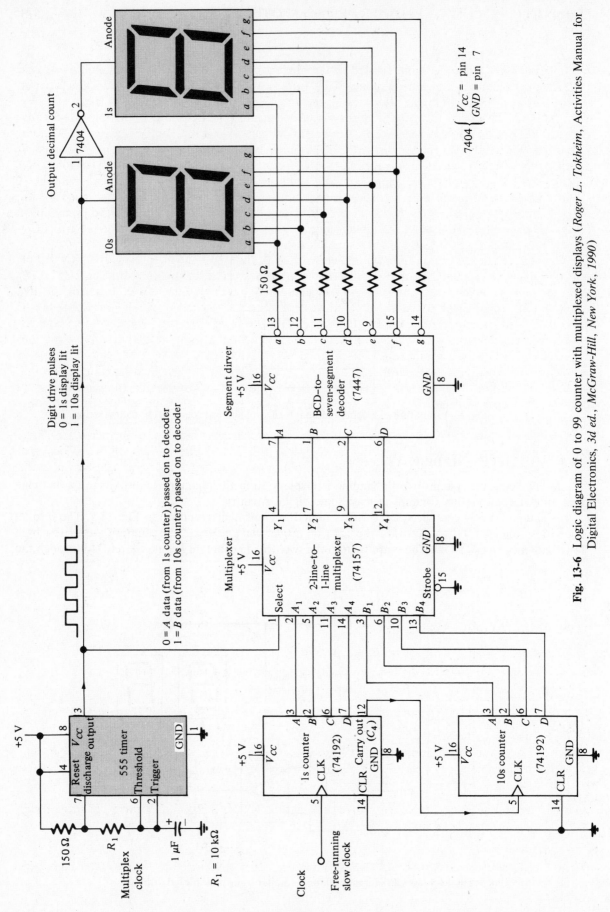

Fig. 13-6 Logic diagram of 0 to 99 counter with multiplexed displays (*Roger L. Tokheim*, Activities Manual for *Digital Electronics, 3d ed., McGraw-Hill, New York, 1990*)

multiplexer (MUX), decoded, and applied to *both* seven-segment LED displays. The multiplex clock (MUX clock) generates a higher-frequency signal (100 Hz). This signal alternately lights the 1s count on the display at the right or the 10s count on the seven-segment LED display at the left.

The block diagram in Fig. 13-5 suggests that the 1s count is passed through the multiplexer and decoded and the 1s display is activated when the MUX clock signal is LOW. When the MUX clock signal goes HIGH, the 10s count is passed through the multiplexer and decoded and the 10s display is activated. In effect, the seven-segment displays are alternately turned on and off about 100 times per second. The human eye interprets that as both seven-segment LED displays being lit continuously.

In this example, *multiplexing* reduces display power consumption and reduces the need for an extra decoder. Multiplexing is widely used with displays to save power. There is less need to multiplex LCD-type displays because they already consume very little power. For this and other reasons, LCD displays are often driven directly and not multiplexed.

The logic diagram in Fig. 13-6 is an implementation of the 0 to 99 counter using TTL ICs. All of the ICs used were examined in some detail earlier in the book except the multiplexer. The 74157 *TTL 2-line-to-1-line multiplexer* serves the purpose of alternately switching either the 1s count or the 10s count onto the input of the decoder. Note that, when the *select line* of the 74157 MUX is LOW, the A data (BCD from 1s counter) is passed to the decoder. At the same time, the 7404 inverter's output is HIGH, which allows the 1s seven-segment display to light. The 10s display is turned off when the MUX clock is LOW because the anode is grounded.

When the select line to the 74157 MUX in Fig. 13-6 goes HIGH, the B data is passed to the decoder. At the same instant, the anode of the 10s seven-segment display is HIGH, which allows it to light. The 1s display is turned off during this time because its anode is grounded by the LOW from the output of the inverter. The 150-Ω resistors limit the current through the display LEDs to a safe level.

The circuit shown in Fig. 13-6 will actually operate. To demonstrate that the displays are being multiplexed, substitute a 150-kΩ resistor for R_1 in the MUX clock circuit. This will slow down the MUX clock so you can see the action of the multiplexer as the displays flash alternately on and off.

SOLVED PROBLEMS

13.8 Refer to Fig. 13-5. When the MUX clock signal is HIGH, the _____ (1s, 10s) count is lit on the _____ (left, right) seven-segment LED display.

Solution:

When the MUX signal shown in Fig. 13-5 is HIGH the 10s count is lit on the left LED display.

13.9 Why are displays multiplexed?

Solution:

Multiplexing of LED displays reduces power consumption and simplifies wiring.

13.10 Refer to Fig. 13-6. Technically, are *both* seven-segment displays ever lit at the same time?

Solution:

Technically, both seven-segment displays shown in Fig. 13-6 are never lit at the same time. To the human eye they both appear to be continuously lit because they are flashing at 100 Hz.

13.11 Refer to Fig. 13-6. What effect would reducing the MUX clock frequency to 5 Hz have on the appearance of the displays?

Solution:

If the frequency of the MUX clock shown in Fig. 13-6 were reduced to 5 Hz, the eye would see the multiplexing action as a flashing of the displays.

13.12 Refer to Fig. 13-6. The logic level on the _____ input to the 74157 MUX determines if the 1s or the 10s count will be passed on to the decoder.

Solution:

The logic level of the select input to the 74157 MUX in Fig. 13-6 determines if the 1s or 10s count will be passed on to the decoder.

13.13 Refer to Fig. 13-6. If the MUX clock is LOW, the 1s count is passed through the decoder to which seven-segment display(s)?

Solution:

When the MUX clock shown in Fig. 13-6 is LOW, the 1s count is passed through the decoder to both displays. However, only the 1s display lights because only its anode is HIGH.

13.14 Refer to Fig. 13-6. What is the job of the 7404 inverter?

Solution:

The inverter shown in Fig. 13-6 activates the anodes of the displays alternately. A HIGH at the anode will activate the display.

13-4 DEMULTIPLEXERS

The operation of a *demultiplexer (DEMUX)* is illustrated in Fig. 13-7. The demultiplexer reverses the operation of the multiplexer (see Fig. 13-1). The single-pole, eight-position rotary switch at the left in Fig. 13-7 shows the fundamental idea of the demultiplexer. Notice that the demultiplexer has a single input and eight outputs. The data at the input can be distributed to one of eight outputs by the mechanical wiper arm on the rotary switch at the left. In the example in Fig. 13-7, the HIGH at the input is routed to output 2 by the rotary switch.

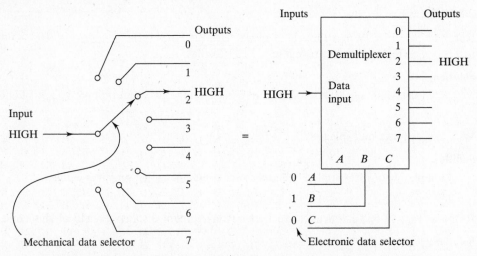

Fig. 13-7 Comparison of a rotary switch and a demultiplexer (data distributor)

A logic symbol for a simplified electronic demultiplexer is drawn at the right in Fig. 13-7. Note the single data input with eight outputs. The demultiplexer also has three data-select inputs (address inputs) for choosing which output is selected. In the example in Fig. 13-7, the HIGH at the input appears at output 2 of the electronic demultiplexer because 010_2 (2 in decimal) is applied to the data-select inputs. The demultiplexer is also called a *decoder* and sometimes a *data distributor*.

The electronic demultiplexer in Fig. 13-7 only allows data to flow from input to output, whereas the rotary switch permits data to flow in both directions. A data distributor, or demultiplexer, can be thought of as being similar to a one-way rotary switch.

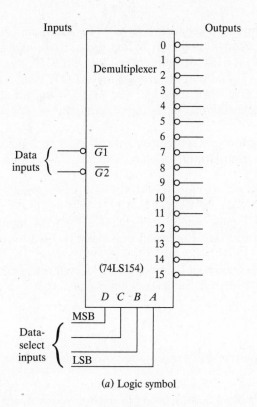

(a) Logic symbol

Function Table																					
Inputs						Outputs															
G1	G2	D	C	B	A	0	1	2	3	4	5	6	7	8	9	10	11	12	13	14	15
L	L	L	L	L	L	L	H	H	H	H	H	H	H	H	H	H	H	H	H	H	H
L	L	L	L	L	H	H	L	H	H	H	H	H	H	H	H	H	H	H	H	H	H
L	L	L	L	H	L	H	H	L	H	H	H	H	H	H	H	H	H	H	H	H	H
L	L	L	L	H	H	H	H	H	L	H	H	H	H	H	H	H	H	H	H	H	H
L	L	L	H	L	L	H	H	H	H	L	H	H	H	H	H	H	H	H	H	H	H
L	L	L	H	L	H	H	H	H	H	H	L	H	H	H	H	H	H	H	H	H	H
L	L	L	H	H	L	H	H	H	H	H	H	L	H	H	H	H	H	H	H	H	H
L	L	L	H	H	H	H	H	H	H	H	H	H	L	H	H	H	H	H	H	H	H
L	L	H	L	L	L	H	H	H	H	H	H	H	H	L	H	H	H	H	H	H	H
L	L	H	L	L	H	H	H	H	H	H	H	H	H	H	L	H	H	H	H	H	H
L	L	H	L	H	L	H	H	H	H	H	H	H	H	H	H	L	H	H	H	H	H
L	L	H	L	H	H	H	H	H	H	H	H	H	H	H	H	H	L	H	H	H	H
L	L	H	H	L	L	H	H	H	H	H	H	H	H	H	H	H	H	L	H	H	H
L	L	H	H	L	H	H	H	H	H	H	H	H	H	H	H	H	H	H	L	H	H
L	L	H	H	H	L	H	H	H	H	H	H	H	H	H	H	H	H	H	H	L	H
L	L	H	H	H	H	H	H	H	H	H	H	H	H	H	H	H	H	H	H	H	L
L	H	X	X	X	X	H	H	H	H	H	H	H	H	H	H	H	H	H	H	H	H
H	L	X	X	X	X	H	H	H	H	H	H	H	H	H	H	H	H	H	H	H	H
H	H	X	X	X	X	H	H	H	H	H	H	H	H	H	H	H	H	H	H	H	H

H = High Level, L = Low Level, X = Don't Care

(b) Function table (*Courtesy of National Semiconductor Corporation*)

Fig. 13-8 The 74LS154 decoder/demultiplexer IC

A commercial demultiplexer is shown in Fig. 13-8. The TTL unit detailed in Fig. 13-8 is described by the manufacturer as a *74LS154 4-line to 16-line decoder/demultiplexer* IC. The logic diagram in Fig. 13-8*a* describes the 74LS154 demultiplexer. The 74LS154 has 16 outputs (0 to 15) with 4 data-select inputs (*D* to *A*). The outputs are all active LOW pins, which means they are normally HIGH and one is pulled LOW when activated. The 74LS154 has two data inputs ($\overline{G1}$ and $\overline{G2}$) which are NORed together to generate the *single* data input. The two data inputs are both active-LOW inputs.

The 74LS154 demultiplexer is sometimes described as a *1-of-16 decoder*. The 74LS154 is a member of the TTL low-power Schottky family. The 74LS154 is a fast decoder with a propagation delay of less than 30 ns.

A truth table (or function table) for the 74LS154 decoder/demultiplexer IC is reproduced in Fig. 13-8*b*. Note that both the data inputs ($\overline{G1}$ and $\overline{G2}$) must be LOW before 1 of the 16 outputs is activated. The data-select inputs can be thought of as address inputs because of the use of the demultiplexer as a memory decoder. For instance, it might be used to select (or address) 1 of 16 RAM chips.

Both TTL and CMOS versions of demultiplexers/decoders are available. Common units include 1-of-4, 1-of-8, 1-of-10, and 1-of-16 decoders/demultiplexers.

SOLVED PROBLEMS

13.15 A _____ (demultiplexer, shift register) reverses the action of a multiplexer.

Solution:

The demultiplexer reverses the action of the multiplexer (compare Figs. 13-1 and 13-7).

13.16 The demultiplexer on the right in Fig. 13-7 could also be referred to as a _____ (1-of-8, 1-of-16) decoder.

Solution:

The demultiplexer on the right in Fig. 13-7 distributes data from a single input to one of eight outputs. It is therefore commonly called a 1-of-8 decoder.

13.17 Demultiplexers are commonly called data _____ (distributors, multivibrators) or _____ (decoders, gates).

Solution:

Demultiplexers are commonly called data distributors or decoders.

13.18 The 74LS154 demultiplexer is a _____ (1-of-8, 1-of-16) decoder with active-_____ (HIGH, LOW) data inputs and active-_____ (HIGH, LOW) outputs.

Solution:

See Fig. 13-8. The 74LS154 demultiplexer is a 1-of-16 decoder with active-LOW data inputs and active-LOW outputs.

13.19 Refer to Fig. 13-8. Both data inputs $\overline{G1}$ and $\overline{G2}$ must be _____ (HIGH, LOW) to activate the selected output on the 74LS154 demultiplexer IC.

Solution:

See Fig. 13-8*b*. Both data inputs ($\overline{G1}$ and $\overline{G2}$) must be LOW to activate the selected output.

13.20 Which output of the 74LS154 demultiplexer will be activated if $\overline{G1}$ and $\overline{G2}$ are both LOW while the data-select inputs are all HIGH.

Solution:

See Fig. 13-8. Output 15 will be activated (LOW) when data inputs ($\overline{G1}$ and $\overline{G2}$) are LOW and all the data-select inputs are HIGH. The address at the data-select inputs is 1111_2, which is decimal 15.

13.21 Which output of the 74LS154 demultiplexer will be activated if $\overline{G1}$ and $\overline{G2}$ are both LOW while the data-select inputs are $D = $ LOW, $C = $ LOW, $B = $ HIGH, and $A = $ HIGH?

Solution:

See Fig. 13-8. Output 3 will be activated (LOW) when data inputs ($\overline{G1}$ and $\overline{G2}$) are LOW and the address at the data-select inputs is 0011_2 (decimal 3).

13-5　LATCHES AND THREE-STATE BUFFERS

Consider the simple digital system shown in Fig. 13-9a. When 7 is pressed on the keyboard, a decimal 7 appears on the display. However, when the key is released, the 7 disappears from the output display. To solve this problem, a *4-bit latch* has been added to the system in Fig. 13-9b so that, when the key is pressed and released, the decimal number will remain lit on the seven-segment display. It can be said that the number 7 is *latched* on the display. The latch could also be referred to as a *buffer* memory.

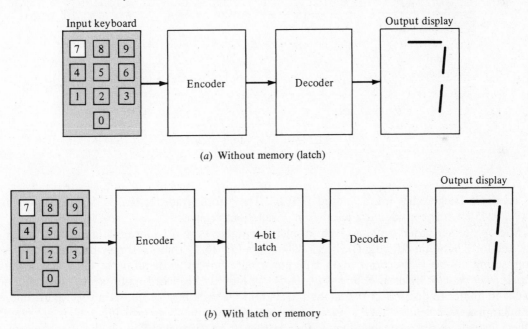

(a) Without memory (latch)

(b) With latch or memory

Fig. 13-9　Block diagram of simple digital systems

A simple latch manufactured in IC form is detailed in Fig. 13-10. This is the TTL 7475 *4-bit transparent latch IC*. A logic diagram for the 7475 latch is shown in Fig. 13-10a with its truth table detailed in Fig. 13-10b. The 7475 IC has four data inputs which accept parallel data. The data at $D_0 - D_3$ pass through the 7475 to both normal and complementary outputs when the *data-enable* inputs are HIGH. With the data-enable inputs HIGH the latch is said to be *transparent* in that any change in data at the data inputs is immediately passed to the data outputs. When the data-enables are activated with a LOW, data is latched (or held) at the outputs. When latched, changes at the data inputs do not cause any changes at the outputs.

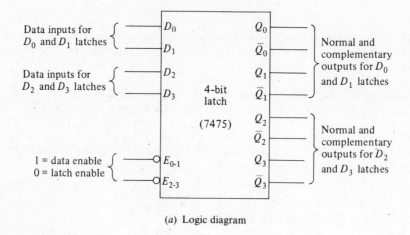

(a) Logic diagram

Mode of operation	Inputs		Outputs	
	E	D	Q	$\bar{Q}$
Data enabled	1	0	0	1
	1	1	1	0
Data latched	0	X	No change	

0 = LOW
1 = HIGH
X = irrelevant

(b) Truth table

(c) Pin diagram

Fig. 13-10 7475 4-bit latch

The 7475 latch comes in a standard DIP IC. The pin diagram for the 7475 IC is drawn in Fig. 13-10c. 7475 latch is considered a parallel-in parallel-out register.

Microprocessor-based systems (such as microcomputers) use a two-way *data bus* to transfer data back and forth between devices. The block diagram in Fig. 13-11 shows a simple microprocessor-based system using a 4-bit bidirectional data bus. For a data bus to work properly, each device must be *isolated from the bus*, by using a *three-state buffer*. A familiar keyboard input is shown with an added three-state buffer to disconnect the latched data from the data bus for all but a very short time when the microprocessor sends a LOW *read* signal. When the buffer's control input C is activated, the latched data drives the data bus lines either HIGH or LOW depending on the data present. The microprocessor then latches this data off the data bus and deactivates the buffer (control C back to HIGH).

The three-state buffer shown in block form in Fig. 13-11 might be implemented using the TTL *74125 quad three-state buffer IC*. A logic symbol for a single *noninverting buffer* is drawn in Fig. 13-12a. A pin diagram of the 74125 IC is in Fig. 13-12b, and a truth table is in Fig. 13-12c. When the control input is LOW, data is passed through the buffer with no inversion. When the control input goes HIGH, the output of the buffer goes to the *high-impedance* state. This is like creating an open between input A and output Y in Fig. 13-12a. Output Y then floats to the voltage level of the data bus line to which it is connected.

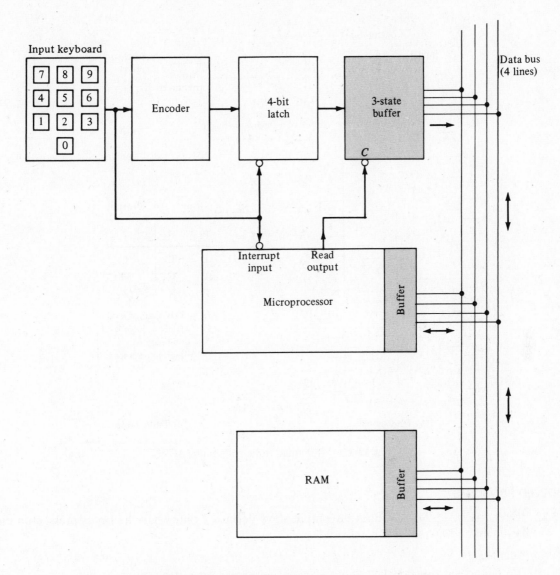

Fig. 13-11 Buffers used to isolate devices from a data bus

Three-state buffers are commonly built into devices designed to interface with a microcomputer bus. Figure 13-11 shows the buffer as part of the microprocessor and RAM (random-access memory or read/write memory). Many devices called *peripheral interface adapters* (*PIAs*) which contain latches, buffers, registers, and control lines are available. These special ICs are available for each specific microprocessor and take care of the input/output needs of the system.

A variety of latches are available in both TTL and CMOS. Latches commonly come in 4- or 8-bit *D* flip-flop versions. Some latches have three-state outputs.

Many buffer ICs are available using either TTL or CMOS technology. TTL buffers come with totem-pole, open-collector, or three-state outputs. Buffers may be of the inverting or noninverting types. Many buffers, such as the 74125 in Fig. 13-12, allow data to pass through the unit only in one direction. A variation of the buffer is the *bus transceiver* which allows *two-way* flow to or from a bus. The buffers identified as part of the microprocessor and RAM in Fig. 13-11 are really two-way buffers or bus transceivers.

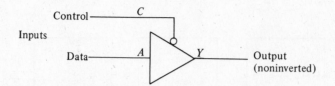

(a) Logic symbol of a three-state buffer

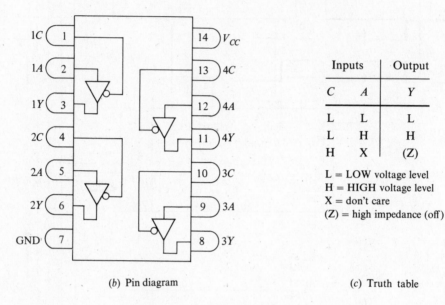

Inputs		Output
C	A	Y
L	L	L
L	H	H
H	X	(Z)

L = LOW voltage level
H = HIGH voltage level
X = don't care
(Z) = high impedance (off)

(b) Pin diagram   (c) Truth table

Fig. 13-12 74125 quad three-state buffer IC

SOLVED PROBLEMS

13.22 Refer to Fig. 13-9a. Why does the output show decimal 7 only when the key is pressed on the keyboard and not when it is released?

Solution:

The system shown in Fig. 13-9a does not contain a latch to hold the data at the inputs to the decoder. To latch data, the system must be modified to the one shown in Fig. 13-9b.

13.23 Refer to Fig. 13-13. The 7475 IC has active _____ (HIGH, LOW) enable inputs.

Solution:

The bubbles at the enable inputs to the 7475 IC shown in Fig. 13-13 mean these are active LOW inputs.

13.24 Refer to Fig. 13-13. List the mode of operation of the 7475 latch for each time period.

Solution:

time period t_1 = data enabled   time period t_5 = data enabled
time period t_2 = data latched   time period t_6 = data latched
time period t_3 = data latched   time period t_7 = data latched
time period t_4 = data enabled

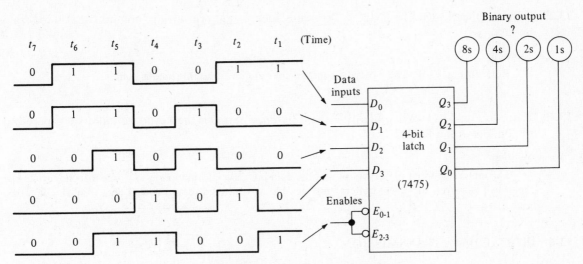

Fig. 13-13 Latch pulse-train problem

13.25 Refer to Fig. 13-13. List the 4-bit binary output at the indicators of the 7475 IC for each time period.

Solution:

time period $t_1 = 0001$ (data enabled)   time period $t_5 = 0111$ (data enabled)
time period $t_2 = 0001$ (data latched)   time period $t_6 = 0111$ (data latched)
time period $t_3 = 0001$ (data latched)   time period $t_7 = 0111$ (data latched)
time period $t_4 = 1000$ (data enabled)

13.26 Each device connected to a data bus (such as the one shown in Fig. 13-11) must be isolated from the bus by a _____ .

Solution:

Devices on a data bus are isolated from the bus by using a three-state buffer. This buffer is often built into the peripheral interface adapter or memory ICs. A two-way buffer is called a bus transceiver.

13.27 Refer to Fig. 13-11. If the 9 were pressed on the keyboard, what might be the sequence of events for the microprocessor to read this number?

Solution:

Refer to Fig. 13-11. Closing the 9 key causes binary 1001 to be latched and the microprocessor *interrupted* (signaled that keyboard is sending data). The microprocessor completes its current task and sends a LOW *read* signal to the three-state buffer. Data (binary 1001) flows through the buffer onto the data bus. The microprocessor latches this data off the data bus and disables the read signal (read output back to HIGH). The outputs of the three-state buffer return to their high-impedance state.

13.28 Refer to Fig. 13-11. If the latched data is binary 1001 and the control C input to the buffer is HIGH, then the outputs of the three-state buffer are at what logic levels?

Solution:

The HIGH on the control C pin of the three-state buffer places the outputs of the buffers in a high-impedance state. That means the buffer outputs will float to whatever logic levels exist on the data bus.

13.29 Refer to Fig. 13-12. The 74125 IC contains four _____ (inverting, noninverting) three-state buffers.

Solution:

Refer to Fig. 13-12. The 74125 IC contains four noninverting three-state buffers.

13.30 Refer to Fig. 13-11. What might be the difference between the keyboard buffers compared to the buffers in the RAM?

Solution:

The buffers between the keyboard and the data bus pass information in only one direction (onto the data bus). However, the RAM buffers must be able to send data to and accept data from the data bus.

13-6 DIGITAL DATA TRANSMISSION

Digital data transmission is the process of sending information from one part of a system to another. Sometimes the locations are close, and sometimes they are many miles apart. Either parallel or serial data transmission can be used. Serial data transmission is more useful when sending information long distances.

Figure 13-14a illustrates the idea of *parallel data transmission*. This is typical inside microprocessor-based systems where entire groups of bits (called *words*) are transferred at the same time. In Fig. 13-14a eight lines are needed to transmit the parallel data. A parallel system is used when speed is important. The disadvantage of parallel transmission is the cost of providing many registers, latches, and conductors for the many bits of data. The data bus shown in Fig. 13-11 is another example of parallel data transmission inside a microcomputer. In the case of the bus system, data can flow in both directions and additional buffering for each device connected to the bus is required.

Figure 13-14b illustrates the idea of *serial data transmission*. There is only one transmission line, and data is sent serially (one bit at a time) over the line. One format for sending asynchronous data serially is shown in Fig. 13-14b. A 7-bit ASCII code (see ASCII code in Fig. 2-11) can be sent by using this serial format. The line is normally HIGH, as shown on the left of the waveform. The LOW *start bit* signals the start of a word. The data bits are transmitted one at a time with the LSB (D_0) first. After the 7 data bits ($D_0 - D_6$), a parity bit for error detection is transmitted. Finally, two HIGH *stop*

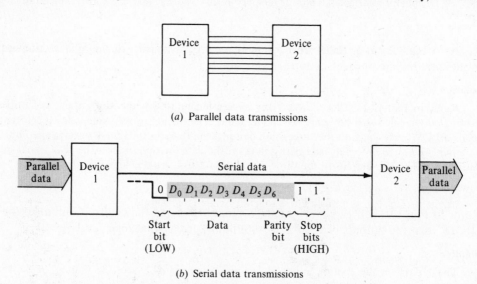

(a) Parallel data transmissions

(b) Serial data transmissions

Fig. 13-14 Digital data transmission methods

bits indicating that the character is complete are sent. These 11 bits transmit one ASCII character representing a letter, number, or control code.

Note that both devices 1 and 2 in the parallel transmission system in Fig. 13-14*a* would require parallel-in parallel-out type registers. In the serial data transmission system shown in Fig. 13-14*b*, device 1 would require a parallel-in serial-out register. Device 2 in Fig. 13-14*b* would require a serial-in parallel-out storage unit to reassemble the data back into parallel format.

Manufacturers produce specialized complex ICs that perform the task of serial data transmission. One such device is the *universal asynchronous receiver-transmitter*, or *UART*. The UART takes care of the parallel-to-serial and serial-to-parallel conversions for the transmitter and receiver. A typical UART is the AY-5-1013 by General Instrument. Other complex ICs that handle serial data transmission are the Motorola 6850 *asynchronous communication interface adapter* (*ACIA*) and the Intel 8251 *universal synchronous-asynchronous receiver-transmitter* (*USART*).

Serial data transmissions can be either asynchronous or synchronous. Asynchronous formats need start and stop bits (see Fig. 13-14*b*). There are also several synchronous serial protocols. Two of them are IBM's *binary synchronous protocol* (*BISYNC*) and IBM's *synchronous data link control* (*SDLC*).

The speed at which serial data is transmitted is referred to as the *baud rate*. As an example, look at Fig. 13-14*b*. It takes 11 bits to send a single character. If 10 characters per second are transmitted, then 110 bits are sent per second. The rate of data transfer will then be 110 *baud* (110 bits per second). Confusion sometimes occurs if baud rate is compared with *data* bits transmitted per second. In the above example, the 10 words transmitted per second contain only 70 data bits. Therefore, 110 baud equals only 70 data bits per second.

Many microcomputer owners use parallel and serial data transmission when interfacing with peripheral equipment. They may use either parallel or serial interfaces for their printers. They may use *modems* (modulator-demodulators) for sending and receiving data over phone lines. Some serial interface devices used with home computers send and receive data at speeds of 9600 baud.

SOLVED PROBLEMS

13.31 Digital data can be transmitted in either parallel or _____ form.

> **Solution:**
>
> Digital data can be transmitted in either parallel or serial form.

13.32 _____ (parallel, serial) data transmission is the process of transferring whole data words at the same time.

> **Solution:**
>
> Parallel data transmission is the process of transferring whole data words at the same time.

13.33 Refer to Fig. 13-14*a*. Device 1 must be a _____-in_____-out type register.

> **Solution:**
>
> Device 1 (Fig. 13-14*a*) must be a parallel-in parallel-out register.

13.34 Refer to Fig. 13-14*b*. Device 2 must be a _____-in_____-out register.

> **Solution:**
>
> Device 2 (Fig. 13-14*b*) must be a serial-in parallel-out device.

13.35 Refer to Fig. 13-11. The data bus system is an example of _____ (parallel, serial) data transmission.

Solution:

A data bus system is an example of parallel data transmission. Bus systems are widely used in microprocessor-based equipment, including microcomputers.

13.36 Refer to the waveform in Fig. 13-14b. The 11 bits transmit one _____ (ASCII, Basic) character representing a letter, number, or control code.

Solution:

The 11 bits shown in Fig. 13-14b serially transmit one ASCII character.

13.37 List at least one complex IC that can handle the task of serial data transmission.

Solution:

Several complex ICs used for serial data transmission are available from manufacturers. Three of these specialized ICs are the universal asynchronous receiver-transmitter (UART), the asynchronous communication interface adapter (ACIA), and the universal synchronous-asynchronous receiver-transmitter (USART).

13.38 A _____ (modem, parallel-in parallel-out register) is the complex device used to send and receive serial data over telephone lines.

Solution:

A modem (modulator-demodulator) is the device used to send and receive data over phone lines.

13-7 PROGRAMMABLE LOGIC ARRAYS

A *programmable logic array (PLA)* is an IC that can be programmed to execute a complex logic function. They are commonly used to implement combinational logic, but some PLAs can be used to implement sequential logic designs. The PLA is a one-package solution to many logic problems that may have many inputs and multiple outputs. Programmable logic arrays are closely related to PROMs and are programmed much like a PROM. A PLA may also be called a *programmable logic device (PLD)*. Both PLA and PLD seem to be generic terms used for these programmable logic units. One popular programmable logic device is the *PAL*® *(programmable array logic)*, available from several manufacturers.

Using PLDs cuts cost because fewer ICs are used to implement a logic circuit. PLDs are faster than using many SSI gate ICs on a printed circuit board. Software tools are available for programming the PLDs, making it easy to add changes in the prototype designs. Other advantages of the PLDs are the lower cost of inventory because they are somewhat generic components and the moderate cost of upgrades and modifications. The PLD is a very reliable component. Proprietary logic designs can be hidden from competitors by using the security fuse provided by the manufacturer.

A logic diagram for a simple PLA device is shown in Fig. 13-15a. Note that this unit has only two inputs and a single output. A typical commercial product may have 12 inputs and 10 outputs, as is the case for the PAL12L10A IC. In Fig. 13-15a note the AND-OR pattern of logic gates which can implement any minterm (sum of products) Boolean expression. The simplified PLA in Fig. 13-15a has intact fuses (fusible links) used for programming the AND gates. The OR gate is not programmed in this unit. The PLA in Fig. 13-15a shows the device as it comes from the manufacturer—with all fuses intact. The PLA in Fig. 13-15a needs to be programmed by burning open selected fuses.

PAL® Registered Trademark of Advanced Micro Devices, Inc.

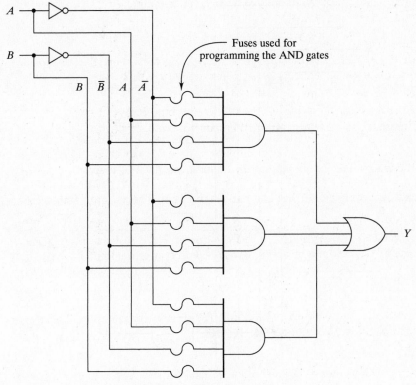

(a) Fuses intact (as from manufacturer)

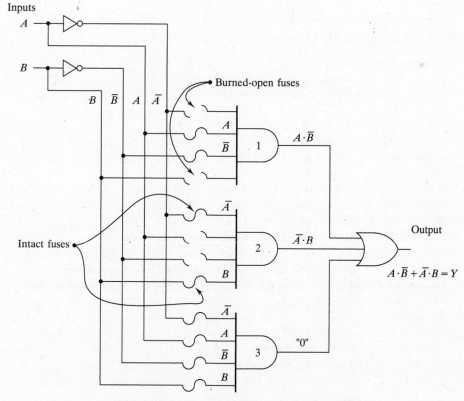

(b) Selected fuses burned open to solve logic problems

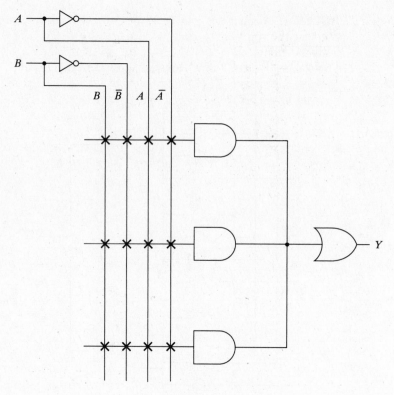

(a) All fuses intact

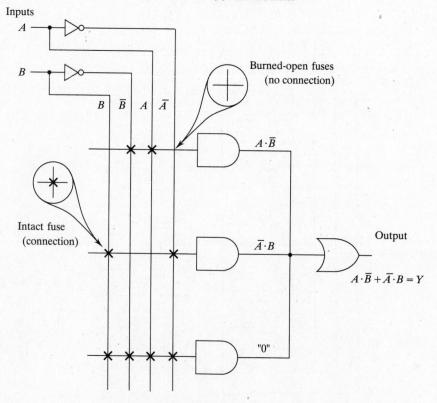

(b) Selected fuses burned open to solve logic problems

Fig. 13-16 Simple PLA using abbreviated notation system (fuse map)

The PLA in Fig 13-15b has been programmed to implement the minterm Boolean expression $A \cdot \overline{B} + \overline{A} \cdot B = Y$. Notice that the top 4-input AND gate (gate 1) has two fusible links burned open, leaving the A and $\overline{B}$ terms connected. Gate 1 ANDs the A and $\overline{B}$ terms. AND gate 2 has two burned-open fuses, leaving the $\overline{A}$ and B inputs connected. Gate 2 ANDs the $\overline{A}$ and B terms. AND gate 3 is *not needed* to implement this Boolean expression. All fuses are left intact as shown in Fig. 13-15b, which means the output of AND gate 3 will always be a logical 0. This logical 0 will have no effect on the operation of the OR gate. The OR gate in Fig. 13-15b logically ORs the $A \cdot \overline{B}$ and $\overline{A} \cdot B$ terms implementing the Boolean expression. In this simplest example, the $A \cdot \overline{B} + \overline{A} \cdot B = Y$ minterm expression was implemented using a programmable logic array. Recall that the Boolean expression $A \cdot \overline{B} + \overline{A} \cdot B = Y$ describes the 2-input XOR function which could probably be implemented cheaper using a 2-input XOR gate SSI IC.

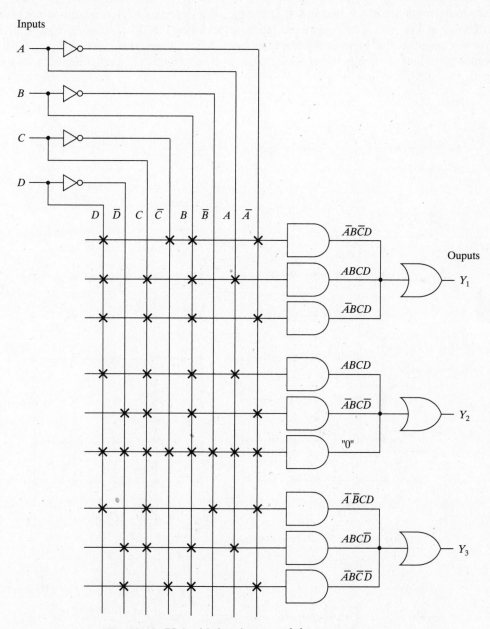

Fig. 13-17 PLA with four inputs and three outputs

An abbreviated notation system used with PLAs is illustrated in Fig. 13-16. Note that all AND and OR gates have only one input, while in reality each AND gate has four inputs, and the OR gate has three inputs (see Fig. 13-15a). The PLA has all fuses intact before programming. This is shown in Fig. 13-15a as a regular logic diagram. Figure 13-16a shows all fuses intact (each $\times$ represents an intact fuse) by using the abbreviated notation system.

The Boolean expression $A \cdot \overline{B} + \overline{A} \cdot B = Y$ is implemented in Fig. 13-15b. The same Boolean expression is implemented in Fig. 13-16b but only using the abbreviated notation system to describe the programming of the PLA. Notice in Fig. 13-16b that an $\times$ at an intersection means an intact fuse while no $\times$ means a burned-open fuse (no connection).

The abbreviated notation system is used because commercial PLAs are much larger than the simplified device drawn in Figs. 13-15 and 13-16. This notation is sometimes called a *fuse map*.

A more complex PAL-type programmable logic device is illustrated in Fig. 13-17. This PLA features four inputs and three outputs. It is common for decoders to have many outputs (such as the 7442 decoder in Fig. 7-7). The programmable logic device in Fig. 13-17 is not a commercial product.

Three combinational logic problems have been solved using the PLA in Fig. 13-17. First the Boolean expression $\overline{A} \cdot B \cdot \overline{C} \cdot D + A \cdot B \cdot C \cdot D + \overline{A} \cdot B \cdot C \cdot D = Y_1$ is implemented using the upper

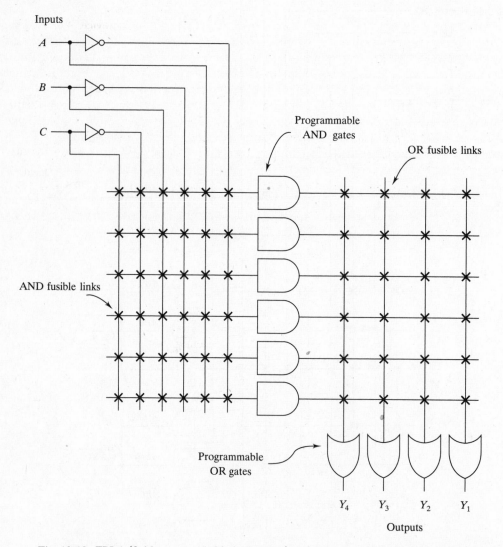

Fig. 13-18 FPLA (field-programmable logic array) with programmable AND and OR arrays

group of AND-OR gates. Remember that an × on the fuse map means an intact fuse while no × means a burned-open fuse. The second Boolean expression $A \cdot B \cdot C \cdot D + \overline{A} \cdot B \cdot C \cdot \overline{D} = Y_2$ is implemented using the middle group of AND-OR gates. Note that the bottom AND gate in the middle group is not needed. Therefore it has all eight fuses intact, which means it generates a logical 0 which has no effect on the output of the OR gate. The third Boolean expression $\overline{A} \cdot \overline{B} \cdot C \cdot D + A \cdot B \cdot C \cdot \overline{D} + \overline{A} \cdot B \cdot \overline{C} \cdot \overline{D} = Y_3$ is implemented using the bottom group of AND-OR gates.

An alternative PLA architecture is shown in Fig. 13-18. This PLA provides both programmable AND and OR arrays. The programmable logic devices studied before only contained programmable AND gates. This type of device is sometimes called a *field-programmable logic array (FPLA)*. Notice in Fig. 13-18 that each fusible link in both the AND and OR arrays is marked with an ×, meaning that all the links are intact (not burned).

One catalog of ICs groups programmable logic devices first by the process technology used to manufacture the units. Second, they are grouped as either one-time programmable or erasable. The erasable units can be either of the UV (ultraviolet) light type or electrically erasable. Third, they are grouped by whether the PLD has combinational logic or registered/latched outputs. Traditionally PLDs have been used to solve complex combinational logic problems. The registered PLDs contain both gates and flip-flops, providing the means of latching output data or of designing sequential logic circuits such as counters.

The PAL10H8 is an example of a small commercial programmable logic device. The pin diagram in Fig. 13-19a shows a block diagram of the PAL10H8 programmable logic array. Notice that the block diagram shows 10 inputs and 8 outputs along with the programmable AND array. A more detailed logic diagram of the PAL10H8 is reproduced in Fig. 13-19b. This detailed logic diagram looks much like the simple programmable logic devices studied earlier. The PAL10H8 IC is a Schottky TTL device with titanium tungsten fusible links. The PAL10H8 has a propagation delay of less than 35 ns. The PAL10H8 requires a standard 5-V dc power supply. The PAL10H8 is available in the 20-pin DIP (shown in Fig. 13-19a) or a 20-lead plastic chip carrier package for surface mounting.

Part number decoding and ordering information supplied by National Semiconductor for the PAL series of programmable logic arrays is shown in Fig. 13-20. Note that the letters PAL indicate the family of devices. In this example, the next number (10) indicates the number of inputs to the AND array. The middle letter (*H* in this example) indicates the type of output. The *H* means the outputs are active-HIGH. The next number (8 in this example) indicates the number of outputs. The trailing

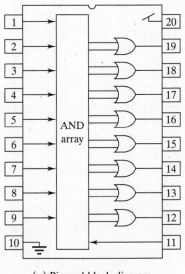

(a) Pin and block diagram

Fig. 13-19 PAL10H8 programmable logic device (*Courtesy of National Semiconductor Corporation*)

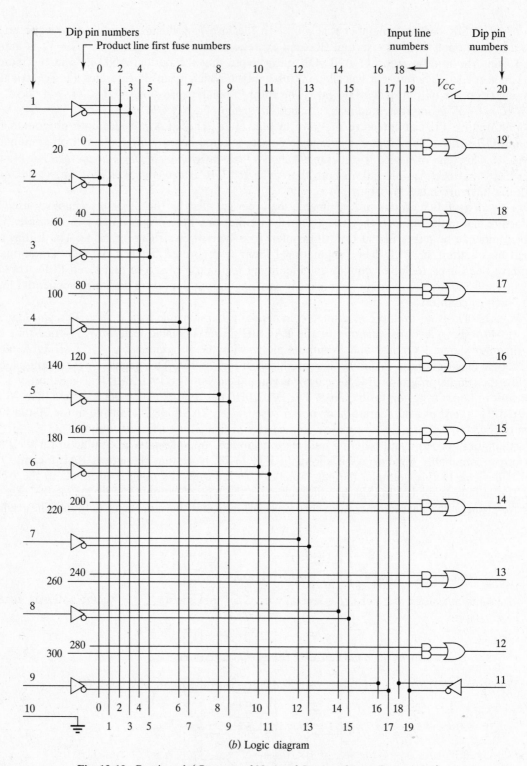

(b) Logic diagram

Fig. 13-19 Continued. (*Courtesy of National Semiconductor Corporation*)

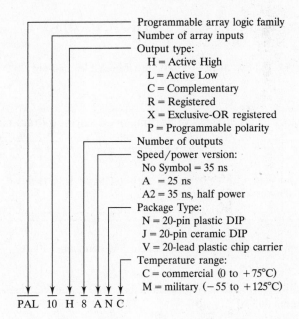

Fig. 13-20 Decoding a PAL part number (*Courtesy of National Semiconductor Corporation*)

letters indicate the speed/power version, package type, and temperature range. Note that both commercial and military versions of the PAL10H8 are available.

SOLVED PROBLEMS

13.39 The letters PLA stand for _____ when dealing with programmable logic ICs.

Solution:

The letters PLA stand for programmable logic array. PLA and PLD (programmable logic device) have become generic terms for these ICs.

13.40 The letters PAL® stand for _____.

Solution:

The letters PAL® stand for programmable array logic.

13.41 Programmable logic devices are commonly used to implement _____ (combinational, fuzzy) logic circuits.

Solution:

PLDs are commonly used to implement combinational logic circuits.

13.42 The letters FPLA stand for _____ when dealing with programmable logic ICs.

Solution:

The letters FPLA stand for field-programmable logic array.

13.43 PLAs and FPLAs are commonly programmed by the _____ (manufacturer, user).

Solution:

PLAs and FPLAs are commonly programmed in the field by the user.

PAL® is a registered trademark of Advanced Micro Devices, Inc.

13.44 The PLA is a close relative of the _____ (PROM, RAM) IC.

Solution:

    The PLA is a close relative of the PROM (programmable read-only memory).

13.45 Programming of most PLDs consists of burning open selected titanium tungsten _____ within the device.

Solution:

    Programming one-time PLDs consists of burning open selected fuses within the device. Some programmable logic devices are erasable.

13.46 What is the fundamental difference between a PAL and an FPLA?

Solution:

    An FPLA (see Fig. 13-18) has both programmable AND and OR gates, while a PAL (see Fig. 13-19) contains only programmable AND gates.

13.47 PLAs are organized to implement _____ (maxterm, minterm) Boolean expressions using an AND-OR pattern of logic gates.

Solution:

    PLAs are organized to implement minterm (sum-of-products) Boolean expressions using an AND-OR pattern of logic gates.

13.48 Refer to Fig. 13-19. The PAL10H8 IC is a programmable logic device with _____ (number) inputs and _____ (number) outputs with a programmable _____ (AND, OR) array.

Solution:

    The PAL10H8 is a PLD with 10 inputs and 8 outputs with a programmable AND array.

13.49 Refer to Fig. 13-20. Explain the meaning of a programmable logic device with a part number of PAL24L10A from National Semiconductor.

Solution:

    Decoding the part number PAL24L10A is as follows:
PAL = programmable array logic family
24 = 24 inputs
L = active-LOW outputs
10 = 10 outputs
A = propagation delay of 25 ns

13.50 Using a simple fuse map like the one pictured in Fig. 13-16a, program this PLA to implement the minterm Boolean expression $\overline{A} \cdot \overline{B} + A \cdot B = Y$.

Solution:

    See Fig. 13-21.

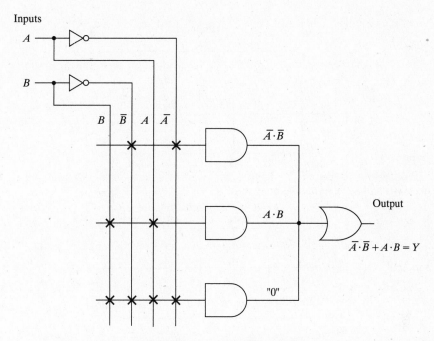

Fig. 13-21 PLA fuse map solution

13-8 MAGNITUDE COMPARATOR

A *magnitude comparator* is a device that compares two binary numbers and outputs a response such as A is equal to B ($A = B$), A is greater than B ($A > B$), or A is less than B ($A < B$). One commercial unit is the *74HC85 4-bit magnitude comparator*. A DIP pin diagram for the 74HC85 magnitude comparator is drawn in Fig. 13-22a. The 74HC85 IC has eight data-comparing inputs. Two 4-bit binary numbers ($A_3 A_2 A_1 A_0$ and $B_3 B_2 B_1 B_0$) are entered into the data-comparing inputs. The 74HC85 IC compares the two 4-bit numbers and generates one of three active-HIGH outputs. The three outputs are either $A > B_{out}$ (pin 5 is HIGH) or $A = B_{out}$ (pin 6 is HIGH) or $A < B_{out}$ (pin 7 is HIGH). Under normal conditions, only one of the three outputs is HIGH for any one comparison. A detailed truth table for the 74HC85 4-bit magnitude comparator is reproduced in Fig. 13-22b.

The 74HC85 is a high-speed CMOS magnitude comparator having a propagation delay of about 27 ns. This 74HC85 IC can operate on a wide range of voltages, from 2 to 6 V. This CMOS unit consumes little power but can drive up to 10 LS-TTL loads.

A single 74HC85 IC compares two 4-bit numbers, but it can easily be expanded to handle 8-, 12-, 16-bit, or more numbers. The cascade inputs are commonly used when expanding the word size of the magnitude comparator. Typical *cascading* of 74HC85 ICs is shown in Fig. 13-23. Note that the cascade inputs of IC_1 are permanently connected as follows: ($A > B_{in}$) = LOW, ($A < B_{in}$) = LOW, and ($A = B_{in}$) = HIGH. The cascade inputs of IC_2 are fed directly from the $A > B_{out}$, $A = B_{out}$, and $A < B_{out}$ outputs of the previous 74HC85 (IC_1). The circuit in Fig. 13-23 compares the magnitude of two 8-bit binary numbers $A_7 A_6 A_5 A_4 A_3 A_2 A_1 A_0$ and $B_7 B_6 B_5 B_4 B_3 B_2 B_1 B_0$. In response to the comparison, IC_2 drives one of three outputs HIGH. As an example in Fig. 13-23, if A_7 to A_0 equals 11111111 and B_7 to B_0 equals 10101010, then the $A > B_{out}$ output from IC_2 is activated and driven HIGH. In this example, all other outputs ($A = B_{out}$ and $A < B_{out}$) remain deactivated at a LOW logic level.

A simple electronic game can be designed using the 74HC85 magnitude comparator. The game is a version of "guess the number." In the classic computer version, a random number is generated within a range, and the player tries to guess the number. The computer responds with such responses

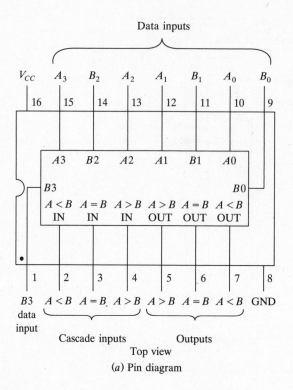

(a) Pin diagram

Comparing inputs				Cascading inputs			Outputs		
A_3, B_3	A_2, B_2	A_1, B_1	A_0, B_0	$A > B$	$A < B$	$A = B$	$A > B$	$A < B$	$A = B$
$A_3 > B_3$	X	X	X	X	X	X	H	L	L
$A_3 < B_3$	X	X	X	X	X	X	L	H	L
$A_3 = B_3$	$A_2 > B_2$	X	X	X	X	X	H	L	L
$A_3 = B_3$	$A_2 < B_2$	X	X	X	X	X	L	H	L
$A_3 = B_3$	$A_2 = B_2$	$A_1 > B_1$	X	X	X	X	H	L	L
$A_3 = B_3$	$A_2 = B_2$	$A_1 < B_1$	X	X	X	X	L	H	L
$A_3 = B_3$	$A_2 = B_2$	$A_1 = B_1$	$A_0 > B_0$	X	X	X	H	L	L
$A_3 = B_3$	$A_2 = B_2$	$A_1 = B_1$	$A_0 < B_0$	X	X	X	L	H	L
$A_3 = B_3$	$A_2 = B_2$	$A_1 = B_1$	$A_0 = B_0$	H	L	L	H	L	L
$A_3 = B_3$	$A_2 = B_2$	$A_1 = B_1$	$A_0 = B_0$	L	H	L	L	H	L
$A_3 = B_3$	$A_2 = B_2$	$A_1 = B_1$	$A_0 = B_0$	X	X	H	L	L	H
$A_3 = B_3$	$A_2 = B_2$	$A_1 = B_1$	$A_0 = B_0$	H	H	L	L	L	L
$A_3 = B_3$	$A_2 = B_2$	$A_1 = B_1$	$A_0 = B_0$	L	L	L	H	H	L

(b) Truth table

Fig. 13-22 74HC85 4-bit magnitude comparator (*Courtesy of National Semiconductor Corporation*)

as "Correct," "Too high," or "Too low." The player can then guess again until reaching the correct number. The player with the fewest guesses wins the game.

A logic diagram of the guess-the-number game is illustrated in Fig. 13-24. To play the game, first press switch SW_1, allowing clock pulses to reach the clock input ($\overline{CP}$) of the 4-bit binary counter (74HC393). When the switch is released, the counter will stop at some random binary count from 0000 to 1111. The random count is applied to the B data-compare inputs of the 4-bit magnitude comparator. Next the player makes a guess (from 0000 to 1111_2), which is applied to the A data-compare inputs of the comparator. The 74HC85 IC compares the magnitudes of the guess and

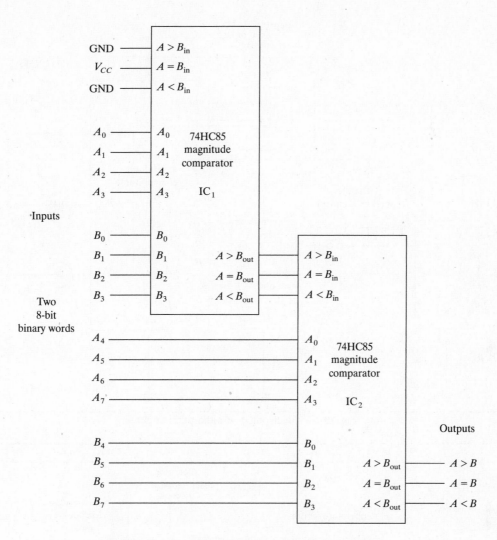

Fig. 13-23 Cascading 74HC85 magnitude comparator

random inputs and generates a HIGH output at one output, lighting one of the LEDs. If the guess is the same as the random number, the $A = B_{out}$ output goes HIGH, causing the green LED to light, and the player wins. If the guess is lower than the random number, the $A < B_{out}$ output goes HIGH, causing the yellow LED to light. If the guess is higher than the random number, the $A > B_{out}$ output goes HIGH, causing the red LED to light. The person's next guess can then be adjusted based on the information gained from the "Too high" or "Too low" outputs of the guess-the-number game.

In the guess-the-number game, the three outputs of the 74HC85 generate information that is used by the player to adjust the next guess. In like manner, magnitude comparators may be used in digital equipment to generate *feedback* to circuitry that can make adjustments in the input. Feedback is a critical item in automated equipment. For instance, if a physical variable (such as temperature, speed, position, time, light intensity, pressure, weight, etc.) is converted into binary form by an A/D converter, this measurement can be sent to one of the data-compare inputs of a magnitude comparator. The other data-compare inputs are set by the operator at the proper level. The outputs of the magnitude comparator will be used to activate circuitry to drive the physical variable toward the proper level.

A simple example of how a magnitude comparator might be used in a control application is shown in Fig. 13-25. In this example the temperature in an oven is to be controlled. The temperature sensor

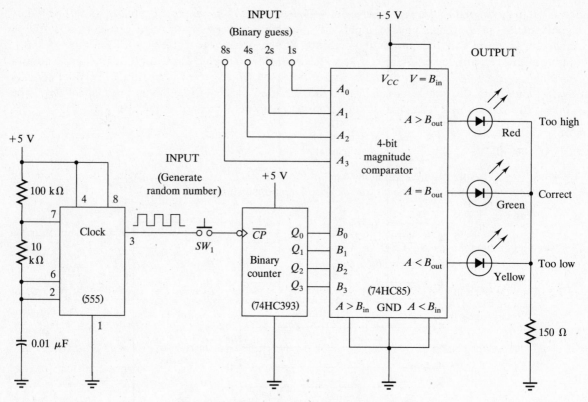

Fig. 13-24 Electronic guess-the-number game

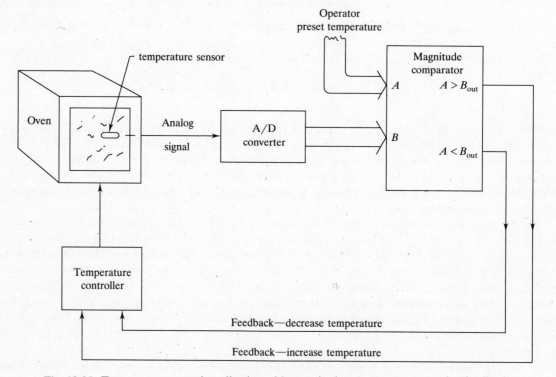

Fig. 13-25 Temperature control application with magnitude comparator generating feedback

sends an analog signal to the A/D converter which generates a *proportional* binary signal. The binary signal enters the B data-compare inputs of a magnitude comparator. The operator sets the A data-compare inputs at the proper temperature. If the oven temperature is too low, the $A > B_{out}$ output of the magnitude comparator is activated with this signal fed back to the temperature control unit. This unit causes the temperature to be increased. If the oven temperature is too high, the $A < B_{out}$ output of the comparator is activated and is fed back to the temperature control unit. The temperature control unit would cause the temperature to be decreased in the oven.

SOLVED PROBLEMS

13.51 A _____ (magnitude, voltage) comparator is a(n) _____ (analog, digital) device that can compare two binary numbers and output a response such as $A = B$, $A > B$, or $A < B$.

Solution:

A magnitude comparator is a digital device that can compare two binary numbers and output a response such as $A = B$, $A > B$, or $A < B$.

13.52 A single 74HC85 IC will compare two _____ (4-, 8-, 16-) bit binary numbers and output one of three responses such as _____, _____, or $A < B$.

Solution:

A single 74HC85 IC will compare two 4-bit binary numbers and output one of three responses such as $A = B$, $A > B$, or $A < B$.

13.53 What is the purpose of the cascading inputs on the 74HC85 IC shown in Fig. 13-22*a*?

Solution:

The 74HC85 ICs can be cascaded (see Fig. 13-23) to make an 8-, 12-, or 16-bit magnitude comparator. If not cascaded, the $A = B_{in}$ input should be tied to V_{CC} while the $A > B_{in}$ and $A < B_{in}$ cascade inputs should be grounded.

13.54 Refer to Fig. 13-23. If the inputs are A_7 to $A_0 = 00110011$ and B_7 to $B_0 = 11110000$, which output will be activated?

Solution:

If A_7 to $A_0 = 00110011$ and B_7 to $B_0 = 11110000$, then the $A < B_{out}$ output of IC_2 will be activated with a HIGH.

13.55 Refer to Fig. 13-24. The 555 timer IC is wired as a(n) _____ (astable, monostable) multivibrator in this game circuit.

Solution:

The 555 timer IC in Fig. 13-24 is wired as an astable multivibrator generating a continuous string of clock pulses.

13.56 Refer to Fig. 13-24. If the binary counter holds the number 0101_2 and your guess is 1000_2, the _____ (color) LED will light, indicating your guess is _____ (correct, too high, too low).

Solution:

If the binary counter holds the number 0101 and your guess is 1000, the red LED will light, indicating your guess is too high.

13.57 Refer to Fig. 13-25. The magnitude comparator in this control application is used to generate digital _____ (feedback, random) signals which cause the temperature controller to turn the oven heating element either on or off.

Solution:

The magnitude comparator in Fig. 13-25 is used to generate digital feedback signals which cause the temperature controller to turn the oven heating element either on or off.

13.58 Refer to Fig. 13-25. If the preset temperature setting is 11110000_2 (at input A) and the temperature signal is 11001111_2 (at input B), then the _____ (decrease, increase) temperature feedback line will be activated.

Solution:

If the preset temperature setting is 11110000_2 and the temperature signal is 11001111_2, then the increase temperature feedback line will be activated.

13.59 Refer to Fig. 13-26. List the *color* of the output LED that is lit for each time period (t_1 to t_5).

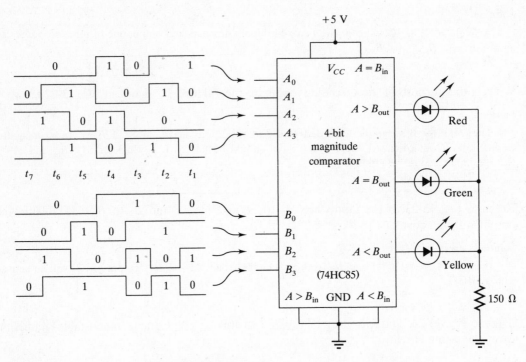

Fig. 13-26 Magnitude comparator pulse-train problem

Solution:

The color of the output LED that is lit for each time period is as follows:
time period t_1 = yellow LED lit
time period t_2 = green LED lit
time period t_3 = red LED lit
time period t_4 = yellow LED lit
time period t_5 = green LED lit

13-9 SCHMITT TRIGGER DEVICES

Waveforms with fast rise times and fast fall times are preferred in digital circuits. A square wave is an example of a good digital signal because it has almost vertical LOW-to-HIGH and HIGH-to-LOW edges. A square wave is said to have fast rise and fall times.

A waveform, such as the sine wave in Fig. 13-27, has a slow rise time and a slow fall time. Using a sine wave to drive a normal gate, counter, or other digital device will cause unreliable operation. In Fig. 13-27 a *Schmitt trigger* inverter is being used to "square up" the sine wave by forming a square wave at the output. The Schmitt trigger inverter is reshaping the waveform. Schmitt trigger devices are used for "squaring up" waveforms. This process is sometimes called *signal conditioning*.

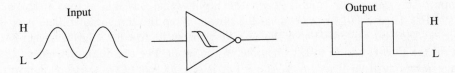

Fig. 13-27 Schmitt trigger inverter used to "square up" waveform

A voltage profile of a common 7404 inverter IC is compared with that of a 7414 Schmitt trigger inverter in Fig. 13-28. Of special interest is the switching threshold of the inverters. The *switching*

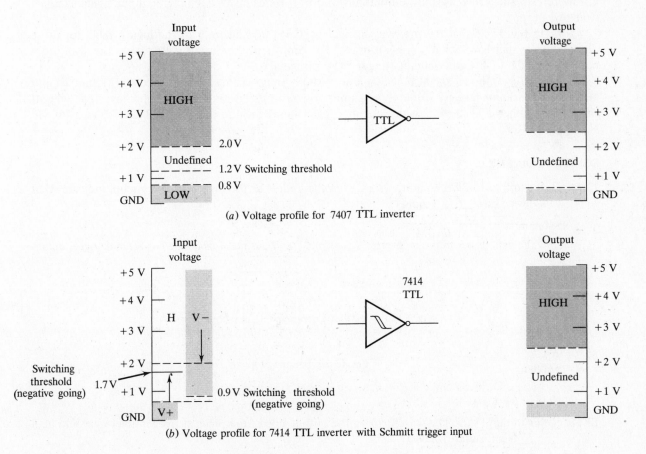

(*a*) Voltage profile for 7407 TTL inverter

(*b*) Voltage profile for 7414 TTL inverter with Schmitt trigger input

Fig. 13-28 Input switching thresholds

threshold is the input voltage at which the outputs of the digital device flip to their opposite state. In examining the input voltage profiles in Fig. 13-28, note that the switching threshold is always within the unshaded forbidden or undefined region of the device.

In Fig. 13-28a you will notice that the switching threshold for the standard 7404 inverter is 1.2 V. For the 7404 inverter, as voltage *increases* from 0 to about 1.1 V, the input is considered to be LOW (output of inverter would be HIGH). As the voltage increases closer to the threshold voltage of 1.2 V, the output would flip to the opposite state (output of inverter to LOW). On the standard 7404 inverter, as voltage *decreases* from near 5 to 1.3 V, the input is considered to be HIGH (output of inverter would remain LOW). As the voltage continues to decrease to the threshold voltage of 1.2 V, the output would flip to the opposite state (output of inverter to HIGH). The important idea on the standard 7404 inverter is that the threshold voltage is the same for both L-to-H and H-to-L transitions of the input. This can cause trouble when the input signal has a slow rise time because several oscillations (H-L-H or L-H-L) can occur at the output when the threshold voltage is crossed.

In Fig. 13-28b you will notice that the switching threshold for the Schmitt trigger 7414 inverter is different for the L-to-H and H-to-L transitions of the input. For the Schmitt trigger 7414 inverter, as the voltage *increases* from 0 to 1.6 V, the input is considered LOW (output of inverter would be HIGH). As the voltage increases to the threshold voltage of 1.7 V, the output would flip to the opposite state (output of inverter would go LOW). On the Schmitt trigger 7414 inverter, as voltage *decreases* from near 5 to 1 V, the input is considered to be HIGH (output remains LOW). As the voltage continues to decrease to the threshold voltage of 0.9 V, the output would flip to the opposite state (output of inverter to HIGH). This difference in threshold voltage for a positive-going (L-to-H) and a negative-going (H-to-L) input signal is called *hysteresis*. Each input of Schmitt trigger devices has hysteresis which increases noise immunity and transforms a slowly changing input signal to a fast changing output.

Note in Fig. 13-28 that the *hysteresis symbol* is placed in the center of the logic symbol for those digital devices that have Schmitt trigger inputs. The profiles of the output voltages are the same from both the standard 7404 and Schmitt trigger 7414 inverters.

Schmitt trigger inputs are also available in CMOS. Some of these include the 4093 Quad 2-input NAND gate Schmitt trigger, 40106 Hex Schmitt trigger inverter, and 74HC14 Hex inverting Schmitt trigger ICs. Other TTL devices with Schmitt trigger inputs include the 74LS132 (74132) and 74LS13 NAND gate ICs.

SOLVED PROBLEMS

13.60 Refer to Fig. 13-29. The hysteresis sign in the middle of the inverter logic symbol indicates that this device has _____ inputs.

  Solution:

   The hysteresis sign in the inverter logic symbol indicates that this device has Schmitt trigger inputs.

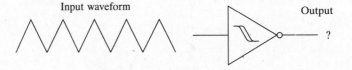

         Fig. 13-29 Sample problem

13.61 Refer to Fig. 13-29. The waveform on the output side of the Schmitt trigger inverter would be a _____ (sign, square) wave.

  Solution:

   The waveform on the output side of the Schmitt trigger inverter would be a square wave.

13.62 The Schmitt trigger inverter in Fig. 13-29 is being used as a signal _____ (conditioner, multiplexer) in this circuit.

Solution:

The Schmitt trigger inverter is being used as a signal conditioner in this circuit. It "squares up" the triangular waveform to form a square wave.

13.63 What is hysteresis when dealing with a Schmitt trigger digital device?

Solution:

See Fig. 13-28b. Hysteresis is the input characteristic of a Schmitt trigger device that sets the switching threshold higher for an L-to-H input (about 1.7 V on the 7414 inverter) and lower for an H-to-L input (about 0.9 V on the 7414 inverter). This greatly improves its noise immunity and its ability to "square up" input signals with slow rise and fall times.

Supplementary Problems

13.64 The 74150 IC is described by the manufacturer as a 16-input _____-_____/_____.
Ans.   data-selector/multiplexer

13.65 The 74150 IC can be used for changing (a) (parallel, serial) input data to (b) (parallel, serial) output data. The 74150 IC can also be used for solving (c) (combinational, sequential) logic problems.   *Ans.* (a) parallel (b) serial (c) combinational

13.66 Draw a block diagram of a 74150 data selector being used to solve the logic problem described by the Boolean expression $\overline{A}B\overline{C}\,\overline{D} + \overline{A}\,BCD + \overline{A}\,\overline{B}\,\overline{C}D + AB\overline{C}D + \overline{A}BCD = Y$.   *Ans.*   See Fig. 13-30.

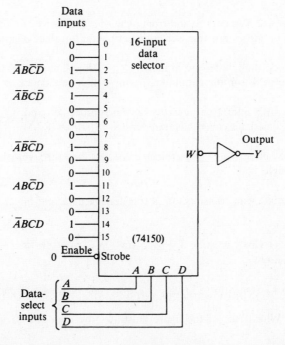

Fig. 13-30 Combinational logic problem solved by using a 74150 data selector

13.67 Refer to Fig. 13-5. The letters MUX on the counter block diagram stand for _____.
Ans. multiplexer

13.68 Refer to Fig. 13-5. When the MUX clock signal is LOW, the _____ (1s, 10s) count is lit on the _____ (left, right) seven-segment LED display.   *Ans.* 1s, right

13.69 To reduce power consumption, _____ (LCD, LED) displays are most often multiplexed.
Ans. LED (light-emitting diode)

13.70 Refer to Fig. 13-5. If the MUX clock frequency were reduced to 1 Hz, what would happen?
Ans. The displays would flash on and off. The eye would see the multiplexing action.

13.71 Refer to Fig. 13-6. A _____ (HIGH, LOW) at the anode terminal would activate the seven-segment display.   *Ans.* HIGH

13.72 A demultiplexer reverses the action of a(n) _____.   *Ans.* multiplexer

13.73 Refer to the demultiplexer at the right in Fig. 13-7. If the data-select inputs are $C = 1$, $B = 1$, and $A = 0$, then output _____ (number) is selected and a HIGH will appear at that output.   *Ans.* 6 or 110_2

13.74 Demultiplexers are also called _____ distributors or _____.   *Ans.* data distributors, decoders

13.75 Refer to Fig. 13-8. Which output of the 74LS154 DEMUX is activated if $G1$ and $G2$ are both LOW and the data-select inputs are $D = 1$, $C = 1$, $B = 0$, and $A = 0$.   *Ans.* output 12 or 1100_2

13.76 Refer to Fig. 13-9a. When the decimal 7 key is pressed and released, what will appear on the seven-segment output display?
Ans. nothing (There is no latch to hold the 7 at the inputs of the decoder.)

13.77 The 7475 is a TTL 4-bit _____ _____ IC.   *Ans.* transparent latch

13.78 Microprocessor-based systems transfer data back and forth on a bidirectional parallel path called a _____ _____.   *Ans.* data bus

13.79 A two-way buffer that will pass data to and from a data bus and serves to isolate a device from the bus is called a(n) _____.   *Ans.* bus transceiver or peripheral interface adapter (PIA)

13.80 If the outputs of a three-state buffer are in their _____ (high-impedance, transmit) state, they will float to whatever logic levels exist on the data bus.   *Ans.* high-impedance or high-Z

13.81 Refer to Fig. 13-11. The interfaces between the microprocessor and _____ (keyboard, RAM) are bidirectional buffers sometimes called bus transceivers.   *Ans.* RAM

13.82 The 74125 three-state buffer will _____ (block data, pass data through) when its control input is LOW.
Ans. pass data through

13.83 _____ (Parallel, Serial) data transmission is transferring data one bit at a time over a single line.
Ans. Serial

13.84 Refer to Fig. 13-14b. Device 1 must be a _____-in _____-out device.
Ans. parallel-in serial-out

13.85 The abbreviation UART stands for _____.   *Ans.* universal asynchronous receiver-transmitter

13.86 Refer to Fig. 13-31. Which part of the figure illustrates the idea of a serial-in parallel-out register?
Ans. part b

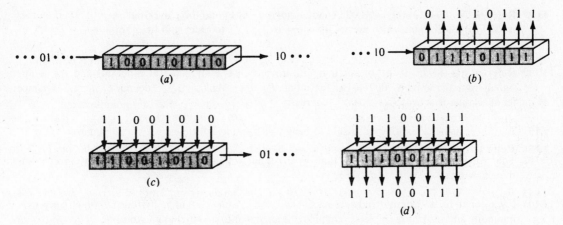

Fig. 13-31 Types of registers

13.87 Refer to Fig. 13-31. Which part of the figure illustrates the idea of a parallel-in serial-out register?
Ans. part *c*

13.88 A data bus, such as that used within a microcomputer, forms a bidirectional path for transmitting _____ (parallel, serial) data.   *Ans.* parallel

13.89 Refer to Fig. 13-11. All devices interfaced with a data bus must use buffers having _____ (three-state, totem-pole) outputs.   *Ans.* three-state

13.90 The letters PLD stand for _____ when dealing with programmable logic.
Ans. programmable logic device

13.91 Refer to Fig. 13-15*a*. This PLA would be _____ (activated, programmed) by burning open selected fuses.   *Ans.* programmed

13.92 Refer to Fig. 13-17. An abbreviated notation system, sometimes called a _____ (fuse, Karnaugh) map, was used to document the programming of this PLA.   *Ans.* fuse

13.93 Refer to Fig. 13-18. This is a fuse map for a _____ (FPLA, PAL).
Ans. FPLA (field-programmable logic array)

13.94 Refer to Fig. 13-20. A programmable logic device with a part number of PAL16L8 has _____ (number) inputs, _____ (number) outputs. The outputs are active _____ (HIGH, LOW) pins.
Ans. 16, 8, LOW

13.95 The 74HC85 is a 4-bit _____ comparator IC.   *Ans.* magnitude

13.96 Refer to Fig. 13-26. Which color LED is lit during time period t_6?
Ans. red (The output $A > B_{out}$ goes HIGH.)

13.97 Refer to Fig. 13-26. During time period t_7, the _____ (color) LED is lit because the _____ output goes HIGH.   *Ans.* green, $A = B_{out}$

13.98 When 74HC85 ICs are connected together to form 8-, 12-, 16-bit magnitude comparators, it is said that they are _____ (cascaded, multiplied).   *Ans.* cascaded (See Fig. 13-23.)

13.99 Refer to Fig. 13-24. If the 74HC393 counter stops at 1101_2 and the player's guess is 0111_2, then the _____ (color) LED will light, which means the guess is _____ (correct, too high, too low).
Ans. yellow, too low

13.100 Refer to Fig. 13-25. If the preset temperature setting is 00011000_2 (at input A) and the temperature signal from the oven is 00011011_2 (at input B), then the _____ (decrease, increase) temperature feedback line is activated.   *Ans.* decrease

13.101 A _____ (sine, square) wave is said to have fast rise and fall times.   *Ans.* square

13.102 The switching _____ (threshold, time) is the input voltage at which the outputs of an inverter flip to their new state.   *Ans.* threshold

13.103 Each input of a Schmitt trigger device has _____ (ac coupling, hysteresis) which increases noise immunity and transforms a slowly changing input signal to a fast-changing output.
Ans. hysteresis

13.104 Schmitt trigger devices are commonly used for signal _____ (conditioning, multiplexing).
Ans. conditioning

Index

SCHAUM'S INTERACTIVE OUTLINE SERIES

Schaum's Outlines and Mathcad™ Combined. . .
The Ultimate Solution.

NOW AVAILABLE! Electronic, interactive versions of engineering titles from the Schaum's Outline Series:

- *Electric Circuits*
- *Electromagnetics*
- *Feedback and Control Systems*
- *Thermodynamics For Engineers*
- *Fluid Mechanics and Hydraulics*

McGraw-Hill has joined with MathSoft, Inc., makers of Mathcad, the world's leading technical calculation software, to offer you interactive versions of popular engineering titles from the Schaum's Outline Series. Designed for students, educators, and technical professionals, the *Interactive Outlines* provide comprehensive on-screen access to theory and approximately 100 representative solved problems. Hyperlinked cross-references and an electronic search feature make it easy to find related topics. In each electronic outline, you will find all related text, diagrams and equations for a particular solved problem together on your computer screen. Every number, formula and graph is interactive, allowing you to easily experiment with the problem parameters, or adapt a problem to solve related problems. The *Interactive Outline* does all the calculating, graphing and unit analysis for you.

These "live" *Interactive Outlines* are designed to help you learn the subject matter and gain a more complete, more intuitive understanding of the concepts underlying the problems. They make your problem solving easier, with power to quickly do a wide range of technical calculations. All the formulas needed to solve the problem appear in real math notation, and use Mathcad's wide range of built in functions, units, and graphing features. This interactive format should make learning the subject matter easier, more effective and even fun.

For more information about *Schaum's Interactive Outlines* listed above and other titles in the series, please contact:

Schaum Division
McGraw-Hill, Inc.
1221 Avenue of the Americas
New York, New York 10020
Phone: 1-800-338-3987

To place an order, please mail the coupon below to the above address or call the 800 number.

-- ✂ --

Schaum's Interactive Outline Series
using Mathcad®

(Software requires 80386/80486 PC or compatibles, with Windows 3.1 or higher, 4 MB of RAM, 4 MB of hard disk space, and 3 1/2" disk drive.)

AUTHOR/TITLE	Interactive Software Only ($29.95 ea) ISBN	Quantity Ordered	Software and Printed Outline ($38.95 ea) ISBN	Quantity Ordered
MathSoft, Inc./DiStefano: Feedback & Control Systems	07-842708-8	_____	07-842709-6	_____
MathSoft, Inc./Edminister: Electric Circuits	07-842710-x	_____	07-842711-8	_____
MathSoft, Inc./Edminister: Electromagnetics	07-842712-6	_____	07-842713-4	_____
MathSoft, Inc./Giles: Fluid Mechanics & Hydraulics	07-842714-2	_____	07-842715-0	_____
MathSoft, Inc./Potter: Thermodynamics For Engineers	07-842716-9	_____	07-842717-7	_____

NAME_____ ADDRESS_____

CITY _____ STATE_____ ZIP_____

ENCLOSED IS ❏ A CHECK ❏ MASTERCARD ❏ VISA ❏ AMEX (✓ ONE)

ACCOUNT #_____ EXP. DATE _____

SIGNATURE_____

MAKE CHECKS PAYABLE TO McGRAW-HILL, INC. PLEASE INCLUDE LOCAL SALES TAX AND $1.25 SHIPPING/HANDLING

Schaum's Outlines and Solved Problems Books
in the
BIOLOGICAL SCIENCES

*SCHAUM OFFERS IN SOLVED-PROBLEM AND QUESTION-AND-ANSWER FORMAT
THESE UNBEATABLE TOOLS FOR SELF-IMPROVEMENT.*

❋ Fried **BIOLOGY** ORDER CODE 022401-3/$12.95
(including 888 solved problems)

❋ Jessop **ZOOLOGY** ORDER CODE 032551-0/$13.95
(including 1050 solved problems)

❋ Kuchel et al. **BIOCHEMISTRY** order code 035579-7/$13.95
(including 830 solved problems)

❋ Meislich et al. **ORGANIC CHEMISTRY, 2/ed** ORDER CODE 041458-0/$13.95
(including 1806 solved problems)

❋ Stansfield **GENETICS, 3/ed** ORDER CODE 060877-6/$12.95
(including 209 solved problems)

❋ Van de Graaff/Rhees **HUMAN ANATOMY AND PHYSIOLOGY** ORDER CODE 066884-1/$12.95
(including 1470 solved problems)

❋ Bernstein **3000 SOLVED PROBLEMS IN BIOLOGY** ORDER CODE 005022-8/$16.95

❋ Meislich et al. **3000 SOLVED PROBLEMS IN ORGANIC CHEMISTRY** ORDER CODE 056424-8/$22.95

Each book teaches the subject thoroughly through Schaum's pioneering solved-problem
format and can be used as a supplement to any textbook. If you want to excel in
any of these subjects, these books will help and they belong on your shelf.

Schaum's Outlines have been used by more than 25,000,000 student's worldwide!

PLEASE ASK FOR THEM AT YOUR LOCAL BOOKSTORE OR USE THE COUPON BELOW TO ORDER.